Springer Tracts in Natural Philosophy

Volume 31

Edited by C. Truesdell

Springer Tracts in Natural Philosophy

Tullio Valent

Boundary Value Problems of Finite Elasticity

Local Theorems on Existence, Uniqueness, and Analytic Dependence on Data

Springer-Verlag
New York Berlin Heidelberg
London Paris Tokyo

Tullio Valent
Dipartimento di Matematica Pura ed Applicata
Università di Padova
Padova 35131
Italy

ISBN-13: 978-1-4612-8326-3   e-ISBN-13: 978-1-4612-3736-5

DOI: 10.1007/978-1-4612-3736-5

AMS Classification: 73C35

Library of Congress Cataloging-in-Publication Data
Valent, Tullio
   Boundary value problems of finite elasticity.
   (Springer tracts in natural philosophy; v. 31)
   Bibliography: p.
   Includes index.
   1. Elasticity. 2. Boundary value problems.
I. Title. II. Series.
QA931.V34   1987   531'.0535   87-9665

Typeset by Asco Trade Typesetting Ltd., Hong Kong.

9 8 7 6 5 4 3 2 1

Springer-Verlag New York Berlin Heidelberg
Springer-Verlag Berlin Heidelberg New York

To Pia, Antonio, and Lucia

Preface

In this book I present, in a systematic form, some local theorems on existence, uniqueness, and analytic dependence on the load, which I have recently obtained for some types of boundary value problems of finite elasticity. Actually, these results concern an n-dimensional $(n \geq 1)$ formal generalization of three-dimensional elasticity. Such a generalization, besides being quite spontaneous, allows us to consider a great many interesting mathematical situations, and sometimes allows us to clarify certain aspects of the three-dimensional case. Part of the matter presented is unpublished; other arguments have been only partially published and in lesser generality. Note that I concentrate on simultaneous local existence and uniqueness; thus, I do not deal with the more general theory of existence. Moreover, I restrict my discussion to compressible elastic bodies and I do not treat unilateral problems. The clever use of the inverse function theorem in finite elasticity made by STOPPELLI [1954, 1957a, 1957b], in order to obtain local existence and uniqueness for the traction problem in hyperelasticity under dead loads, inspired many of the ideas which led to this monograph.

Chapter I aims to give a very brief introduction to some general concepts in the mathematical theory of elasticity, in order to show how the boundary value problems studied in the sequel arise.

Chapter II is very technical; it supplies the framework for all subsequent developments. Theorems on continuity, differentiability, and analyticity for composition operators are established in this chapter; they will suggest the later choices of the spaces for solutions and data. From

Theorem 6.1 it follows, for example, that to study our nonlinear problems using the implicit function theorem, the Sobolev spaces connected with a weak formulation of its (formally) linearized problems do not work. Thus we need appropriate regularity theorems for linear boundary value problems with a rather mild smoothness of some coefficients. The main object of Chapter III is precisely to provide such regularity theorems.

Subsequent chapters are devoted to obtaining theorems of existence, uniqueness, and analytic dependence on the load, near special deformations, for boundary value problems of place (in Chapter IV) and traction (in Chapters V and VI) in finite elastostatics. Loads independent of the deformation (dead loads) and loads depending on the deformation (live loads) are both considered. For the problem of place under dead loads some "semiglobal" results are also given. Evidently, a reasonable dependence of the load on the deformation, while not creating serious difficulties for the boundary condition of place, gives rise to a very wide variety of boundary value problems, with difficulties of every kind when we deal with the traction problem. On the other hand, for the traction problem, any physically realistic load depends (nontrivially) on the deformation.

In Chapter V, I present an abstract method of attacking the traction problem with general loads when certain conditions are satisfied: this method leads to an abstract theorem of existence, uniqueness, and analytic dependence on a parameter (Theorem 5.1). A first application of this theorem is given in the second part of Chapter V in treating the case of dead loads. Two more applications of that abstract theorem are made in Chapter VI, where a very important class of traction problems is studied: namely, those in which the prescribed surface traction is parallel to the normal of the boundary of the unknown deformed equilibrium configuration. Within this class there are boundary value problems to which the abstract method of Chapter V does not apply. One of these is particularly interesting and realistic: that is, the boundary value problem arising from the study of the equilibrium of a heavy elastic body submerged in a quiet, homogeneous, heavy liquid.

Most of Chapter VI is devoted to (the n-dimensional version of) this boundary value problem. The main result of the book is a theorem of existence, uniqueness, and analytic dependence on a parameter for this problem, near suitable deformations (see Theorem 4.17). I believe that some key ideas devised in proving this theorem may suggest a way of attacking boundary value problems of traction different from those discussed here. Moreover, I note that in traction problems, the particular deformations near to which I find existence, uniqueness, and analytic dependence on a parameter are unstressed; but, bearing in mind the analysis of BHARATHA & LEVINSON [1978], CAPRIZ & PODIO-GUIDUGLI [1979], and WAN & MARSDEN [1983], we can realize how the meth-

odologies presented here can be adapted in order to study traction
boundary value problems near stressed deformations.

Of course, the methods and results of this book have a quite different
character from those based on the calculus of variations and the search
for suitable constitutive assumptions (such as polyconvexity of the stored-
energy function assumed by BALL [1977]). Rather, they may be useful as
a first step in a global approach to boundary value problems of finite
elasticity.

I conclude by expressing my gratitude to Professor C. TRUESDELL for
inviting me to write this monograph.

Padova   TULLIO VALENT
January 1987

Contents

Contents

CHAPTER I

A Brief Introduction to Some General Concepts in Elasticity

This chapter is not intended to be an introduction to elasticity. It is merely a very quick survey of those definitions and axioms that relate strictly to the mathematical problems studied in the remainder of the book. For a deep and exhaustive treatment of the matter we refer, e.g., to TRUESDELL & TOUPIN [1960], TRUESDELL & NOLL [1965], WANG & TRUESDELL [1973], TRUESDELL [1974, 1977], GURTIN [1981], MARSDEN & HUGHES [1983], and CIARLET [1986]. What we will say has a physical meaning if $n = 3$. Nevertheless, since the definitions, axioms, and problems considered in this book generalize spontaneously to an n-dimensional setting ($n \geq 1$), we prefer to present the theory directly in the n-dimensional form while still using the terminology relative to the three-dimensional case.

§1. Some Notations

Throughout this book (except in the statement of Lemma 2.1' of Chapter II) Ω is a nonempty, bounded, open subset of $\mathbb{R}^n$ with $n \geq 1$. $\bar{\Omega}$ is the closure of Ω and $\partial \Omega$ is the boundary of Ω. We denote by $M_{m \times n}$ the set of real $m \times n$ matrices, and we write M_n instead of $M_{n \times n}$. The unit element of the ring M_n is indicated by I.

We use the following notations: $\text{Sym}_n = \{Z \in M_n : Z = Z^T\}$, $\text{Skew}_n = \{Z \in M_n : Z + Z^T = 0\}$, $M_n^+ = \{Z \in M_n : \det Z > 0\}$, and $O_n^+ = \{Z \in M_n^+ : Z^{-1} = Z^T\}$; where $\det Z$ is the determinant of Z, Z^T is the transpose of

Z, and Z^{-1} is the inverse of Z. For any $Z \in M_n$, tr Z denotes the trace of Z and cof Z denotes the matrix of cofactors of Z, so that

$$\text{cof } Z = (\det Z)(Z^{-1})^{\text{T}}$$

whenever $\det Z \neq 0$.

For $Z = (Z_{ih})_{i=1,\ldots,m;\, h=1,\ldots,p} \in M_{m \times p}$ and $W = (W_{hj})_{h=1,\ldots,p;\, j=1,\ldots,n} \in M_{p \times n}$ we set

$$ZW = \left(\sum_{h=1}^{p} Z_{ih} W_{hj} \right)_{i=1,\ldots,m;\, j=1,\ldots,n},$$

so that $Zy = (\sum_{j=1}^{n} Z_{ij} y_j)_{i=1,\ldots,m}$ for $Z = (Z_{ij})_{i=1,\ldots,m;\, j=1,\ldots,n} \in M_{m \times n}$ and $y = (y_j)_{j=1,\ldots,n} \in \mathbb{R}^n$.

Moreover, for $Z = (Z_{ij})_{i=1,\ldots,m;\, j=1,\ldots,n}$ and $W = (W_{ij})_{i=1,\ldots,m;\, j=1,\ldots,n}$ belonging to $M_{m \times n}$ we set

$$Z \cdot W = \sum_{\substack{i=1,\ldots,m \\ j=1,\ldots,n}} Z_{ij} W_{ij} \quad \text{and} \quad |Z| = (Z \cdot Z)^{1/2}.$$

For $y = (y_i)_{i=1,\ldots,n}$ and $z = (z_j)_{j=1,\ldots,n}$ belonging to $\mathbb{R}^n$ we denote by $y \otimes z$ and $y \wedge z$ the matrices defined by $y \otimes z = (y_i z_j)_{i,j=1,\ldots,n}$ and $y \wedge z = y \otimes z - z \otimes y$. Furthermore, if S, T are functions defined on Ω with values in $M_{m \times p}$ and $M_{p \times n}$, respectively, then ST denotes the function defined on Ω by putting $(ST)(x) = S(x)T(x)$, $\forall x \in \Omega$. If S, T are functions defined on Ω with values in $M_{m \times n}$, then $S \cdot T$ and $|S|$ are the functions defined on Ω by putting $(S \cdot T)(x) = S(x) \cdot T(x)$ and $|S|(x) = |S(x)|$, $\forall x \in \Omega$. Finally, if u, v are $\mathbb{R}^n$-valued functions defined on Ω, then $u \otimes v$ and $u \wedge v$ are the functions defined on Ω by putting $(u \otimes v)(x) = u(x) \otimes v(x)$ and $(u \wedge v)(x) = u(x) \wedge v(x)$, $\forall x \in \Omega$. Sometimes, to make our notation simpler, we identify a constant function with its value; thus, if u is an $\mathbb{R}^n$-valued function defined on Ω and $y \in \mathbb{R}^n$, $Z \in M_n$, then we indicate by $y \cdot u$, $u \cdot y$, and Zu the functions defined on Ω by putting $(y \cdot u)(x) = (u \cdot y)(x) = y \cdot u(x)$, $(Zu)(x) = Z(u(x))$, $\forall x \in \Omega$.

If U is an open subset of $\mathbb{R}^n$, the gradient of a function $v: U \to \mathbb{R}^m$, $m \geq 1$, is denoted by Dv, i.e.,

$$Dv = (D_j v_i)_{i=1,\ldots,m;\, j=1,\ldots,n},$$

where D_j is the jth partial derivative and $v_i = v \cdot e_i$ with $(e_i)_{i=1,\ldots,n}$ the canonical base of $\mathbb{R}^n$, while the divergence of a function $S: U \to M_{m \times n}$ is denoted by div S, so that

$$\text{div } S = \left(\sum_{j=1}^{n} D_j S_{ij} \right)_{i=1,\ldots,m},$$

where $S_{ij} = S \cdot (e_i \otimes e_j)$.

For any subset $\mathscr{P}$ of $\mathbb{R}^n$, the identity function of $\mathscr{P}$ into $\mathbb{R}^n$ is denoted by

$$\iota_{\mathscr{P}}.$$

We say that a *part* $\mathscr{P}$ of Ω is *regular*, provided $\mathscr{P}$ is open, its boundary in $\mathbb{R}^n$ is piecewise smooth, and such that the Gauss–Green formula

$$\int_{\mathscr{P}} Dv \, dx = \int_{\partial\mathscr{P}} v \otimes v \, d\sigma$$

holds for every C^1 function $v\colon \bar{\mathscr{P}} \to \mathbb{R}^n$, where v is the outward, unit normal to $\partial\mathscr{P}$ and σ is the hypersurface measure on $\partial\mathscr{P}$. (Of course, the left-hand side is an integral with respect to the Lebesgue measure on $\mathbb{R}^n$.)

Throughout this chapter we suppose that Ω is regular and that $\Omega =$ interior of $\bar\Omega$. For any Lebesgue measurable subset $\mathscr{P}$ of $\mathbb{R}^n$,

$$\mathrm{vol}(\mathscr{P})$$

will denote the Lebesgue measure of $\mathscr{P}$.

§2. Deformations and Motions

We will always regard the bounded, open subset Ω of $\mathbb{R}^n$ as a fixed *reference configuration* of a body. By a *deformation* of $\bar\Omega$ we mean an one-to-one C^1 function $\phi\colon \bar\Omega \to \mathbb{R}^n$ such that

$$\det D\phi > 0.$$

The function $u\colon \bar\Omega \to \mathbb{R}^n$, defined by $u = \phi - \iota_{\bar\Omega}$, is called the *displacement* of $\bar\Omega$ relative to the deformation ϕ.

Using the invariance of domain theorem (cf., e.g., MEISTERS & OLECH [1963], p. 64) and the inverse function theorem (see Appendix I), we realize that, if ϕ is a deformation of $\bar\Omega$, then $\phi(\Omega)$ is an open subset of $\mathbb{R}^n$, $\phi(\partial\Omega) = \partial\phi(\Omega)$, $\phi(\bar\Omega) = \overline{\phi(\Omega)}$, and the inverse of ϕ is a C^1 function on $\phi(\bar\Omega)$. Thus a deformation of $\bar\Omega$ is an orientation-preserving C^1 diffeomorphism of $\bar\Omega$ onto a subset of $\mathbb{R}^n$.

A *motion* (of the body) is a mapping $t \mapsto \phi_t$ from $\mathbb{R}$ into the set of deformations of $\bar\Omega$. We suppose that, for each $x \in \bar\Omega$, the mapping $t \mapsto \phi_t(x)$ is of class C^1: its derivative at t will be denoted by $\dot\phi_t(x)$ and the mapping $x \mapsto \dot\phi_t(x)$ will be denoted by $\dot\phi_t$. In other words, we suppose that $t \mapsto \phi_t$ is of class C^1 when the linear space of functions from $\bar\Omega$ into $\mathbb{R}^n$ (where $t \mapsto \phi_t$ takes its values) is thought of as equipped with the topology of pointwise convergence, and we will denote its derivative at t by $\dot\phi_t$. Analogously, if the second derivative at t of the mapping $t \mapsto \phi_t$ exists, it will be denoted by $\ddot\phi_t$.

A deformation ϕ of $\bar\Omega$ is said to be *rigid* if $|\phi(x') - \phi(x'')| = |x' - x''|$,

$\forall x'$, $x'' \in \bar{\Omega}$. It is well known that ϕ is rigid if and only if for every x', $x'' \in \bar{\Omega}$ we have $\phi(x') - \phi(x'') = R(x' - x'')$ with $R \in \mathbb{O}_n^+$. A motion $t \mapsto \phi_t$ is said to be rigid when ϕ_t is a rigid deformation of $\bar{\Omega}$ for every t.

If $\text{vol}(\phi(\mathscr{P})) = \text{vol}(\mathscr{P})$ for any Borel set $\mathscr{P}$ of Ω, then the deformation ϕ of $\bar{\Omega}$ is said to be *isochoric*. A motion $t \mapsto \phi_t$ is said to be isochoric when $\text{vol}(\phi_t(\mathscr{P})) = \text{vol}(\mathscr{P})$ for any Borel set $\mathscr{P}$ of Ω and time t. Given a deformation ϕ of $\bar{\Omega}$ different from $\imath_{\bar{\Omega}}$, we will denote by x_ϕ the points of $\phi(\bar{\Omega})$. Moreover, for the sake of clarity, *to designate a function defined on $\phi(\Omega)$ or $\partial\phi(\Omega)$, we use the subscript ϕ*.

$$* \quad * \quad * \quad * \quad *$$

We now recall, without proofs, some well-known facts about the change of variable formulas for integrals.

If $\mathscr{P}$ is a Borel set of Ω and $f_\phi \colon \phi(\mathscr{P}) \to \mathbb{R}$ is a Borel function, then f_ϕ is (Lebesgue) integrable if and only if $f_\phi \circ \phi \det D\phi$ is (Lebesgue) integrable on $\mathscr{P}$; moreover,

$$\int_{\phi(\mathscr{P})} f_\phi \, dx_\phi = \int_{\mathscr{P}} f_\phi \circ \phi \det D\phi \, dx. \tag{2.1}$$

Note that, by (2.1), a deformation ϕ of $\bar{\Omega}$ is isochoric if and only if $\det D\phi = 1$.

If $\mathscr{P}$ is a regular part of Ω (cf. §1), v the outward unit normal to $\partial\mathscr{P}$, and $f_\phi \colon \partial\phi(\mathscr{P}) \to \mathbb{R}$ a Borel function, then f_ϕ is integrable on $\partial\phi(\mathscr{P})$ with respect to the hypersurface measure σ_ϕ on $\partial\phi(\mathscr{P})$ if and only if $f_\phi \circ \phi|(\text{cof } D\phi)v|$ is integrable on $\partial\mathscr{P}$ with respect to the hypersurface measure σ on $\partial\mathscr{P}$; moreover,

$$\int_{\partial\phi(\mathscr{P})} f_\phi \, d\sigma_\phi = \int_{\partial\mathscr{P}} f_\phi \circ \phi|(\text{cof } D\phi)v| \, d\sigma. \tag{2.2}$$

Remark that, as evidently $(Zy) \cdot z = y \cdot (Z^T z)$, $\forall(y, z, Z) \in \mathbb{R}^n \times \mathbb{R}^n \times \mathbb{M}_n$, we have

$$((\text{cof } Z)y) \cdot (Zz) = (\det Z)(y \cdot z), \qquad \forall(y, z, Z) \in \mathbb{R}^n \times \mathbb{R}^n \times \mathbb{M}_n.$$

Using this equality and bearing in mind that (for any regular point $x \in \partial\mathscr{P}$) the differential of ϕ at x carries the tangent hyperplane to $\partial\mathscr{P}$ at x onto the tangent hyperplane to $\partial\phi(\mathscr{P})$ at $\phi(x)$, we easily realize that

$$v_\phi(\phi(x)) = \frac{(\text{cof } D\phi(x))v(x)}{|\text{cof } D\phi(x))v(x)|}, \tag{2.3}$$

where v_ϕ is the outward unit normal to $\partial\phi(\mathscr{P})$. Combining (2.2) with (2.3) we immediately obtain

$$\int_{\partial\phi(\mathscr{P})} f_\phi v_\phi \, d\sigma_\phi = \int_{\partial\mathscr{P}} f_\phi \circ \phi(\text{cof } D\phi)v \, d\sigma \tag{2.4}$$

for every Borel function f_ϕ: $\partial\phi(\mathscr{P}) \to \mathbb{R}$ integrable on $\partial\phi(\mathscr{P})$ with respect to the hypersurface measure σ_ϕ on $\partial\phi(\mathscr{P})$.

For any continuous function S_ϕ: $\phi(\bar{\Omega}) \to \mathbb{M}_n$, we denote by S the $\mathbb{M}_n$- valued function defined on $\bar{\Omega}$ by

$$S = (S_\phi \circ \phi) \operatorname{cof} D\phi, \tag{2.5}$$

so that, by (2.4),

$$\int_{\partial\phi(\mathscr{P})} S_\phi v_\phi \, d\sigma_\phi = \int_{\partial\mathscr{P}} Sv \, d\sigma.$$

Later, we shall need the following equality

$$\operatorname{div} S = \det D\phi(\operatorname{div} S_\phi) \circ \phi, \tag{2.6}$$

which is an easy consequence of the equality

$$\operatorname{div}(\operatorname{cof} D\phi) = 0. \tag{2.7}$$

Equation (2.7) is known as the Piola identity. For a proof of (2.7) see, e.g., MARSDEN & HUGHES [1983].

§3. Mass. Force

To any deformation ϕ: $\bar{\Omega} \to \mathbb{R}^n$ we associate a positive measure m_ϕ on the σ-algebra of Borel sets of $\phi(\Omega)$, and assume that

$$m_\phi(\phi(\mathscr{P})) = m(\mathscr{P}) \tag{3.1}$$

for every Borel set $\mathscr{P}$ of Ω where m stands for $m_{i_{\bar{n}}}$. By m_ϕ we mean the mass of the body in the configuration $\phi(\bar{\Omega})$; condition (3.1), written for every deformation ϕ of $\bar{\Omega}$ and every Borel set $\mathscr{P}$ of Ω, expresses *conservation of mass*.

We suppose that m is absolutely continuous with respect to the Lebesgue measure: let μ be the density of m with respect to the Lebesgue measure, so that

$$m(\mathscr{P}) = \int_{\mathscr{P}} \mu \, dx$$

for every Borel set $\mathscr{P}$ of Ω. If μ_ϕ: $\phi(\Omega) \to \mathbb{R}$ is the function related to μ by

$$\mu = \mu_\phi \circ \phi \det D\phi, \tag{3.2}$$

then, in view of (2.1), we have

$$\int_{\phi(\mathscr{P})} \mu_\phi \, dx_\phi = \int_{\mathscr{P}} \mu \, dx$$

for every Borel set $\mathscr{P}$ of Ω. Therefore the measure m_ϕ is absolutely continuous with respect to the Lebesgue measure on $\phi(\Omega)$ with density μ_ϕ. μ will be called the *mass density in the reference configuration* and μ_ϕ the *mass density in the configuration* $\phi(\bar{\Omega})$.

$$* \quad * \quad * \quad * \quad *$$

The mechanical actions exerted on a part of a body by its environment during a motion are described by *forces*. Classically, the forces are of two types: body forces and contact forces. Body forces act on the points in a part $\mathscr{P}$ of a body that has a volume, while contact forces act on the points of $\mathscr{P}$ that lie on its boundary.

During a motion $t \mapsto \phi_t$ of a body, the *body forces* may be defined by assigning for each t an $\mathbb{R}^n$-valued measure β_{ϕ_t} on the σ-algebra of Borel sets of $\phi_t(\Omega)$, while the *contact forces* (or *surface forces*) may be defined by assigning, for each t and each regular part $\mathscr{P}$ of Ω, an $\mathbb{R}^n$-valued measure $\sigma_{\phi_t, \mathscr{P}}$ on the σ-algebra of Borel sets of $\partial\phi_t(\mathscr{P})$: if B is a ball of $\mathbb{R}^n$ centered at any point $x \in \partial\phi_t(\mathscr{P})$ and Σ is the Borel set $B \cap \partial\phi_t(\mathscr{P})$ of $\partial\phi_t(\mathscr{P})$, then $\sigma_{\phi_t, \mathscr{P}}(\Sigma)$ represents the contact action exerted on $\phi_t(\mathscr{P})$, through Σ, by the environment of $\phi_t(\mathscr{P})$.

As to forces, we will accept the following usual assumptions.

(i) β_{ϕ_t} is absolutely continuous with respect to the measure m_{ϕ_t}; thus, if b_{ϕ_t} is the density of β_{ϕ_t} with respect to the measure m_{ϕ_t}, then for any Borel set $\mathscr{P}$ of Ω

$$\beta_{\phi_t}(\phi_t(\mathscr{P})) = \int_{\phi_t(\mathscr{P})} \mu_{\phi_t}(x_{\phi_t}) b_{\phi_t}(x_{\phi_t}) \, dx.$$

(ii) $\sigma_{\phi_t, \mathscr{P}}$ is absolutely continuous with respect to the hypersurface measure on $\partial\phi_t(\mathscr{P})$.

(iii) (*Cauchy's Axiom*). Let $x_{\phi_t} \in \phi_t(\Omega)$, and let e be a unit vector of $\mathbb{R}^n$. The density at x_{ϕ_t} of $\sigma_{\phi_t, \mathscr{P}}$ (with respect to the hypersurface measure on $\partial\phi_t(\mathscr{P})$) is the same for any regular part $\mathscr{P}$ of Ω such that $x_{\phi_t} \in \partial\phi_t(\mathscr{P})$, and such that e is the outward unit normal to $\partial\phi_t(\mathscr{P})$ at x_{ϕ_t}.

In view of (ii) and (iii), the contact forces during a motion $t \mapsto \phi_t$ can be defined by assigning, for each t, an $\mathbb{R}^n$-valued function

$$(x_{\phi_t}, e) \mapsto s_{\phi_t}(x_{\phi_t}, e)$$

defined on $(\phi_t(\Omega) \times \{e \in \mathbb{R}^n : |e| = 1\}) \cap \{(x_{\phi_t}, v(x_{\phi_t})) : x_{\phi_t} \in \partial\phi_t(\Omega), x_{\phi_t} \text{ is a regular point}\}$ and such that, for each t and each regular part $\mathscr{P}$ of Ω, the function $x \mapsto s_{\phi_t}(x_{\phi_t}, v(x_{\phi_t}))$, where $v(x_{\phi_t})$ is the outward unit normal to $\partial\phi_t(\mathscr{P})$ at x_{ϕ_t}, is integrable on $\partial\phi_t(\mathscr{P})$ with respect to the hypersurface measure. Usually, the functions s_{ϕ_t} and b_{ϕ_t} are called the *density of surface force* and the *density of body force* (at time t and relative to the motion $t \mapsto \phi_t$), respectively. For a careful analysis of these matters we refer to TRUESDELL [1977].

§4. Euler's Axiom. Cauchy's Theorem

If $t \mapsto \phi_t$ is a motion and $\mathcal{P}$ is a Borel set of Ω, we put

$$
\begin{cases}
m_0(\mathcal{P}, t) = \displaystyle\int_{\phi_t(\mathcal{P})} \mu_{\phi_t}(x_{\phi_t}) \dot{\phi}_t(\phi_t^{-1}(x_{\phi_t}))\, dx_{\phi_t}, \\[3mm]
m_1(\mathcal{P}, t) = \displaystyle\int_{\phi_t(\mathcal{P})} x_{\phi_t} \wedge \mu_{\phi_t}(x_{\phi_t}) \dot{\phi}_t(\phi_t^{-1}(x_{\phi_t}))\, dx_{\phi_t},
\end{cases}
$$

where ϕ_t^{-1} is the inverse of the deformation $\phi_t \colon \bar{\Omega} \to \phi_t(\bar{\Omega}) \subseteq \mathbb{R}^n$. The elements $m_0(\mathcal{P}, t)$ and $m_1(\mathcal{P}, t)$ of $\mathbb{R}^n$ are called the *linear momentum* and the *angular momentum* (with respect to 0) of $\mathcal{P}$ at time t, respectively.

Continuum mechanics is based on the following axiom which relates the motion of a body to the forces sustaining it.

Euler's Axiom. *Let $t \mapsto \phi_t$ be a motion such that for any regular part $\mathcal{P}$ of Ω, the functions $t \mapsto m_0(\mathcal{P}, t)$ and $t \mapsto m_1(\mathcal{P}, t)$ are differentiable. Then, for every regular part $\mathcal{P}$ of Ω and every t,*

$$
\dot{m}_0(\mathcal{P}, t) = \int_{\phi_t(\mathcal{P})} \mu_{\phi_t}(x_{\phi_t}) b(x_{\phi_t})\, dx_{\phi_t} + \int_{\partial \phi_t(\mathcal{P})} s_{\phi_t}(x_{\phi_t}, \nu(x_{\phi_t}))\, d\sigma_{\phi_t}(x_{\phi_t}), \quad (4.1)
$$

and

$$
\dot{m}_1(\mathcal{P}, t) = \int_{\phi_t(\mathcal{P})} x_{\phi_t} \wedge \mu_{\phi_t}(x_{\phi_t}) b_{\phi_t}(x_{\phi_t})\, dx_{\phi_t}
$$

$$
+ \int_{\partial \phi_t(\mathcal{P})} x_{\phi_t} \wedge s_{\phi_t}(x_{\phi_t}, \nu(x_{\phi_t}))\, d\sigma_{\phi_t}(x_{\phi_t}), \quad (4.2)
$$

where $\nu(x_{\phi_t})$ denotes the outward unit normal to $\partial \phi_t(\mathcal{P})$ at point x_{ϕ_t} and the superimposed dot denotes derivative with respect to t.

Equation (4.1) expresses the *balance of linear momentum*, while (4.2) expresses the *balance of angular momentum*.

Remark 4.1. *If the motion $t \mapsto \phi_t$ is sufficiently smooth (e.g., if the second derivative with respect to t of the function $(t, x) \mapsto \phi_t(x)$ exists and is continuous), then*

$$
\dot{m}_0(\mathcal{P}, t) = \int_{\phi_t(\mathcal{P})} \mu_{\phi_t}(x_{\phi_t}) \ddot{\phi}_t(\phi_t^{-1}(x_{\phi_t}))\, dx_{\phi_t}, \quad (4.3)
$$

$$
\dot{m}_1(\mathcal{P}, t) = \int_{\phi_t(\mathcal{P})} x_{\phi_t} \wedge \mu_{\phi_t}(x_{\phi_t}) \ddot{\phi}_t(\phi_t^{-1}(x_{\phi_t}))\, dx_{\phi_t}. \quad (4.4)
$$

Proof. In view of (2.1) and (3.2) we obtain

$$
m_0(\mathcal{P}, t) = \int_{\mathcal{P}} \mu_{\phi_t}(\phi_t(x)) \det D\phi_t(x) \dot{\phi}_t(x)\, dx = \int_{\mathcal{P}} \mu(x) \dot{\phi}_t(x)\, dx,
$$

and thus

$$\dot{m}_0(\mathscr{P}, t) = \int_{\mathscr{P}} \mu(x)\ddot{\phi}_t(x)\, dx = \int_{\phi_t(\mathscr{P})} \mu_{\phi_t}(x_{\phi_t})\ddot{\phi}_t(\phi_t^{-1}(x_{\phi_t}))\, dx_{\phi_t};$$

likewise, we get

$$\dot{m}_1(\mathscr{P}, t) = \int_{\phi_t(\mathscr{P})} x_{\phi_t} \wedge \mu_{\phi_t}(x_{\phi_t})\ddot{\phi}_t(\phi_t^{-1}(x_{\phi_t}))\, dx_{\phi_t}$$

$$+ \int_{\phi_t(\mathscr{P})} \mu_{\phi_t}(x_{\phi_t})\dot{\phi}_t(\phi_t^{-1}(x_{\phi_t})) \wedge \dot{\phi}_t(\phi_t^{-1}(x_{\phi_t}))\, dx_{\phi_t}$$

$$= \int_{\phi_t(\mathscr{P})} \mu_{\phi_t} \wedge \mu_{\phi_t}(x_{\phi_t})\ddot{\phi}_t(\phi_t^{-1}(x_{\phi_t}))\, dx_{\phi_t}. \quad \square$$

The following lemma has central importance in continuum mechanics.

Lemma 4.2. *Let U be an open subset of $\mathbb{R}^n$ and let $f: U \times \{e \in \mathbb{R}^n: |e| = 1\} \to \mathbb{R}^m$, with $m \geq 1$, be a function such that for each $e \in \mathbb{R}^n$ with $|e| = 1$, $x \mapsto f(x, e)$ is continuous. Assume further that for every compact polyhedron $\mathscr{P}$ in U*

$$\left| \int_{\partial\mathscr{P}} f(x, v(x))\, d\sigma(x) \right| \leq c(\mathscr{P})\, \text{vol}(\mathscr{P})$$

with $v(x)$ an oriented unit normal to $\partial\mathscr{P}$ at x, and $\mathscr{P} \mapsto c(\mathscr{P})$ a positive, real-valued function which is bounded on every decreasing sequence of compact polyhedra in U. Then, for each $x \in U$, the function $e \mapsto f(x, e)$ is the restriction to $\{e \in \mathbb{R}^n: |e| = 1\}$ of a linear function from $\mathbb{R}^n$ to $\mathbb{R}^m$, that is

$$f(x, e) = F(x)e$$

with $F(x)$ an $m \times n$ real matrix.

For a proof of Lemma 4.2, in the case $n = 3$, we refer to TRUESDELL [1977] and GURTIN & MARTINS [1976], where a more general result is proved; the generalization of the proofs given in the case $n = 3$ to the case when n is any integer ≥ 1 is easy.

Using Lemma 4.2 we now prove without difficulty the next theorem.

Theorem 4.3 (Cauchy). *Let $t \mapsto \phi_t$ be a motion such that the second derivative with respect to t of the function $(t, x) \mapsto \phi_t(x)$ exists and is continuous. Suppose that μ_{ϕ_t} and b_{ϕ_t} are continuous and that, for each $e \in \mathbb{R}^n$ with $|e| = 1$, the function $x \mapsto s_{\phi_t}(x_{\phi_t}, e)$ is of class C^1 in $\phi_t(\Omega)$. Then:*

(i) *Equation (4.1) is satisfied for every regular part $\mathscr{P}$ of Ω if and only if there is a C^1 function*

$$S_{\phi_t}: \phi_t(\Omega) \to \mathbb{M}_n$$

such that

$$s_{\phi_t}(x_{\phi_t}, e) = S_{\phi_t}(x_{\phi_t})e, \qquad \forall x_{\phi_t} \in \phi_t(\Omega) \tag{4.5}$$

and

$$\text{div } S_{\phi_t} + \mu_{\phi_t} b_{\phi_t} = \mu_{\phi_t} \ddot{\phi}_t \circ \phi_t^{-1} \quad \text{in } \phi_t(\Omega). \tag{4.6}$$

(ii) *Whenever (4.1) is satisfied for every regular part $\mathcal{P}$ of Ω, (4.2) holds for every regular part $\mathcal{P}$ of Ω if and only if for each $x_{\phi_t} \in \phi_t(\Omega)$ the matrix $S_{\phi_t}(x_{\phi_t})$ is symmetric.*

Proof. First, observe that if the closure of $\mathcal{P}$ in $\mathbb{R}^n$ is contained in Ω, then (by Remark 4.1 and the hypotheses made on μ_{ϕ_t}, b_{ϕ_t}, and the motion $t \mapsto \phi_t$), (4.1) yields

$$\left| \int_{\partial \phi_t(\mathcal{P})} s_{\phi_t}(x_{\phi_t}, \nu(x_{\phi_t})) \, d\sigma_{\phi_t}(x_{\phi_t}) \right|$$

$$\leq \left(\sup_{x \in \phi_t(\mathcal{P})} |\mu_{\phi_t}(x_{\phi_t})(b_{\phi_t}(x_{\phi_t}) - \ddot{\phi}_t(\phi_t^{-1}(x_{\phi_t}))|\right) \text{vol}(\phi_t(\mathcal{P}))$$

with

$$\sup_{x \in \phi_t(\mathcal{P})} |\mu_{\phi_t}(x_{\phi_t})(b_{\phi_t}(x_{\phi_t}) - \ddot{\phi}_t(\phi_t^{-1}(x_{\phi_t})))| < +\infty.$$

Hence, if (4.1) is satisfied for every regular part $\mathcal{P}$ of Ω, then for every polyhedron $\mathcal{P}$ in $\phi_t(\Omega)$ we have

$$\left| \int_{\partial \mathcal{P}} s_{\phi_t}(x_{\phi_t}, \nu(x_{\phi_t})) \, d\sigma_{\phi_t}(x_{\phi_t}) \right| \leq c(\mathcal{P}) \text{ vol}(\mathcal{P}),$$

where $\nu(x_{\phi_t})$ is the outward unit normal to $\partial \mathcal{P}$ at x_{ϕ_t} (for any regular point x_{ϕ_t} of $\partial \mathcal{P}$) and

$$c(\mathcal{P}) = \sup_{x_{\phi_t} \in \mathcal{P}} |\mu_{\phi_t}(x_{\phi_t})(b_{\phi_t}(x_{\phi_t}) - \ddot{\phi}_t(\phi_t^{-1}(x_{\phi_t})))|.$$

Therefore, in view of Lemma 4.2, there is a function $S_{\phi_t}: \phi_t(\Omega) \to \mathbb{M}_n$ such that (4.5) holds.

Note that, since

$$S_{\phi_t}(x_{\phi_t}) = \sum_{i=1}^{n} S_{\phi_t}(x_{\phi_t})e_i \otimes e_i = \sum_{i=1}^{n} s_{\phi_t}(x_{\phi_t}, e_i) \otimes e_i,$$

where $(e_i)_{i=1,\ldots,n}$ is the canonical base of $\mathbb{R}^n$ and since the functions $x_{\phi_t} \mapsto s_{\phi_t}(x_{\phi_t}, e_i)$ are (by hypothesis) of class C^1, then $S_{\phi_t}: \phi_t(\Omega) \to \mathbb{M}_n$ is a C^1 function. Thus, to complete the proof of part (i) it suffices to recall (4.3) and remark that, by the divergence theorem,

$$\int_{\partial \phi_t(\mathcal{P})} S_{\phi_t}(x_{\phi_t})\nu(x_{\phi_t}) \, d\sigma_{\phi_t}(x_{\phi_t}) = \int_{\phi_t(\mathcal{P})} \text{div } S_{\phi_t}(x_{\phi_t}) \, dx_{\phi_t}.$$

Finally, property (ii) is easily proved using (4.4) and the fact that

$$\int_{\partial \phi_t(\mathscr{P})} x_{\phi_t} \otimes S_{\phi_t}(x_{\phi_t}) v(x_{\phi_t})\, d\sigma_{\phi_t}(x_{\phi_t}) = \int_{\phi_t(\mathscr{P})} x_{\phi_t} \otimes \operatorname{div} S_{\phi_t}(x_{\phi_t})\, dx_{\phi_t}$$

$$+ \int_{\phi_t(\mathscr{P})} S_{\phi_t}(x_{\phi_t})\, dx_{\phi_t};$$

the last equality can be verified without difficulty by once more using the divergence theorem. □

The function $S_{\phi_t}: \phi_t(\Omega) \to \mathbb{M}_n$ is called the *Cauchy stress* at the deformation ϕ_t (related to the motion $t \mapsto \phi_t$). Henceforth, we will assume that the motion and the functions μ_{ϕ_t}, b_{ϕ_t}, and s_{ϕ_t} have the smoothness specified in the statement of Theorem 4.3. Note that if there is a deformation ϕ such that $\phi_t = \phi$ for each t belonging to an interval J, then for each $t \in J$, (4.6) yields

$$\operatorname{div} S_\phi + \mu_\phi b_\phi = 0 \quad \text{in } \phi(\Omega). \tag{4.7}$$

Equations (4.6) and (4.7) are known as the *Cauchy equations* of motion and equilibrium, respectively.

The disadvantage of equations (4.6) and (4.7) is that $\phi_t(\Omega)$ and $\phi(\Omega)$ are, in general, unknown. The introduction of the function

$$S_t: \Omega \to \mathbb{M}_n,$$

related to S_{ϕ_t} by

$$S_t = S_{\phi_t} \circ \phi_t \operatorname{cof} D\phi_t, \tag{4.8}$$

and the function $S: \Omega \to \mathbb{M}_n$ related to S_ϕ by (2.5), allows one to go back to the reference configuration Ω without formal changes in equations (4.6) and (4.7). Indeed, in view of (3.2), it follows from (4.6) and (4.7), after multiplication by $\det D\phi_t$ [respectively $\det D\phi$], that

$$\operatorname{div} S_t + \mu b_{\phi_t} \circ \phi_t = \mu \ddot{\phi}_t \quad \text{in } \Omega, \tag{4.9}$$

$$\operatorname{div} S + \mu b_\phi \circ \phi = 0 \quad \text{in } \Omega. \tag{4.10}$$

S is known as the (first) *Piola–Kirchhoff stress.*

Unlike S_{ϕ_t}, S_t is not symmetric. In view of (4.8), the symmetry of S_{ϕ_t} implies that

$$S_t(D\phi_t)^{\mathrm{T}} \quad \text{is symmetric.} \tag{4.11}$$

Note that if $s_{\phi_t}: \partial \phi_t(\Omega) \to \mathbb{R}^n$ is the density of the measure $\sigma_{\phi_t, \Omega}$ expressing the density of the contact forces exerted on $\phi_t(\Omega)$ by its environment through $\partial \phi_t(\Omega)$, then, for any regular point $\bar{x}_{\phi_t}$ of $\partial \phi_t(\Omega)$, we have (by Cauchy's axiom)

$$s_{\phi_t}(\bar{x}_{\phi_t}, v_{\phi_t}(\bar{x}_{\phi_t})) = s_{\phi_t}(\bar{x}_{\phi_t}). \tag{4.12}$$

For each sufficiently small number $\lambda > 0$ we have $\bar{x}_{\phi_t} - \lambda v_{\phi_t}(\bar{x}_{\phi_t}) \in \Omega$, and hence (in view of Theorem 4.3)

$$s_{\phi_t}(\bar{x}_{\phi_t} - \lambda v_{\phi_t}(\bar{x}_{\phi_t}), v_{\phi_t}(\bar{x}_{\phi_t})) = S_{\phi_t}(\bar{x}_{\phi_t} - \lambda v_{\phi_t}(\bar{x}_{\phi_t}))v_{\phi_t}(\bar{x}_{\phi_t}).$$

The left-hand side of this equality is also defined for $\lambda = 0$. Suppose that the function $\lambda \mapsto s_{\phi_t}(\bar{x}_{\phi_t} - \lambda v_{\phi_t}(\bar{x}_{\phi_t}), v_{\phi_t}(\bar{x}_{\phi_t}))$ is (left-continuous) at $\lambda = 0$:

$$\lim_{\lambda \to 0^-} S_{\phi_t}(\bar{x}_{\phi_t} - \lambda v_{\phi_t}(\bar{x}_{\phi_t}))v_{\phi_t}(\bar{x}_{\phi_t}) = s_{\phi_t}(\bar{x}_{\phi_t}, v_{\phi_t}(\bar{x}_{\phi_t})).$$

Then (4.12) takes the form

$$\lim_{\lambda \to 0^-} S_{\phi_t}(\bar{x}_{\phi_t} - \lambda v_{\phi_t}(\bar{x}_{\phi_t}))v_{\phi_t}(\bar{x}_{\phi_t}) = \sigma_{\phi_t}(\bar{x}_{\phi_t}). \tag{4.13}$$

Now assume that the function $S_{\phi_t}: \phi_t(\Omega) \to \mathbb{M}_n$ has a continuous extension to $\overline{\phi_t(\Omega)}$. Then, in view of (4.13), we have

$$S_{\phi_t} v_{\phi_t} = \sigma_t \tag{4.14}$$

at any regular point of $\partial(\Omega)$. Note that, taking into account (2.3) and (4.8), (4.14) becomes

$$S_t v = \sigma_t \tag{4.15}$$

at any regular point of $\partial\Omega$, with

$$\sigma_t = |(\text{cof } D\phi_t)v| s_{\phi_t} \circ \phi_t.$$

§5. Constitutive Assumptions. Elastic Body

The basic axioms of §3 and §4 concern all bodies and motions, independently of the physical materials of which those bodies are composed. To take into account the influence of the material properties of a body on its mechanical behavior, it is necessary to introduce further assumptions which are usually called *constitutive assumptions*. They are restrictions upon the forces or the motions or both.

There are various types of constitutive assumptions. Following GURTIN [1981], Chapter 6, we distinguish three types of constitutive assumptions:

(i) *Restrictions upon the possible motions the body may undergo* (or, *internal constraints*). The simplest and strongest example of internal constraint consists in the assumption that only rigid motions are possible. Another important example is the assumption that only isochoric motions are possible; this restriction, known as the *incompressibility constraint*, is usually assumed for liquids such as water under normal conditions of flow.

(ii) *Restrictions upon the form of the Cauchy stress.* A typical example is the assumption that the Cauchy stress S_ϕ is a pressure, i.e., $S_\phi = -\pi_\phi I$, with π_ϕ a positive real-valued function defined on $\phi(\Omega)$. Such an assumption is made for most fluids when viscosity is negligible.

(iii) *Equations relating the stress to the motion.* The equation defining an elastic body is, for us, the most important example of this third type of constitutive assumption.

Again, following GURTIN [1981], we give some definitions. Any pair $((\phi_t)_{t \in \mathbb{R}}, (S_{\phi_t})_{t \in \mathbb{R}})$, with $(\phi_t)_{t \in \mathbb{R}}$ a motion and S_{ϕ_t} a C^1 function of $\phi_t(\Omega)$ into Sym_n, will be called a *dynamical process*. This definition is clearly suggested by (Cauchy's) Theorem 4.3. By a *material body* we mean a body with a mass and a set $\mathscr{C}$ of dynamical processes. $\mathscr{C}$ will be called the *constitutive class* of the material body: it is the set of those dynamical processes that satisfy all constitutive assumptions.

We list a few examples of material bodies

(a) *Incompressible material body.* This is the case when $\mathscr{C}$ is the set of those dynamical processes $((\phi_t)_{t \in \mathbb{R}}, (S_{\phi_t})_{t \in \mathbb{R}})$ such that the motion $(\phi_t)_{t \in \mathbb{R}}$ is isochoric (see §2). In view of (2.1), for every motion $t \mapsto \phi_t$ of an incompressible material body, we have, at each t,

$$\det D\phi_t = 1$$

and consequently, by (3.2),

$$\mu_{\phi_t}(\phi_t(x)) = \mu(x).$$

(b) *Ideal fluid.* A material body is called an ideal fluid if μ is constant and the constitutive class is the set of dynamical processes $((\phi_t)_{t \in \mathbb{R}}, (S_{\phi_t})_{t \in \mathbb{R}})$ such that ϕ_t is isochoric for each t and that

$$S_{\phi_t} = -\pi_{\phi_t} I$$

with π_{ϕ_t} a positive real-valued function defined on $\phi_t(\Omega)$.

(c) *Elastic body.* A material body is called an elastic body if its constitutive class $\mathscr{C}$ is defined by a function

$$s: \Omega \times \mathbb{M}_n^+ \to \text{Sym}_n,$$

in the sense that $\mathscr{C}$ is the set of dynamical processes $((\phi_t)_{t \in \mathbb{R}}, (S_{\phi_t})_{t \in \mathbb{R}})$ such that

$$S_{\phi_t}(\phi_t(x)) = s(x, D\phi_t(x)), \qquad \forall (x, t) \in \Omega \times \mathbb{R}. \tag{5.1}$$

From (5.1) it follows that the function s is completely determined by the knowledge of the Cauchy stress S_{ϕ_t} corresponding to the motions $t \mapsto \phi_t$, ϕ_t being an affine function independent of t. The function s is called the *response function* of the material. Note that (5.1) implies that, for each $(x, t) \in \Omega \times \mathbb{R}$, the value of the Cauchy stress at the point $\phi_t(x)$ depends only on the gradient $D\phi_t$ of ϕ_t (and, therefore, it is independent, in particular, of the past deformation history).

It is useful to remark that if $a: \bar{\Omega} \times \mathbb{M}_n^+ \to \mathbb{M}_n^+$ is the function defined by putting

$$a(x, Z) = s(x, Z) \operatorname{cof} Z, \qquad \forall (x, Z) \in \bar{\Omega} \times \mathbb{M}_n^+, \tag{5.2}$$

then from (2.5) and (5.1) it immediately follows that

$$S_t(x) = a(x, D\phi_t(x)), \qquad \forall (x, t) \in \Omega \times \mathbb{R}. \tag{5.3}$$

Observe that (4.11) yields

$$a(x, D\phi_t(x))(D\phi_t(x))^{\mathrm{T}} \in \mathrm{Sym}_n, \qquad \forall (x, t) \in \Omega \times \mathbb{R}.$$

This holds, in particular, for $\phi_t(x) = Zx$, $\forall (x, t) \in \Omega \times \mathbb{R}$ where Z is an element of $\mathbb{M}_n^+$; thus, we have

$$a(x, Z)Z^{\mathrm{T}} \in \mathrm{Sym}_n, \qquad \forall (x, Z) \in \Omega \times \mathbb{M}_n^+. \tag{5.4}$$

An elastic body is said to be *hyperelastic* if there is a function $w: \Omega \times \mathbb{M}_n^+ \to \mathbb{R}$ such that, for $i, j = 1, \ldots, n$,

$$a_{ij}(x, Z) = D_{Z_{ij}} w(x, Z), \qquad \forall (x, Z) \in \Omega \times \mathbb{M}_n^+,$$

where $a_{ij}(x, Z)$ is the (i, j)th element of the matrix $a(x, Z)$, and w is called the *stored-energy function*.

§6. Frame-Indifference of the Material Response

By an (orientation-preserving) *change in observer* we mean a bijection $(y, t) \mapsto (y^*, t^*)$ of $\mathbb{R}^n \times \mathbb{R}$ onto itself such that $y \mapsto y^*$ is an orientation-preserving isometry on $\mathbb{R}^n$ and $t \mapsto t^*$ is an order-preserving isometry on $\mathbb{R}$. As is well known (see NOLL [1964]), such a function must have the form

$$\begin{cases} t^* = t + \alpha, \\ y^* = \tau_t + R_t y, \end{cases} \tag{6.1}$$

with $\alpha \in \mathbb{R}$ independent of (y, t), $\tau_t \in \mathbb{R}^n$ independent of y, and $R_t \in \mathbb{O}_n^+$ independent of y.

Two *motions* $t \mapsto \phi_t$ and $t \mapsto \phi_t^*$ will be said to be *related by a change in observer* if

$$\phi_{t+\alpha}^*(x) = \tau_t + R_t \phi_t(x), \qquad \forall (x, t) \in \Omega \times \mathbb{R}, \tag{6.2}$$

with $\alpha \in \mathbb{R}$ independent of (x, t), $\tau_t \in \mathbb{R}^n$ independent of x, and $R_t \in \mathbb{O}_n^+$ independent of x. Note that if $\rho_t: \mathbb{R}^n \to \mathbb{R}^n$ is the rigid deformation of $\mathbb{R}^n$ defined by putting

$$\rho_t(y) = \tau_t + R_t y, \qquad \forall y \in \mathbb{R}^n,$$

then (6.2) can be written in the form

$$\phi_{t+\alpha}^* = \rho_t \circ \phi_t.$$

Let the motions $t \mapsto \phi_t$ and $t \mapsto \phi_t^*$ be related by (6.2), let s_{ϕ_t} and $s_{\phi_t^*}$ be the corresponding densities of surface forces at time t, and let S_{ϕ_t} and $S_{\phi_t^*}$ be the Cauchy stresses associated with s_{ϕ_t} and $s_{\phi_t^*}$. For each $e \in \mathbb{R}^n$

with $|e| = 1$, and for each time t we put

$$e_t^* = R_t e.$$

We have, for every $(x, t) \in \Omega \times \mathbb{R}$ and for every $e \in \mathbb{R}^n$ with $|e| = 1$,

$$\begin{cases} s_{\phi_t}(\phi_t(x), e) = S_{\phi_t}(\phi_t(x))e, \\ s_{\phi_{t+\alpha}^*}(\phi_{t+\alpha}^*(x), e_t^*) = S_{\phi_{t+\alpha}^*}(\phi_{t+\alpha}^*(x))e_t^*. \end{cases} \tag{6.3}$$

Reflecting on the physical meaning of $s_{\phi_t}(\phi_t(x), e)$ and $s_{\phi_{t+\alpha}^*}(\phi_{t+\alpha}^*(x), e_t^*)$, and recalling (6.2), one should expect that

$$s_{\phi_t}(\phi_t(x), e) = R_t^{\mathrm{T}} s_{\phi_{t+\alpha}^*}(\phi_{t+\alpha}^*(x), e_t^*). \tag{6.4}$$

Note that, in view of (6.3), (6.4) holds for every $(x, t) \in \Omega \times \mathbb{R}$ and every $e \in \mathbb{R}^n$ with $|e| = 1$, if and only if

$$S_{\phi_{t+\alpha}^*}(\phi_{t+\alpha}^*(x)) = R_t S_{\phi_t}(\phi_t(x)) R_t^{\mathrm{T}}, \qquad \forall(x, t) \in \Omega \times \mathbb{R}. \tag{6.5}$$

This discussion lead us, following GURTIN [1981], p. 144, to the definition of dynamical process related by a change in observer. We will say that two *dynamical processes* $((\phi_t)_{t \in \mathbb{R}}, (S_{\phi_t})_{t \in \mathbb{R}})$ and $((\phi_t^*)_{t \in \mathbb{R}}, (S_{\phi_t^*})_{t \in \mathbb{R}})$ are *related by a change in observer* if (6.2) and (6.5) hold with $\alpha \in \mathbb{R}$ independent of (x, t), $\tau_t \in \mathbb{R}^n$ independent of x, and $R_t \in \mathbb{O}_n^+$ independent of x. Moreover, we will say that the *response of a material body* is *independent of the observer* (or *frame-indifferent*) if its constitutive class $\mathscr{C}$ contains every dynamical process related to any element of $\mathscr{C}$ by a change in observer.

It is easily seen that the response of an ideal fluid (cf. §5) is independent of the observer. As for the elastic bodies we emphasize the following remark.

Remark 6.1. *The response of an elastic body is independent of the observer if and only if*

$$s(x, RZ) = Rs(x, Z)R^{\mathrm{T}}, \qquad \forall(x, Z, R) \in \Omega \times \mathbb{M}_n^+ \times \mathbb{O}_n^+, \tag{6.6}$$

where s is the response function of the body.

Proof. Suppose that the response of an elastic body is independent of the observer, i.e., suppose that ((5.1), (6.2), (6.5)) imply

$$S_{\phi_{t+\alpha}^*}(\phi_{t+\alpha}^*(x)) = s(x, D\phi_{t+\alpha}^*(x)), \qquad \forall(x, t) \in \Omega \times \mathbb{R}. \tag{6.7}$$

Then, for any pair $((\phi_t)_{t \in \mathbb{R}}, (\phi_t^*)_{t \in \mathbb{R}})$ of motions related by (6.2) we have

$$s(x, D\phi_{t+\alpha}^*(x)) = R_t s(x, D\phi_t(x)) R_t^{\mathrm{T}}, \qquad \forall(x, t) \in \Omega \times \mathbb{R}. \tag{6.8}$$

Therefore, observing that

$$D\phi_{t+\alpha}^*(x) = R_t D\phi_t(x), \qquad \forall(x, t) \in \Omega \times \mathbb{R},$$

we obtain, for every motion $t \mapsto \phi_t$ and every function $t \mapsto R_t$ of $\mathbb{R}$ into $\mathbb{O}_n^+$,

$$s(x, R_t D\phi_t(x)) = R_t s(x, D\phi_t(x)) R_t^{\mathrm{T}}, \qquad \forall(x, t) \in \Omega \times \mathbb{R}. \tag{6.9}$$

In particular, (6.9) holds when $\phi_t(x) = Zx$, $\forall(x, t) \in \Omega \times \mathbb{R}$ and $R_t = R$, $\forall t \in \mathbb{R}$, with (Z, R) any element of $\mathbb{M}_n^+ \times \mathbb{O}_n^+$; therefore (6.6) holds.

Conversely, suppose that (6.6) is satisfied. Then (6.9) holds for every motion $t \mapsto \phi_t$ and every function $t \mapsto R_t$ of $\mathbb{R}$ into $\mathbb{O}_n^+$. Consequently, (6.8) holds for every motion $t \mapsto \phi_t^*$ related to the motion $t \mapsto \phi_t$ by (6.2). Hence ((5.1), (6.2), (6.5)) imply (6.7); thus, the response of the elastic body is independent of the observer. $\square$

Note that, since evidently

$$\operatorname{cof}(RZ) = R \operatorname{cof} Z, \qquad \forall(R, Z) \in \mathbb{O}_n^+ \times \mathbb{M}_n^+,$$

from (5.2), it easily follows that

$$a(x, RZ) - Ra(x, Z) = (s(x, RZ)R - Rs(x, Z)) \operatorname{cof} Z.$$

Hence (6.6) and (6.9) are equivalent.

Then Remark 6.1 gives

Remark 6.2. *The response of an elastic body is independent of the observer if and only if*

$$a(x, RZ) = Ra(x, Z), \qquad \forall(x, R, Z) \in \Omega \times \mathbb{O}_n^+ \times \mathbb{M}_n^+, \tag{6.10}$$

where a is the function related to s by (5.2).

Another interesting consequence of Remark 6.1 is

Remark 6.3. *If the response of an elastic body is independent of the observer, then the response function $s: \Omega \times \mathbb{M}_n^+ \to \operatorname{Sym}_n$ is completely determined by its restriction to $\Omega \times \{Z \in \operatorname{Sym}_n: Z$ is positive definite$\}$.*

In fact, from (6.6) it immediately follows that

$$s(x, Z) = R_Z s(x, U_Z) R_Z^{\mathsf{T}},$$

where R_Z and U_Z are the elements of $\mathbb{O}_n^+$ and Sym_n, respectively, such that U_Z is positive definite and†

$$Z = R_Z U_Z.$$

We will always assume that the response of an elastic body is independent of the observer.

---

† Since $\det Z > 0$, the matrix $Z^{\mathsf{T}} Z$ is symmetric and positive definite; therefore, there exists a unique element U_Z of Sym_n such that U_Z is positive definite and

$$U_Z^2 = Z^{\mathsf{T}} Z.$$

By putting $R = Z U_Z^{\frac{1}{2}}$, it is easily verified that $R_Z \in \mathbb{O}_n^+$; moreover, we have $Z = R_Z U_Z$. Note that, if $Z = R_Z U_Z$ with $R_Z \in \mathbb{O}_n^+$, then the factorization $Z = R_Z U_Z$, with $R_Z \in \mathbb{O}_n^+$ and U_Z symmetric and positive definite, is unique.

CHAPTER II

Composition Operators in Sobolev and Schauder Spaces. Theorems on Continuity, Differentiability, and Analyticity

This chapter lays the basis for our next developments. It mainly concerns the study on continuity, differentiability, and analyticity of composition operators (i.e., operators of the type $\sigma \mapsto F(\sigma)$ where σ is an $\mathbb{R}^N$-valued function defined on Ω, and $F(\sigma)$ is the real-valued function defined on Ω by setting $F(\sigma)(x) = f(x, \sigma(x))$, $\forall x \in \Omega$ with $f: \Omega \times \mathbb{R}^N \to \mathbb{R}$ a given function). Various theorems are established in Sobolev spaces and in Schauder spaces.

In the case of Sobolev spaces, a property of (pointwise) multiplication, proved in §2, plays a central role in obtaining our results (see Lemma 2.1 and Corollary 2.3). Theorems 4.5 and 4.6 concern the differentiability of operators of the type $(f, \sigma) \mapsto F(\sigma)$, where the meaning of the symbols is as given previously. The analyticity of $\sigma \mapsto F(\sigma)$ from $W^{m+r,p}(\Omega, \mathbb{R}^N)$ into $W^{m,p}(\Omega)$ with $m \geq 1$, $r \geq 0$, $(m + r)p > n$, and from $C^{m,\lambda}(\overline{\Omega}, \mathbb{R}^N)$ into $C^{m,\lambda}(\overline{\Omega})$ with $m \geq 0$, is proved assuming that $f \in C^{\infty}(\overline{\Omega} \times \mathbb{R}^N)$ and that the functions $(x, y) \mapsto D_x^{\alpha} f(x, y)$, $|\alpha| \leq m$, are analytic in y uniformly with respect to x (cf. §5).

We complete the chapter by proving a theorem on failure of differentiability (Theorem 6.1), which will lead us, in §4, to define the admissibility of a linearization with respect to a pair of Banach spaces, one for solutions and one for data.

§1. Some Facts About Sobolev and Schauder Spaces

As we have said in Chapter I, §1, Ω will always denote a nonempty, bounded, open subset of $\mathbb{R}^n$, except in the statement of Lemma 2.1′ where Ω does not need to be bounded. Let m be an integer ≥ 0 and let

p, λ be real numbers with $p \geq 1$ and $0 < \lambda \leq 1$. $C^m(\Omega)$ denotes the set of C^m real-valued functions in Ω (i.e., real-valued functions defined on Ω and having continuous derivatives up to order m in Ω). $C^\infty(\Omega)$ denotes the set of C^∞ real-valued functions in Ω (i.e., real-valued functions defined in Ω and having continuous derivatives of all orders in Ω). $\mathscr{D}(\Omega)$ denotes the (real, locally convex, linear) space of those elements of $C^\infty(\Omega)$ which have compact support, equipped with Schwartz's inductive limit topology. The dual $\mathscr{D}'(\Omega)$ of $\mathscr{D}(\Omega)$ is the space of (real) distributions on Ω.

A function $v: B \to \mathbb{R}$, with B a subset of $\mathbb{R}^n$, is said to be *of class C^m* [respectively C^∞] if for each $x \in B$ there is an open neighborhood V_x of x in $\mathbb{R}^n$ and a C^m [respectively C^∞] function $v_x: V_x \to \mathbb{R}$ such that $v_x|_{V_x \cap B} = v|_{V_x \cap B}$. Then we also say that v is a C^m function [respectively C^∞ function] in B. It is well known that if B is a closed subset of an open subset A of $\mathbb{R}^n$ and $v: B \to \mathbb{R}$ is of class C^m [respectively C^∞], then there is a function of class C^m [respectively C^∞] in A whose restriction to B coincides with v (see, e.g., DIEUDONNÉ [1970], p. 18). We will denote by $C^m(B)$ [respectively $C^\infty(B)$] the set of real-valued functions defined on B and of class C^m [respectively C^∞]. It is also known that a function $v: \bar\Omega \to \mathbb{R}$ is of class C^m [respectively C^∞] if and only if $v \in C^m(\Omega)$ [respectively $C^\infty(\Omega)$] and for $|\alpha| \leq m$ [respectively for all α] $D^\alpha v$ has a continuous extension to $\bar\Omega$ (see, again, DIEUDONNÉ [1970], p. 19). Here α is the multi-index $(\alpha_1, \ldots, \alpha_n)$ with $\alpha_1, \ldots, \alpha_n$ nonnegative integers, $|\alpha| = \sum_{j=1}^n \alpha_j$, and $D^\alpha = D_1^{\alpha_1} \ldots D_n^{\alpha_n}$. The (real) linear space $C^m(\bar\Omega)$ is a Banach space with respect to the norm $|\cdot|_m$ defined by

$$|v|_m = \sum_{|\alpha| \leq m} \sup_{x \in \Omega} |D^\alpha v(x)|.$$

The closure of $\mathscr{D}(\Omega)$ in $C^m(\bar\Omega)$ will be denoted by $C_0^m(\bar\Omega)$. Note that $C_0^m(\bar\Omega) = \{v \in C^m(\bar\Omega): D^\alpha v|_{\partial\Omega} = 0 \text{ for } |\alpha| \leq m\}$.

The Schauder space $C^{m,\lambda}(\bar\Omega)$ denotes the (real Banach) space of functions $v \in C^m(\bar\Omega)$ such that, for $|\alpha| = m$, $D^\alpha v$ satisfies a Hölder condition of exponent λ in Ω (i.e., the subset $\{|D^\alpha v(x') - D^\alpha v(x'')|/|x' - x''|^\lambda: x', x'' \in \Omega, x' \neq x''\}$ of $\mathbb{R}$ is bounded), with the norm $\|\cdot\|_{m,\lambda}$ defined by

$$\|v\|_{m,\lambda} = |v|_m + \sum_{|\alpha|=m} \sup_{\substack{x', x'' \in \Omega \\ x' \neq x''}} \frac{|D^\alpha v(x') - D^\alpha v(x'')|}{|x' - x''|^\lambda}.$$

Those bounded, open subsets Ω of $\mathbb{R}^n$ such that

$$\sup_{\substack{x', x'' \in \Omega \\ x' \neq x''}} \frac{|v(x') - v(x'')|}{|x' - x''|} \leq c_\Omega \sum_{|\alpha|=1} \sup_{x \in \Omega} |D^\alpha v(x)|, \qquad \forall v \in C^1(\bar\Omega), \qquad (1.1)$$

with c_Ω a number > 0 independent of v, will have a special interest for us.

We easily realize that (1.1) is true, e.g., if Ω has the property that any pair of points $x', x'' \in \Omega$ can be joined by a rectifiable arc in Ω having length not exceeding some fixed multiple of $|x' - x''|$, or, more generally, if Ω is a finite union of open subsets of $\mathbb{R}^n$ having the above property, and such that the distance between any pair of them is > 0. If (1.1) holds then, evidently,

$$C^{m+1}(\bar{\Omega}) \subseteq C^{m,1}(\bar{\Omega}) \subseteq C^{m,\lambda}(\bar{\Omega}).$$

An important consequence of (1.1), the proof of which is easy, is that the (pointwise) multiplication $(u, v) \mapsto uv$ is a continuous mapping from $C^{m,\lambda}(\bar{\Omega}) \times C^{m,\lambda}(\bar{\Omega})$ into $C^{m,\lambda}(\bar{\Omega})$, i.e., that

$$u, v \in C^{m,\lambda}(\bar{\Omega}) \;\Rightarrow\; uv \in C^{m,\lambda}(\bar{\Omega}), \qquad \|uv\|_{m,\lambda} \leqq c_{m,\lambda} \|u\|_{m,\lambda} \|v\|_{m,\lambda}, \quad (1.2)$$

where $c_{m,\lambda}$ is a number > 0 independent of u and v. Thus $C^{m,\lambda}(\bar{\Omega})$ is a Banach algebra, provided (1.1) holds.

We say that Ω is of class C^m [respectively $C^{m,\lambda}$], with $m \geqq 1$, if $\bar{\Omega}$ is a submanifold with boundary $\mathbb{R}^n$ of class C^m [respectively $C^{m,\lambda}$], i.e., if for each $x \in \partial\Omega$ there is an open neighborhood U_x of x in $\mathbb{R}^n$ and a homeomorphism τ_x of U_x onto the ball $\{\xi \in \mathbb{R}^n : |\xi| < 1\}$ of $\mathbb{R}^n$, such that $\tau_x(\bar{\Omega} \cap U_x) = \{\xi \in \mathbb{R}^n : |\xi| < 1, \xi_n \geqq 0\}$ and such that $\tau_x \in (C^m(\bar{U}_x))^n$, $\tau_x^{-1} \in (C^m(\{\xi \in \mathbb{R}^n : |\xi| < 1\}))^n$ [respectively $\tau_x \in (C^{m,\lambda}(\bar{U}_x))^n$, $\tau_x^{-1} \in (C^{m,\lambda}(\{\xi \in \mathbb{R}^n : |\xi| < 1\}))^n$]. If Ω is of class $C^{m,\lambda}$, we denote by $C^{m,\lambda}(\partial\Omega)$ the (real linear) space of functions $g: \partial\Omega \to \mathbb{R}$ such that $g \circ \tau_x^{-1} \in C^{m,\lambda}(B)$ where $B = \{\xi \in \mathbb{R}^n : |\xi| < 1, \xi_n = 0\}$. On $C^{m,\lambda}(\partial\Omega)$ we consider the norm $\|\cdot\|_{m,\lambda,\partial\Omega}$ defined by

$$\|g\|_{m,\lambda,\partial\Omega} = \sup_{x \in \partial\Omega} \|g \circ \tau_x^{-1}\|_{C^{m,\lambda}(B)}$$

for a fixed choice of the family $(\tau_x)_{x \in \partial\Omega}$. It is easy to check that different families give equivalent norms.

Remark 1.1. If Ω is of class C^1, then (1.1) holds and hence $C^{m,\lambda}(\bar{\Omega})$ is a Banach algebra.

Proof. Let Ω be of class C^1. Then Ω is a finite union of open subsets of $\mathbb{R}^n$ which are either convex or diffeomorphic to a convex subset of $\mathbb{R}^n$ (open semiball). Hence Ω has a finite number of connected components, each of which is a finite union of open subsets of $\mathbb{R}^n$ which are either convex or diffeomorphic to a convex subset of $\mathbb{R}^n$. Moreover, the distance between any two connected components of Ω is > 0. It is easily seen that if x', x'' are two points of a connected component K of Ω and $l(x', x'')$ is the infimum bound of the lenghts of rectifiable arcs in K joining x' and x'', then $l(x', x'') \leqq c|x' - x''|$ where c is a number > 0 independent of x' and x''. Thus, by the discussion following (1.1), we can conclude that (1.1) holds. $\square$

Remark 1.2. Let Ω be such that (1.1) holds and let $\phi: \bar{\Omega} \to \mathbb{R}^n$ be a function of class C^1. If for a rigid deformation ϕ_0 of $\bar{\Omega}$

$$\sum_{|\alpha|=1} \sup_{x \in \Omega} |D^\alpha(\phi - \phi_0)(x)| < \frac{1}{c_\Omega} \tag{1.3}$$

then ϕ is one-to-one.

Proof. It suffices to observe that, combining (1.3) with (1.1) written for $v = \phi - \phi_0$, we have

$$|\phi(x') - \phi(x'') - R_0(x' - x'')| < |x' - x''|$$

whenever $x' \neq x''$, where R_0 ($\in \mathbb{O}_n^+$) is the value of the gradient of ϕ_0 (at any $x \in \bar{\Omega}$). $\quad \square$

$L^p(\Omega)$ denotes the (real Banach) space of (classes of) real-valued, measurable functions v defined on Ω such that $|v|^p$ is Lebesgue-integrable, while $L^\infty(\Omega)$ denotes the (real Banach) space of (classes of) real-valued, measurable functions that are bounded almost everywhere. $W^{m,p}(\Omega)$ [respectively $W^{m,\infty}(\Omega)$] denotes the (real Banach) space of elements v of $L^p(\Omega)$ [respectively $L^\infty(\Omega)$] such that, for $|\alpha| \leq m$, the weak derivative $D^\alpha v$ belongs to $L^p(\Omega)$ [respectively $L^\infty(\Omega)$] equipped with the norm $\|\cdot\|_{m,p}$ defined by

$$\|v\|_{m,p} = \left(\sum_{|\alpha| \leq m} \|D^\alpha v\|_{0,p}^p \right)^{1/p} \quad \left[\text{respectively } \|v\|_{m,\infty} = \sup_{|\alpha| \leq m} \|D^\alpha v\|_{0,\infty} \right],$$

where $\|\cdot\|_{0,p}$ is the usual norm of $L^p(\Omega)$ and $\|v\|_{0,\infty} = \text{ess sup}_{x \in \Omega} |v(x)|$. By definition, $W_0^{m,p}(\Omega)$ is the closure of $\mathscr{D}(\Omega)$ in $W^{m,p}(\Omega)$. $W^{-m,p}(\Omega)$ denotes the real, linear space of distributions on Ω of the type $\sum_{|\alpha| \leq m} D^\alpha f_\alpha$ with $f_\alpha \in L^p(\Omega)$. If X is a real topological linear space (in particular, a real normed space) we indicate by X' its dual, i.e., the real linear space of continuous, linear functions of X into $\mathbb{R}$.

We quickly recall some facts about the Sobolev spaces $W^{m,p}(\Omega)$, the proof of which can be found in ADAMS [1975] and NEČAS [1967].

The linear operator which maps any element of $(W_0^{m,p}(\Omega))'$ onto its restriction to $\mathscr{D}(\Omega)$ is a bijection of $(W_0^{m,p}(\Omega))'$ onto $W^{-m,p'}(\Omega)$, where p' is related to p by $(1/p) + (1/p') = 1$ provided $p > 1$ and $p' = \infty$ if $p = 1$.

$\mathscr{D}(\mathbb{R}^n)$ is dense in $W^{m,p}(\Omega)$; $C^\infty(\Omega) \cap W^{m,p}(\Omega)$ is dense in $W^{m,p}(\Omega)$, where $C^\infty(\Omega) \cap W^{m,p}(\Omega)$ denotes the set of elements of $W^{m,p}(\Omega)$ that contain a function belonging to $C^\infty(\Omega)$.

If Ω is of class C^m and v is an oriented unit normal to $\partial\Omega$, then the following implications are true:

$$\begin{cases} v \in W_0^{m,p}(\Omega) \cap C^{m-1}(\bar{\Omega}) \;\Rightarrow\; \dfrac{\partial^j v}{\partial v^j} = 0, \quad 0 \leq j \leq m - 1 \quad \text{on } \partial\Omega, \\[4mm] v \in C^m(\bar{\Omega}), \quad \dfrac{\partial^j v}{\partial v^j} = 0, \quad 0 \leq j \leq m - 1 \quad \text{on } \partial\Omega \;\Rightarrow\; v \in W_0^{m,p}(\Omega); \end{cases}$$

for this reason when $v \in W_0^{m,p}(\Omega)$ v is said to verify, in a generalized sense, the boundary conditions $\partial^j v / \partial v^j = 0$, $j = 0, \ldots, m - 1$, on $\partial\Omega$.

The following (*Poincaré's inequality*)

$$\|v\|_{0,p} \leq c \sum_{|\alpha|=1} \|D^\alpha v\|_{0,p}$$

holds for all $v \in W_0^{1,p}(\Omega)$. Ω is said to have the *cone property* if there are numbers $\alpha > 0$ and $h > 0$ such that for each $x \in \Omega$ we can construct a right spherical cone with vertex x, opening α, and height h such that it lies in Ω. It is easily seen that Ω has the cone property, provided it is of class C^1.

From the well-known *Sobolev embedding theorem* (see ADAMS [1975], Theorem 5.23) it follows that if Ω has the cone property and m, k are integers with $k \geq 1$ and $m \geq 0$, then the following continuous embeddings hold:

$$W^{m+k,p}(\Omega) \subset W^{m,q}(\Omega), \quad \forall q \in \mathbb{R} \quad \text{such that } 1 \leq q \leq np/(n - kp),$$

$$\text{provided } kp < n; \quad (1.4)$$

$$W^{m+k,p}(\Omega) \subset W^{m,q}(\Omega), \quad \forall q \in \mathbb{R} \quad \text{such that } q \geq 1, \quad \text{provided } kp = n; \quad (1.5)$$

$$W^{m+k,p}(\Omega) \subset C_B^m(\Omega), \quad \text{provided } kp > n; \quad (1.6)$$

where $C_B^m(\Omega) = \{v \in C^m(\Omega): D^\alpha v$ is bounded for $|\alpha| \leq m\}$ with the norm $|\cdot|_m$ and where the inclusion $W^{m+k,p}(\Omega) \subset C_B^m(\Omega)$ should be interpreted in the sense that each element of $W^{m+k,p}(\Omega)$ is an equivalence class containing a function belonging to $C_B^m(\Omega)$. Moreover, the embeddings (1.5) and (1.6) are compact† and the embedding (1.4) is compact for all $q \in \mathbb{R}$ with $1 \leq q < np/(n - kp)$. Finally, when Ω is of class $C^{0,1}$ we have the following compact embedding:

$$W^{m+k,p}(\Omega) \subset C^m(\bar{\Omega}), \quad \text{provided } kp > n;$$

$$W^{m+k,p}(\Omega) \subset C^{m,\lambda}(\bar{\Omega}) \quad \text{for } 0 < \lambda < k - (n/p), \quad \text{provided } kp > n \geq (k - 1)p.$$

We now recall a result known as *Lions's lemma* (see, e.g., LIONS & MAGENES [1972]). Let $(X_1, \|\cdot\|_1)$, $(X_2, \|\cdot\|_2)$, $(X_3, \|\cdot\|_3)$ be Banach spaces with $X_1 \subseteq X_2 \subseteq X_3$; if the embedding $X_1 \to X_2$ is compact and the embedding $X_2 \to X_3$ is continuous, then for each number $\varepsilon > 0$ there is a number $c(\varepsilon) > 0$ such that for all $x \in X_1$,

$$\|x\|_2 \leq \|x\|_1 + c(\varepsilon)\|x\|_3.$$

Without claiming to make it plausible, we now give a definition of the spaces $W^{s,p}(\mathbb{R}^n)$ with $s \in \mathbb{R}$.

Let $p > 1$. If $\vartheta \in \mathbb{R}$, $0 < \vartheta < 1$, we denote by $W^{\vartheta,p}(\mathbb{R}^n)$ the (real, linear) space consisting of those $v \in L^p(\mathbb{R}^n)$ such that $\|v\|_{\vartheta,p} < +\infty$, where

$$\|v\|_{\vartheta,p} = \left(\int_{\mathbb{R}^n} |v|^p \, dx + \int_{\mathbb{R}^n} \int_{\mathbb{R}^n} \frac{|v(x') - v(x'')|^p}{|x' - x''|^{n+\vartheta p}} \, dx' \, dx'' \right)^{1/p}.$$

$W^{\vartheta,p}(\mathbb{R}^n)$ is a (real) Banach space with respect to the norm $\|\cdot\|_{\vartheta,p}$.

---

† If X, Y are Banach spaces a linear mapping $f: X \to Y$ is said to be *compact* provided, for any relatively compact subset B of X, $f(B)$ is a bounded subset of Y. It is not difficult to recognize that if $f: X \to Y$ is compact and $(x_k)_{k \in \mathbb{N}}$ is a sequence in X which converges weakly to 0, then the sequence $(f(x_k))_{k \in \mathbb{N}}$ converges to 0 in Y.

For any nonnegative real number s we put

$$W^{s,p}(\mathbb{R}^n) = \{v \in W^{[s],p}(\mathbb{R}^n): D^\alpha v \in W^{\vartheta,p}(\mathbb{R}^n) \text{ for } |\alpha| = s\},$$

where $[s]$ and ϑ are the numbers defined by the following properties: $[s]$ is an integer ≥ 0, $\vartheta \in [0, 1[$, and $s = [s] + \vartheta$. On $W^{s,p}(\mathbb{R}^n)$ we consider the norm $\|\cdot\|_{s,p,\mathbb{R}^n}$ defined by

$$\|v\|_{s,p,\mathbb{R}^n}$$

$$= \left(\sum_{|\alpha| \leq [s]} \int_{\mathbb{R}^n} |D^\alpha v(x)|^p \, dx + \sum_{|\alpha|=[s]} \int_{\mathbb{R}^n} \int_{\mathbb{R}^n} \frac{|D^\alpha v(x') - D^\alpha v(x'')|^p}{|x' - x''|^{n+\vartheta p}} \, dx' \, dx'' \right)^{1/p}.$$

With respect to such a norm $W^{s,p}(\mathbb{R}^n)$ is a (real) Banach space. Finally, if s is a negative real number, we denote by $W^{s,p}(\mathbb{R}^n)$ the dual of $W^{-s,p'}(\mathbb{R}^n)$.

Let us now define $W^{s,p}(\partial\Omega)$ when Ω is of class C^1. As Ω is bounded and of class C^1, there is an open covering $(U_j)_{j=1,\ldots,r}$ of $\partial\Omega$ and, for $j = 1, \ldots, r$, a C^1 diffeomorphism τ_j of U_j onto the ball $\{\xi \in \mathbb{R}^n: |\xi| < 1\}$ such that $\tau_j(\bar{\Omega} \cap U_j) = \{\xi \in \mathbb{R}^n: |\xi| < 1, \xi_n \geq 0\}$. Let $(\varphi_j)_{j=1,\ldots,r}$ be a partition of unity for $\partial\Omega$ subordinated to the covering $(U_j)_{j=1,\ldots,r}$; namely, let $\varphi_j \in \mathscr{D}(\mathbb{R}^n)$ be such that $\sum_{j=1}^r \varphi_j(x) = 1$, $\forall x \in \partial\Omega$, the support of φ_j is contained in U_j and $\varphi_j \geq 0$. For $s \in \mathbb{R}$, $s \geq 0$ and $p \in \mathbb{R}$, $p > 1$ we consider the (real) linear space consisting of those functions $v: \partial\Omega \to \mathbb{R}$ such that $\Phi_j(v) \in W^{s,p}(\mathbb{R}^{n-1})$, $\forall j = 1, \ldots, r$ where $\Phi_j(v)$ is the real-valued function defined on $\mathbb{R}^{n-1}$ by putting

$$\Phi_j(v)(\xi_1, \ldots, \xi_{n-1}) = \begin{cases} (\varphi_j v)(\tau_j^{-1}(\xi_1, \ldots, \xi_{n-1}, 0)) & \text{if } \xi_1^2 + \cdots + \xi_{n-1}^2 < 1, \\ 0 & \text{if } \xi_1^2 + \cdots + \xi_{n-1}^2 \geq 1. \end{cases}$$

The quotient of this linear space with respect to the equivalence relation, defined by agreeing that v is equivalent to u if and only if $\Phi_j(v) = \Phi_j(u)$, $\forall j = 1, \ldots, r$, will be denoted by $W^{s,p}(\partial\Omega)$. $W^{s,p}(\partial\Omega)$ is a Banach space with respect to the norm $\|\cdot\|_{s,p,\partial\Omega}$ defined by putting

$$\|\hat{v}\|_{s,p,\partial\Omega} = \left(\sum_{j=1}^r \|\Phi_j(v)\|_{s,p,\mathbb{R}^{n-1}}^p \right)^{1/p},$$

where $\hat{v}$ denotes the equivalence class containing v. We can verify that the linear space $W^{s,p}(\partial\Omega)$ and the topology defined on it by the norm $\|\cdot\|_{s,p,\partial\Omega}$ does not depend either on the particular atlas $(U_j, \tau_j)_{j=1,\ldots,r}$ of $\partial\Omega$ or on the partition of unity $(\varphi_j)_{j=1,\ldots,r}$.

The convenience of the introduction of spaces $W^{s,p}(\partial\Omega)$ is mostly justified by the following *trace theorem* (see ADAMS [1975], p. 216).

Let Ω be of class C^m ($m \geq 1$) and, for any $v \in C^\infty(\bar{\Omega})$, let

$$\gamma(v) = (\gamma_0(v), \ldots, \gamma_{m-1}(v)),$$

where $\gamma_0(v) = v|_{\partial\Omega}$ and $\gamma_j(v) = \partial^j v/\partial\nu^j$, $j = 1, \ldots, m - 1$. The mapping

$v \mapsto \gamma(v)$ *extends by continuity to an open, continuous, linear mapping of* $W^{m,p}(\Omega)$ *onto*

$$\prod_{j=0}^{m-1} W^{m-j-1/p,p}(\partial\Omega).$$

We conclude this section by noting that if r is any integer ≥ 1 and $\mathcal{F}(X)$ is a set of real-valued functions defined on the set X, we will denote by $\mathcal{F}(X, \mathbb{R}^r)$ the set of $\mathbb{R}^r$-valued functions defined on X whose scalar components belong to $\mathcal{F}(X)$. Likewise, if $\mathcal{G}(\Omega)$ is a subset of $\mathcal{D}'(\Omega)$, then $\mathcal{G}(\Omega, \mathbb{R}^r)$ will denote the set of $\mathbb{R}^r$-valued distributions in Ω whose scalar components belong to $\mathcal{G}(\Omega)$.

Moreover, if v belongs, respectively, to $W^{m,p}(\Omega, \mathbb{R}^r)$, to $W^{m,s}(\partial\Omega, \mathbb{R}^r)$, to $C^{m,\lambda}(\bar\Omega, \mathbb{R}^r)$, and to $C^{m,\lambda}(\partial\Omega, \mathbb{R}^r)$, we will put $\|v\|_{m,p} = \sum_{i=1}^{r} \|v_i\|_{m,p}$, $\|v\|_{m,s,\partial\Omega} = \sum_{i=1}^{r} \|v_i\|_{m,s,\partial\Omega}$, $\|v\|_{m,\lambda} = \sum_{i=1}^{r} \|v_i\|_{m,\lambda}$, and $\|v\|_{m,\lambda,\partial\Omega} = \sum_{i=1}^{r} \|v_i\|_{m,\lambda,\partial\Omega}$, respectively, where v_i is the ith scalar component of v (with respect to the canonical base of $\mathbb{R}^r$).

Finally, if $\mathcal{X}$ is a Banach space continuously embedded in $C^1(\bar\Omega, \mathbb{R}^r)$, we will put

$$\mathcal{X}^+ = \{\phi \in \mathcal{X}: \det D\phi > 0\}.$$

§2. A Property of Multiplication in Sobolev Spaces

The following lemma is a crucial tool in proving the properties of continuity, differentiability, and analyticity of composition operators (Nemytsky operators) acting between Sobolev spaces. It concerns the conditions on the numbers p, q, r, m for (the pointwise) multiplication to be a continuous mapping of $W^{m,p}(\Omega) \times W^{m,q}(\Omega)$ into $W^{m,r}(\Omega)$.

Lemma 2.1. *Suppose that* Ω *has the cone property and let* p, q, r *be real numbers* ≥ 1 *such that* $p \geq r$, $q \geq r$ *and*

$$\frac{m}{n} > \frac{1}{p} + \frac{1}{q} - \frac{1}{r} \tag{2.1}$$

(with m *an integer* ≥ 0). *Then, if* $u \in W^{m,p}(\Omega)$ *and* $v \in W^{m,q}(\Omega)$, *we have* $uv \in W^{m,r}(\Omega)$ *and there is a number* $c > 0$, *independent of* u *and* v, *such that*

$$\|uv\|_{m,r} \leq c \|u\|_{m,p} \|v\|_{m,q},$$

Proof. We proceed by induction on m. We begin by remarking that the statement of the lemma is true when $m = 0$, namely, that if $u \in L^p(\Omega)$ and $v \in L^q(\Omega)$ with $p \geq r$, $q \geq r$ and $1/p + 1/q < 1/r$, then $uv \in L^r(\Omega)$ and $\|uv\|_{0,r} \leq c \|u\|_{0,p} \|v\|_{0,q}$ with c a number > 0 independent of u and v. This is an easy consequence of Hölder's inequality and of the fact that (as Ω has finite volume) $s_1 \leq s_2 \Rightarrow L^{s_1}(\Omega) \subseteq L^{s_2}(\Omega)$.

We now suppose that the statement of the lemma is true for an m (≥ 0) and we prove that, consequently, it is true when m is replaced by $m + 1$. Let p_1, q_1, r_1 be real numbers ≥ 1 such that $p_1 \geq r_1$, $q_1 \geq r_1$, and

$$\frac{m + 1}{n} > \frac{1}{p_1} + \frac{1}{q_1} - \frac{1}{r_1}, \tag{2.2}$$

and let $u_1 \in W^{m+1, p_1}(\Omega)$, $v_1 \in W^{m+1, q_1}(\Omega)$. We first consider the case when $mp_1 \leq n$ and $mq_1 \leq n$ with $m \neq 1$. Putting

$$\tilde{p}_1 = \frac{np_1}{n - p_1}, \qquad \tilde{q}_1 = \frac{nq_1}{n - q_1},$$

by the Sobolev embedding theorem we have

$$u \in W^{m, \tilde{p}_1}(\Omega) \qquad v \in W^{m, \tilde{q}_1}(\Omega);$$

moreover,

$$D_i u \in W^{m, p_1}(\Omega), \qquad D_i v \in W^{m, q_1}(\Omega), \qquad i = i, \ldots, n.$$

Observe that, since

$$\frac{1}{\tilde{p}_1} = \frac{1}{p_1} - \frac{1}{n}, \qquad \frac{1}{\tilde{q}_1} = \frac{1}{q_1} - \frac{1}{n},$$

(2.2) implies

$$\frac{m}{n} > \frac{1}{\tilde{p}_1} + \frac{1}{q_1} - \frac{1}{r_1}, \qquad \frac{m}{n} > \frac{1}{p_1} + \frac{1}{\tilde{q}_1} - \frac{1}{r_1}.$$

Thus, in view of the induction hypothesis, we have

$$v D_i u \in W^{m, r_1}(\Omega), \qquad u D_i v \in W^{m, r_1}(\Omega)$$

and

$$\|v D_i u\|_{m, r_1} \leq c_1 \|v\|_{m, \tilde{q}_1} \|D_i u\|_{m, p_1}, \qquad \|u D_i v\|_{m, r_1} \leq c_1 \|u\|_{m, \tilde{p}_1} \|D_i v\|_{m, q_1},$$

where c_1 is a number > 0 independent of u, v, and i. Then, by the Sobolev embedding theorem, there is a number $c_2 > 0$, independent of u, v, and i, such that

$$\|v D_i u\|_{m, r_1} \leq c_2 \|u\|_{m+1, p_1} \|v\|_{m+1, q_1}, \qquad \|u D_i v\|_{m, r_1} \leq c_2 \|u\|_{m+1, p_1} \|v\|_{m+1, q_1}. \tag{2.3}$$

In a similar way we can show that $uv \in W^{m, r_1}(\Omega)$ and that a number $c_3 > 0$, independent of u and v, exists such that

$$\|uv\|_{m, r_1} \leq c_3 \|u\|_{m+1, p_1} \|v\|_{m+1, q_1}. \tag{2.4}$$

We now prove that (2.3) and (2.4) also hold in the four cases: $mp_1 > n$, $mq_1 > n$, $p_1 = n$ with $m = 1$, and $q_1 = n$ with $m = 1$. If either $mp_1 > n$ or $mq_1 > n$ all hypotheses of the statement of the lemma are obviously satisfied so that, by the induction assumption, multiplication is a continuous mapping from $W^{m, p_1}(\Omega) \times W^{m, q_1}(\Omega)$ into $W^{m, r_1}(\Omega)$ and hence (2.3) and (2.4) hold. Then let $p_1 = n$ [respectively $q_1 = n$] and $m = 1$. Note that when $q_1 > r_1$ [respectively $p_1 > r_1$] all hypotheses of the statement of the lemma are satisfied, and so (2.3) and (2.4) hold. When $q_1 = r_1$ [respectively $p_1 = r_1$] the hypotheses of the statement of our lemma are satisfied provided p_1 [respectively q_1] is replaced by $\tilde{p}_1$ [respectively $\tilde{q}_1$], where $\tilde{p}_1$ [respectively $\tilde{q}_1$] is any number $> p_1$ [respectively $> q_1$]; thus, since (by the Sobolev embedding theorem) $W^{2,p_1}(\Omega)$ [respectively $W^{2,q_1}(\Omega)$] can be continuously embedded in $W^{1,\tilde{p}_1}(\Omega)$ [respectively $W^{1,\tilde{q}_1}(\Omega)$], from the induction hypothesis it follows that multiplication is a continuous mapping from $W^{2,p_1}(\Omega) \times W^{1,q_1}(\Omega)$ [respectively $W^{1,p_1}(\Omega) \times W^{2,q_1}(\Omega)$] into $W^{1,r_1}(\Omega)$, and so (2.3) and (2.4) also hold in this case.

Now, the induction is completed by proving that

$$D_i(uv) = vD_iu + uD_iv, \tag{2.5}$$

because, combining (2.3), (2.4), and (2.5), we obtain $uv \in W^{m+1,r_1}(\Omega)$ and

$$\|uv\|_{m+1,r_1} \leqq c_4 \|u\|_{m+1,p_1} \|v\|_{m+1,q_1},$$

with c_4 a number > 0 independent of u and v.

Accordingly, let us prove (2.5). We recall that $C^\infty(\Omega) \cap W^{m,s}(\Omega)$ is dense in $W^{m,s}(\Omega)$ for any real number $s \geqq 1$. Hence there is a sequence $(u_k)_{k \in \mathbb{N}}$ in $C^\infty(\Omega) \cap W^{m+1,p_1}(\Omega)$ and a sequence $(v_k)_{k \in \mathbb{N}}$ in $C^\infty(\Omega) \cap W^{m+1,q_1}(\Omega)$ such that

$$\lim_{k\to\infty} \|u_k - u\|_{m+1,p_1} = 0, \qquad \lim_{k\to\infty} \|v_k - v\|_{m+1,q_1} = 0. \tag{2.6}$$

In view of (2.3) we have

$$\|v_kD_iu_k - vD_iu\|_{m,r_1} \leqq \|v_kD_i(u_k - u)\|_{m,r_1} + \|(v_k - v)D_iu\|_{m,r_1}$$
$$\leqq \|u_k - u\|_{m+1,p_1}\|v_k\|_{m+1,q_1} + \|u\|_{m+1,p_1}\|v_k - v\|_{m+1,q_1}$$

and

$$\|u_kD_iv_k - uD_iv\|_{m,r_1} \leqq \|u_kD_i(v_k - v)\|_{m,r_1} + \|(u_k - u)D_iv\|_{m,r_1}$$
$$\leqq \|u_k\|_{m+1,p_1}\|v_k - v\|_{m+1,q_1}$$
$$+ \|u_k - u\|_{m+1,p_1}\|v\|_{m+1,q_1},$$

while, by (2.4), we obtain

$$\|u_kv_k - uv\|_{m,r_1} \leqq \|u_k(v_k - v)\|_{m,r_1} + \|(u_k - u)v\|_{m,r_1}$$
$$\leqq \|u_k\|_{m+1,p_1}\|v_k - v\|_{m+1,q_1} + \|u_k - u\|_{m+1,p_1}\|v\|_{m+1,q_1}.$$

Then, from (2.6), it follows that

$$\lim_{k\to\infty} \|v_kD_iu_k + u_kD_iv_k) - (vD_iu + uD_iv)\|_{m,r_1} = 0,$$

$$\lim_{k\to\infty} \|u_kv_k - uv\|_{m,r_1} = 0.$$

Therefore, by Hölder's inequality, we have

$$\int_\Omega (vD_iu + uD_iv)\varphi\, dx + \int_\Omega uvD_i\varphi\, dx = \lim_{k\to\infty} \left(\int_\Omega (v_kD_iu_k + u_kD_iv_k)\varphi\, dx \right.$$
$$\left. + \int_\Omega u_kv_kD_i\varphi\, dx \right)$$

for all $\varphi \in \mathcal{D}(\Omega)$.

Consequently,

$$\int_\Omega (vD_iu + uD_iv)\varphi\, dx + \int_\Omega uvD_i\varphi\, dx = 0, \qquad \forall \varphi \in \mathcal{D}(\Omega), \qquad (2.7)$$

because in view of the equality $D_i(u_kv_k) = v_kD_iu_k + u_kD_iv_k$ we have

$$\int_\Omega (v_kD_iu_k + u_kD_iv_k)\varphi\, dx + \int_\Omega u_kv_kD_i\varphi\, dx = \int_\Omega D_i(u_kv_k\varphi)\, dx$$
$$= 0, \qquad \forall \varphi \in \mathcal{D}(\Omega).$$

Since (2.7) is equivalent to (2.5), our proof is complete.   □

Evidently, from Lemma 2.1 derives the well-known fact (see ADAMS [1975], Theorem 5.23) that $W^{m,p}(\Omega)$ is a Banach algebra provided Ω has the cone property. We emphasize the following consequence of Lemma 2.1: if Ω has the cone property then (the pointwise) *multiplication is a continuous mapping of $W^{m,p}(\Omega) \times W^{m,q}(\Omega)$ into $W^{m,q}(\Omega)$ provided $q \leqq p$ and $mp > n$.*

We remark that the proof of Lemma 2.1 does not require the boundedness of Ω, but only the fact that the volume of Ω is finite. In the general case, when the volume of Ω is not necessarily finite, the following result holds.

Lemma 2.1'. *Let Ω be an open subset of $\mathbb{R}^n$ having the cone property, and let m be an integer ≥ 1 and p, q, r be real numbers ≥ 1 such that $p \geq r$, $q \geq r$, and $m/n > 1/p + 1/q - 1/r$. If the volume of Ω is infinite assume further that $mp \leq n$ when $q \neq r$, that $mq \leq n$ when $p \neq r$, and that $(m-1)/n \leq 1/p + 1/q - 1/r$ when $p \neq r$, $q \neq r$. Then, if $u \in W^{m,p}(\Omega)$ and $v \in W^{m,q}(\Omega)$, we have $uv \in W^{m,r}(\Omega)$ and there is a number $c > 0$, independent of u and v, such that $\|uv\|_{m,r} \leq c\|u\|_{m,p}\|v\|_{m,q}$.*

For the proof of this result we refer to VALENT [1985b].

The following theorem is a consequence (and, at the same time) a generalization of Lemma 2.1; it provides conditions on the numbers p, q, r, m, h, k for multiplication to be a continuous mapping from $W^{m+h,p}(\Omega) \times W^{m+k,q}(\Omega)$ into $W^{m,r}(\Omega)$.

Theorem 2.2. *Suppose that Ω has the cone property, and let m, h, k be integers ≥ 0 and p, q, r be real numbers ≥ 1 such that*

$$\frac{m+h+k}{n} > \frac{1}{p} + \frac{1}{q} - \frac{1}{r}. \tag{2.8}$$

Moreover, suppose that $np/(n-hp) \geq r$ if $hp < n$ and that $nq/(n-kq) \geq r$ if $kq < n$. Then, for $u \in W^{m+h,p}(\Omega)$ and $v \in W^{m+k,q}(\Omega)$, we have $uv \in W^{m,r}(\Omega)$ and there is a number $c > 0$, independent of u and v, such that

$$\|uv\|_{m,r} \leq c \|u\|_{m+h,p} \|v\|_{m+k,q}.$$

Proof. As we shall see, Theorem 2.2 is an almost immediate consequence of Lemma 2.1 and the Sobolev embedding theorem. We can suppose either $h \neq 0$ or $k \neq 0$, because if $h = k = 0$ Theorem 2.2 reduces to Lemma 2.1. Recall that, by the Sobolev embedding theorem, $W^{m+h,p}(\Omega)$ [respectively $W^{m+k,q}(\Omega)$] can be continuously embedded in $W^{m,\tilde{p}}(\Omega)$ [respectively $W^{m,\tilde{q}}(\Omega)$] where $\tilde{p} = \infty$ [respectively $\tilde{q} = \infty$] when $hp > n$ [respectively $kq > n$], $\tilde{p}$ [respectively $\tilde{q}$] is any real number ≥ 1 when $hp = n$ [respectively $kq = n$] and

$$\tilde{p} = \frac{np}{n-hp} \qquad \left[\text{respectively } \tilde{q} = \frac{nq}{n-kq} \right] \tag{2.9}$$

when $hp < n$ [respectively $kq < n$].

Observe that condition (2.8) can be written in the form

$$\frac{m}{n} > \frac{n-hp}{np} + \frac{n-kq}{nq} - \frac{1}{r}. \tag{2.10}$$

Hence, provided $hp < n$ and $kq < n$, Lemma 2.1 assures that multiplication is a continuous mapping from $W^{m,\tilde{p}}(\Omega) \times W^{m,\tilde{q}}(\Omega)$ into $W^{m,r}(\Omega)$ with $\tilde{p}$ and $\tilde{q}$ defined by (2.9); therefore, multiplication is a continuous mapping from $W^{m+h,p}(\Omega) \times W^{m+k,q}(\Omega)$ into $W^{m,r}(\Omega)$.

We reach the same conclusion if $hp \geq n$ and $kq < n$ (and hence if $kq \geq n$ and $hp < n$). Indeed, if $hp = n$ and $kq < n$, in view of (2.10) there is a real number $\tilde{p} > 1$ such that

$$\frac{m}{n} > \frac{1}{\tilde{p}} + \frac{1}{\tilde{q}} - \frac{1}{r} \qquad \text{with} \qquad \tilde{q} = \frac{nq}{n-kq};$$

therefore by Lemma 2.1 multiplication is a continuous mapping from $W^{m,\tilde{p}}(\Omega) \times W^{m,nq/(n-kq)}(\Omega)$ into $W^{m,r}(\Omega)$, and thus, the desired conclusion follows from the fact that $W^{m+h,p}(\Omega) \times W^{m+k,q}(\Omega)$ can be continuously embedded in $W^{m,\tilde{p}}(\Omega) \times W^{m,nq/(n-kq)}(\Omega)$; if $hp > n$ and $kq < n$ the conclusion holds, because $W^{m+h,p}(\Omega) \times W^{m+k,q}(\Omega)$ can be continuously embedded in $C_B^m(\Omega) \times W^{m,nq/(n-kq)}(\Omega)$, multiplication is (clearly) a continuous mapping from $C_B^m(\Omega) \times W^{m,nq/(n-kq)}(\Omega)$ into $W^{m,nq/(n-kq)}(\Omega)$ and $W^{m,nq/(n-kq)}(\Omega)$ is continuously contained in $W^{m,r}(\Omega)$.

Finally, if $hp \geqq n$ and $kq > n$, let $\tilde{p}$ and $\tilde{q}$ be real numbers $\geqq r$ such that $m/n > 1/\tilde{p} + 1/\tilde{q} - 1/r$; by Lemma 2.1, multiplication is a continuous mapping from $W^{m,\tilde{p}}(\Omega) \times W^{m,\tilde{q}}(\Omega)$ into $W^{m,r}(\Omega)$, and thus multiplication is a continuous mapping from $W^{m+h,p}(\Omega) \times W^{m+k,q}(\Omega)$ into $W^{m,r}(\Omega)$, because $W^{m+h,p}(\Omega) \times W^{m+k,q}(\Omega)$ can be continuously embedded in $W^{m,\tilde{p}}(\Omega) \times W^{m,\tilde{q}}(\Omega)$. □

Clearly, the next result is a consequence of Theorem 2.2.

Corollary 2.3. *Suppose that Ω has the cone property and let m, k be integer $\geqq 0$. If $(m + k)p > n$ then, for $u \in W^{m,p}(\Omega)$ and $v \in W^{m+k,p}(\Omega)$, we have $uv \in W^{m,p}(\Omega)$ and there is a number $c > 0$, independent of u and v, such that*

$$\|uv\|_{m,p} \leqq c \|u\|_{m,p} \|v\|_{m+k,p}.$$

§3. On Continuity of Composition Operators in Sobolev and Schauder Spaces

Let N be an integer $\geqq 1$ and let $(x, y) \mapsto f(x, y)$ be a real-valued function defined on $\Omega \times \mathbb{R}^n$. For any function $\sigma: \Omega \to \mathbb{R}^N$ let $F(\sigma): \Omega \to \mathbb{R}$ be the function defined by setting, for all $x \in \Omega$,

$$F(\sigma)(x) = f(x, \sigma(x)). \tag{3.1}$$

Recall that if m is an integer $\geqq 0$ and U is an open subset of $\mathbb{R}^N$, then $C^m(\bar{\Omega} \times U)$ [respectively $C^\infty(\bar{\Omega} \times U)$] is the set of real-valued functions defined on $\bar{\Omega} \times U$ which are the restrictions to $\bar{\Omega} \times U$ of some C^m functions [respectively C^∞ functions] from $\mathbb{R}^n \times U$ into $\mathbb{R}$ (see §1).

Theorem 3.1. *Assume that Ω has the cone property, and suppose that $f \in C^m(\bar{\Omega} \times U)$ where U is an open subset of $\mathbb{R}^N$ and m is an integer $\geqq 1$. Then $\sigma \mapsto F(\sigma)$ is a continuous mapping from the subset $\{\sigma \in W^{m+r,p}(\Omega, \mathbb{R}^N): \sigma(\bar{\Omega}) \subset U\}$ of $W^{m+r,p}(\Omega, \mathbb{R}^N)$ into $W^{m,p}(\Omega)$ for any integer $r \geqq 0$ such that $(m + r)p > n$.*

Proof (by induction on m). If $\sigma(\bar{\Omega}) \subset U$, we denote by $F_{x_i}(\sigma)$ and $F_{y_j}(\sigma)$ $(i = 1, \ldots, n; j = 1, \ldots, N)$, the real-valued functions defined on Ω by putting, for all $x \in \Omega$,

$$F_{x_i}(\sigma)(x) = D_{x_i} f(x, \sigma(x)), \qquad F_{y_j}(\sigma)(x) = D_{y_j} f(x, \sigma(x)). \tag{3.2}$$

We begin with the case $m = 1$. Accordingly, let $f \in C^1(\bar{\Omega} \times U)$ and $(1 + r)p > n$. By the Sobolev embedding theorem each $v \in W^{1+r,p}(\Omega)$ is an equivalence class of functions containing a continuous and bounded function, which we still denote by v and there is a number $c_{1+r,p} > 0$, independent of v, such that

$$\|v\|_{0,\infty} \le c_{1+r,p} \|v\|_{1+r,p}, \qquad \forall v \in W^{1+r,p}(\Omega), \tag{3.3}$$

where $\|\cdot\|_{0,\infty}$ is the norm of $L^\infty(\Omega)$. Then, if $\sigma \in W^{1+k,p}(\Omega, \mathbb{R}^N)$ and $\sigma(\bar{\Omega}) \subset U$, the equivalence classes $F(\sigma)$, $F_{x_i}(\sigma)$, and $F_{y_j}(\sigma)$ can be identified with uniformly continuous (and hence bounded) functions.

Let $\sigma = (\sigma_j)_{j=1,\dots,N} \in W^{1+r,p}(\Omega, \mathbb{R}^N)$, $\sigma(\bar{\Omega}) \subset U$ and let $(\sigma^k)_{k \in \mathbb{N}}$ be a sequence in $C^\infty(\Omega, \mathbb{R}^N) \cap W^{1+r,p}(\Omega, \mathbb{R}^N)$ with $\sigma(\bar{\Omega}) \subset U$ which converges to σ in $W^{1+r,p}(\Omega, \mathbb{R}^N)$, and therefore, by (3.1), in $L^\infty(\Omega, \mathbb{R}^N)$. We have

$$D_i F(\sigma^k) = F_{x_i}(\sigma^k) + \sum_{j=1}^N F_{y_j}(\sigma^k) D_i \sigma_j^k. \tag{3.4}$$

Since $(\sigma^k)_{k \in \mathbb{N}}$ converges to σ in $L^\infty(\Omega, \mathbb{R}^N)$, it is not difficult to see that $(F(\sigma^k))_{k \in \mathbb{N}}$, $(F_{x_i}(\sigma^k))_{k \in \mathbb{N}}$, and $(F_{y_j}(\sigma^k))_{k \in \mathbb{N}}$ converge in $L^\infty(\Omega)$ to $F(\sigma)$, $F_{x_i}(\sigma)$, and $F_{y_j}(\sigma)$, respectively; therefore, $(F_{x_i}(\sigma^k) + \sum_{j=1}^N F_{y_j}(\sigma^k) D_i \sigma_j^k)_{k \in \mathbb{N}}$ converges in $L^p(\Omega)$ to $F_{x_i}(\sigma) + \sum_{j=1}^N F_{y_j}(\sigma) D_i \sigma_j$.

Consequently, by Hölder's inequality we have, for any $\varphi \in \mathcal{D}(\Omega)$,

$$\int_\Omega \left(F_{x_i}(\sigma) + \sum_{j=1}^N F_{y_j}(\sigma) D_i \sigma_j \right) \varphi \, dx + \int_\Omega F(\sigma) D_i \varphi \, dx$$

$$= \lim_{k \to \infty} \left[\int_\Omega (F_{x_i}(\sigma^k) + \sum_{j=1}^N F_{y_j}(\sigma^k) D_i \sigma_j^k) \varphi \, dx + \int_\Omega F(\sigma^k) D_i \varphi \, dx \right]. \tag{3.5}$$

Because of (3.4) we have, for any $k \in \mathbb{N}$ and any $\varphi \in \mathcal{D}(\Omega)$,

$$\int_\Omega (F_{x_i}(\sigma^k) + \sum_{j=1}^N F_{y_j}(\sigma^k) D_i \sigma_j^k) \varphi \, dx + \int_\Omega F(\sigma^k) D_i \varphi \, dx = 0$$

and thus by (3.5) we obtain

$$\int_\Omega (F_{x_i}(\sigma) + \sum_{j=1}^N F_{y_j}(\sigma) D_i \sigma_j) \varphi \, dx + \int_\Omega F(\sigma) D_i \varphi \, dx = 0. \qquad \forall \varphi \in \mathcal{D}(\Omega),$$

which means

$$D_i F(\sigma) = F_{x_i}(\sigma) + \sum_{j=1}^N F_{y_j}(\sigma) D_i \sigma_j. \tag{3.6}$$

Since each of the equivalence classes $F(\sigma)$, $F_{x_i}(\sigma)$, and $F_{y_j}(\sigma)$ contains a continuous and bounded function, (3.4) yields $F(\sigma) \in W^{1,p}(\Omega)$. To prove that the mapping $\sigma \mapsto F(\sigma)$ from $\{\sigma \in W^{1+r,p}(\Omega, \mathbb{R}^N): \sigma(\bar{\Omega}) \subset U\}$ into $W^{1,p}(\Omega)$ is continuous we need only remark that if a sequence $(\sigma^k)_{k \in \mathbb{N}}$, with $\sigma^k(\bar{\Omega}) \subset U$, converges to σ in $W^{1+r,p}(\Omega, \mathbb{R}^N)$, then by (3.3) $(\sigma^k)_{k \in \mathbb{N}}$ converges to σ in $L^\infty(\Omega, \mathbb{R}^N)$ and hence the sequences $(F(\sigma^k))_{k \in \mathbb{N}}$, $(F_{x_i}(\sigma^k))_{k \in \mathbb{N}}$, and $(F_{y_j}(\sigma^k))_{k \in \mathbb{N}}$ converge in $L^\infty(\Omega)$, respectively, to $F(\sigma)$, $F_{x_i}(\sigma)$, and $F_{y_j}(\sigma)$, so that $(D_i F(\sigma^k))_{k \in \mathbb{N}}$ converges to $D_i F(\sigma)$ in $L^p(\Omega)$ in view of (3.6) and thus $(F(\sigma^k))_{k \in \mathbb{N}}$ converges to $F(\sigma)$ in $W^{1,p}(\Omega)$.

As a next step, we suppose that the statement of the theorem is true for some $m \geq 1$ and we show that, consequently, it holds when m is replaced by $m + 1$. In order to do this, assume that $f \in C^{m+1}(\bar{\Omega} \times U)$, that $(m + 1 + r)p > n$, and that $\sigma \in W^{m+1+r,p}(\Omega, \mathbb{R}^N)$ with $\sigma(\bar{\Omega}) \subset U$; and we prove that $F(\sigma) \in W^{m+1,p}(\Omega)$ and that $\sigma \mapsto F(\sigma)$ is a continuous mapping from $\{\sigma \in W^{m+1+r,p}(\Omega, \mathbb{R}^N): \sigma(\bar{\Omega}) \subset U\}$ into $W^{m+1,p}(\Omega)$. Observe that the induction hypothesis makes $\sigma \mapsto F(\sigma)$ a continuous mapping from $\{\sigma \in W^{m+1+r,p}(\Omega, \mathbb{R}^N): \sigma(\bar{\Omega}) \subset U\}$ into $W^{m,p}(\Omega)$. Thus it remains to prove that $\sigma \mapsto D_i F(\sigma)$ is a continuous mapping from $\{\sigma \in W^{m+1+r,p}(\Omega, \mathbb{R}^N): \sigma(\bar{\Omega}) \subset U\}$ into $W^{m,p}(\Omega)$. Accordingly, we first note that (by the Sobolev embedding theorem) each $v \in W^{m+1+r,p}(\Omega, \mathbb{R}^N)$ can be identified with a continuous function and there is a number $c_{m+1+r,p} > 0$, independent of v, such that $\|v\|_{0,\infty} \leq c_{m+1+r,p} \|v\|_{m+1+r,p}$, $\forall v \in W^{m+1+r,p}(\Omega, \mathbb{R}^N)$, so that, by arguments similar to the ones used in the case $m = 1$, we can show that (3.6) holds.

It is now convenient to distinguish the cases $rp > n$, $rp = n$, and $rp < n$. If $rp > n$, from the (induction) assumption, it follows that F_{x_i} and F_{y_j} are continuous mappings from $\{\sigma \in W^{m+r,p}(\Omega, \mathbb{R}^N): \sigma(\bar{\Omega}) \subset U\}$ into $W^{m,p}(\Omega)$; therefore, in view of (3.6), $\sigma \mapsto D_i F(\sigma)$ is a continuous mapping from $\{\sigma \in W^{m+1+r,p}(\Omega, \mathbb{R}^N): \sigma(\Omega) \subset U\}$ into $W^{m,p}(\Omega)$, because (by the Sobolev embedding theorem) $W^{m+r,p}(\Omega)$ can be continuously embedded in $C_B^m(\Omega)$ and the pointwise multiplication is (evidently) a continuous mapping of $C_B^m(\Omega) \times W^{m,p}(\Omega)$ into $W^{m,p}(\Omega)$. Now let $rp = n$ and let $q \in \mathbb{R}$ be such that $q \geq p$ and $mq > n$. Thus $(m + r)q > n$ and hence, by the (induction) assumption, F_{x_i} and F_{y_j} are continuous mappings from $\{\sigma \in W^{m+r,q}(\Omega, \mathbb{R}^N): \sigma(\bar{\Omega}) \subset U\}$ into $W^{m,q}(\Omega)$. Since $(m + r)p > n$, there is a real number $s \geq 1$ such that

$$(m + r)q > n + \frac{n}{qs}. \tag{3.7}$$

Note that, as $rp = n$, by the Sobolev embedding theorem $W^{m+r,p}(\Omega)$ can be continuously embedded in $W^{m,s}(\Omega)$. Since (3.7) is equivalent to $m/n > 1/q + 1/s - 1/p$, Lemma 2.1 ensures that the pointwise multiplication is a continuous mapping from $W^{m,q}(\Omega) \times W^{m,s}(\Omega)$ into $W^{m,p}(\Omega)$. Consequently, from (3.6) it follows that $\sigma \mapsto D_i F(\sigma)$ is a continuous mapping from $\{\sigma \in W^{m+1+r,p}(\Omega, \mathbb{R}^N): \sigma(\bar{\Omega}) \subset U\}$ into $W^{m,p}(\Omega)$.

Finally, let us consider the case $rp < n$. Note that, in this case, the condition $(m + 1 + r)p > n$ is equivalent to the condition

$$(m + r)\frac{np}{n - p} > n. \tag{3.8}$$

Note also that $np/(n - p) > p$ and that the induction hypothesis, combined with (3.8), implies that F_{x_i} and F_{y_j} are continuous mappings from $\{\sigma \in W^{m+r,np/(n-p)}(\Omega, \mathbb{R}^N): \sigma(\bar{\Omega}) \subset U\}$ into $W^{m,np/(n-p)}(\Omega)$. Moreover, by the Sobolev embedding theorem $W^{m+r,p}(\Omega)$ can be continuously embedded in $W^{m,np/(n-rp)}(\Omega)$ and, since $(m + 1 + r)p > n$, from Lemma 2.1 it follows that the pointwise multiplication is a continuous mapping from $W^{m,\,np(n-rp)}(\Omega) \times W^{m,\,np/(n-p)}(\Omega)$ into $W^{m,\,p}(\Omega)$. This implies, by (3.6), that $\sigma \mapsto D_i F(\sigma)$ is a continuous mapping from $\{\sigma \in W^{m+1+r,\,p}(\Omega, \mathbb{R}^N): \sigma(\Omega) \subset U\}$ into $W^{m,\,p}(\Omega)$. Thus the proof is complete. $\square$

$$* \quad * \quad * \quad * \quad *$$

Theorem 3.2. *If* $f \in C^2(\bar{\Omega} \times U)$ *where* U *is an open subset of* $\mathbb{R}^N$, *then for any* $\lambda \in \,]0, 1]$, $\sigma \mapsto F(\sigma)$ *is a continuous mapping from the subset* $\{\sigma \in C^{0,\,\lambda}(\bar{\Omega}, \mathbb{R}^N): \sigma(\bar{\Omega}) \subset U\}$ *of* $C^{0,\,\lambda}(\bar{\Omega}, \mathbb{R}^N)$ *into* $C^{0,\,\lambda}(\bar{\Omega})$.

Proof. Let $f \in C^2(\bar{\Omega} \times U)$. We first prove that $F(\sigma) \in C^{0,\,\lambda}(\bar{\Omega})$ for each $\sigma \in C^{0,\,\lambda}(\bar{\Omega}, \mathbb{R}^N)$ such that $\sigma(\bar{\Omega}) \subset U$. Accordingly, we fix $\sigma \in C^{0,\,\lambda}(\bar{\Omega}, \mathbb{R}^N)$. Since $F(\sigma)$ is, evidently, a continuous function from $\bar{\Omega}$ into $\mathbb{R}$, we need only show that the set

$$\left\{ \frac{f(x', \sigma(x')) - f(x'', \sigma(x''))}{|x' - x''|^\lambda}: x', x'' \in \bar{\Omega}, x' \neq x'' \right\}$$

is bounded. Since $f \in C^2(\bar{\Omega} \times U)$ and $\sigma(\bar{\Omega})$ is closed in U, there is a C^2 function $\tilde{f}: \mathbb{R}^n \times \mathbb{R}^N \to \mathbb{R}$ such that $\tilde{f}|_{\bar{\Omega} \times \sigma(\bar{\Omega})} = \tilde{f}|_{\bar{\Omega} \times \sigma(\bar{\Omega})}$. If $\text{co}(\bar{\Omega})$ and $\text{co}(\sigma(\bar{\Omega}))$ denote the convex hulls of $\bar{\Omega}$ and $\sigma(\bar{\Omega})$, respectively, then a Taylor expansion of order 1 yields, for all $x', x'' \in \bar{\Omega}$ with $x' \neq x''$,

$$\frac{f(x', \sigma(x')) - f(x'', \sigma(x''))}{|x' - x''|^\lambda}$$

$$\leq \sum_{i=1}^n \left\{ \sup |D_{x_i}\tilde{f}(x', x'')| \frac{|x_i' - x_i''|}{|x' - x''|^\lambda}: (x', x'') \in \text{co}(\bar{\Omega}) \times \text{co}(\sigma((\bar{\Omega}))) \right\}$$

$$+ \sum_{j=1}^N \sup \left\{ |D_{y_j}\tilde{f}(x', x'')| \frac{|\sigma_j(x') - \sigma_j(x'')|}{|x' - x''|^\lambda}: (x', x'') \in \text{co}(\bar{\Omega}) \times \text{co}(\sigma(\bar{\Omega})) \right\}$$

$$\leq \sum_{i=1}^n \sup\{|D_{x_i}\tilde{f}(x', x'')|: (x', x'') \in \text{co}(\Omega) \times \text{co}(\sigma(\bar{\Omega}))\} |x' - x''|^{1-\lambda}$$

$$+ \sum_{j=1}^N \sup\{|D_{y_j}\tilde{f}(x', x'')|: (x', x'') \in \text{co}(\bar{\Omega}) \times \text{co}(\sigma(\Omega))\} \frac{|\sigma(x') - \sigma(x'')|}{|x' - x''|^\lambda}$$

$$\leq c \sum_{i=1}^n \sup\{|D_{x_i}\tilde{f}(x', x'')|: (x', x'') \in \text{co}(\bar{\Omega}) \times \text{co}(\sigma(\bar{\Omega}))\}$$

$$+ \|\sigma\|_{0,\,\lambda} \sum_{j=1}^N \sup\{|D_{y_j}\tilde{f}(x', x'')|: (x', x'') \in \text{co}(\bar{\Omega}) \times \text{co}(\sigma(\bar{\Omega}))\},$$

where $c = \sup_{x',\,x'' \in \bar{\Omega}} |x' - x''|^{1-\lambda}$. We remark that, up to this time, we have used only the fact that $f \in C^1(\bar{\Omega} \times U)$.

We now prove that $\sigma \mapsto F(\sigma)$ is a continuous mapping from $C^{0,\,\lambda}(\bar{\Omega}, \mathbb{R}^N)$ into $C^{0,\,\lambda}(\bar{\Omega})$. We begin by observing that if $\tau, \sigma \in C^{0,\,\lambda}(\bar{\Omega}, \mathbb{R}^N)$, $\sigma(\bar{\Omega}) \subset U$, and $\|\tau - \sigma\|_{0,\,\lambda} < d_\sigma$ where d_σ is the distance from $\sigma(\bar{\Omega})$ to $\mathbb{R}^N \backslash U$, then a

Taylor expansion of order 1 gives

$$
\begin{cases}
f(x', \tau(x')) - f(x', \sigma(x')) = \displaystyle\sum_{j=1}^{N} D_{y_j} f(x', y')(\tau_j(x') - \sigma_j(x')), \\[2mm]
f(x'', \tau(x'')) - f(x'', \sigma(x'')) = \displaystyle\sum_{j=1}^{N} D_{y_j} f(x'', y'')(\tau_j(x'') - \sigma_j(x'')),
\end{cases}
$$

and

$$
|f(x, \tau(x)) - f(x, \sigma(x))| \leq \sum_{j=1}^{N} \sup_{y \in \sigma(\bar{\Omega})_{d_{\sigma/2}}} |D_{y_j} f(x, y)| |\tau(x) - \sigma(x)|,
$$

where $\sigma(\bar{\Omega})_{d_{\sigma/2}}$ denotes the neighborhood of order $d_{\sigma/2}$ of $\sigma(\bar{\Omega})$ and y', y'' are suitable points belonging to the segment $(\tau(x'), \sigma(x'))$, the former, and to the segment $(\tau(x''), \sigma(x''))$, the latter. Therefore

$$
\frac{|(f(x', \tau(x')) - f(x'', \sigma(x''))) - (f(x'', \tau(x'')) - f(x'', \sigma(x'')))|}{|x' - x''|^{\lambda}}
$$

$$
= \frac{1}{|x' - x''|^{\lambda}} \left(\sum_{j=1}^{N} D_{y_j} f(x', y')(\tau_j(x') - \sigma_j(x')) \right.
$$

$$
\left. - \sum_{j=1}^{N} D_{y_j} f(x'', y'')(\tau_j(x'') - \sigma_j(x'')) \right)
$$

$$
\leq \sum_{j=1}^{N} \left(\frac{|D_{y_j} f(x', y') - D_{y_j} f(x'', y'')|}{|x' - x''|^{\lambda}} |\tau_j(x') - \sigma_j(x'')| \right.
$$

$$
\left. + |D_{y_j} f(x'', y'')| \frac{|(\tau_j(x') - \sigma_j(x')) - (\tau_j(x'') - \sigma_j(x''))|}{|x' - x''|^{\lambda}} \right).
$$

Then, since $D_{y_j} f \in C^1(\bar{\Omega} \times U)$, the function $F_{y_j}(\sigma): x \mapsto D_{y_j} f(x, \sigma(x))$ belongs to $C^{0,\lambda}(\bar{\Omega})$. Consequently, since $|y' - \sigma(x')| \leq |\tau(x') - \sigma(x')|$ and $|y'' - \sigma(x'')| \leq |\tau(x'') - \sigma(x'')|$, we have

$$
\sup_{x', x'' \in \bar{\Omega}, x' \neq x''} \frac{|D_{y_j} f(x', y') - D_{y_j} f(x'', y'')|}{|x' - x''|^{\lambda}} \leq \|F_{y_j}(\sigma)\|_{0,\lambda} + 1
$$

for $\|\tau - \sigma\|_{0,\lambda}$ small enough. Hence, if $\|\tau - \sigma\|_{0,\lambda}$ is sufficiently small, we have

$$
\|F(\tau) - F(\sigma)\|_{0,\lambda} \leq \left[N + \sum_{j=1}^{N} \left(\|F_{y_j}(\sigma)\|_{0,\lambda} \right. \right.
$$

$$
\left. \left. + 2 \sup_{x \in \bar{\Omega}, y \in \sigma(\bar{\Omega})_{d_{\sigma/2}}} |D_{y_j} f(x, y)| \right) \right] \|\tau - \sigma\|_{0,\lambda}.
$$

Thus we have proved that $F: \{\sigma \in C^{0,\lambda}(\bar{\Omega}, \mathbb{R}^N): \sigma(\bar{\Omega}) \subset U\} \to C^{0,\lambda}(\bar{\Omega})$ is continuous at the (arbitrary) point σ. $\square$

Theorem 3.3. *Assume that Ω is such that* (1.1) *holds and suppose that $f \in C^{1+\sup(1,m)}(\bar{\Omega} \times U)$ with U an open subset of $\mathbb{R}^N$ and $m \geqq 0$. Then, for any $\lambda \in {]}0, 1]$, $\sigma \mapsto F(\sigma)$ is a continuous mapping from the subset $\{\sigma \in C^{m,\lambda}(\bar{\Omega}, \mathbb{R}^N): \sigma(\bar{\Omega}) \subset U\}$ of $C^{m,\lambda}(\bar{\Omega}, \mathbb{R}^N)$ into $C^{m,\lambda}(\bar{\Omega}, \mathbb{R}^N)$.*

Proof (by induction on m). The statement of Theorem 3.3 is true when $m = 0$, because of Theorem 3.2. We suppose that it is true for an m and we prove that, consequently, it remains true on replacing m with $m + 1$. Accordingly, let $f \in C^{m+2}(\bar{\Omega} \times U)$. In view of the induction hypothesis the mappings F_{x_i} and F_{y_j} act continuously from the subset $\{\sigma \in C^{m,\lambda}(\bar{\Omega}, \mathbb{R}^N): \sigma(\bar{\Omega} \subset U\}$ of $C^{m,\lambda}(\bar{\Omega}, \mathbb{R}^N)$ into $C^{m,\lambda}(\bar{\Omega})$; hence, bearing in mind that, by (1.1), $C^{m,\lambda}(\bar{\Omega})$ is a Banach algebra and observing that (3.6) holds, we easily see that F continuously maps the subset $\{\sigma \in C^{m+1,\lambda}(\bar{\Omega}, \mathbb{R}^N): \sigma(\bar{\Omega}) \subset U\}$ of $C^{m+1,\lambda}(\bar{\Omega}, \mathbb{R}^N)$ into $C^{m+1,\lambda}(\bar{\Omega})$.   □

§4. On Differentiability of Composition Operators in Sobolev and Schauder Spaces

This section is devoted to the differentiability of the operators $\sigma \mapsto F(\sigma)$ and $(f, \sigma) \mapsto F(\sigma)$, with $F(\sigma)$ defined in §3.

Theorem 4.1. *Assume that Ω has the cone property. If $f \in C^{m+1}(\bar{\Omega} \times U)$ with U an open subset of $\mathbb{R}^N$ and $m \geq 1$, then for any integer $r \geq 0$ such that $(m + r)p > n$, $\sigma \mapsto F(\sigma)$ is a C^1 mapping from the subset $\{\sigma \in W^{m+r,p}(\Omega, \mathbb{R}^N): \sigma(\bar{\Omega}) \subset U\}$ of $W^{m+r,p}(\Omega, \mathbb{R}^N)$ into $W^{m,p}(\Omega)$ and its differential at any σ is the (continuous, linear) mapping $F'(\sigma)$ of $W^{m+r,p}(\Omega, \mathbb{R}^N)$ into $W^{m,p}(\Omega)$ defined by putting, for all $\tau \in W^{m+r,p}(\Omega, \mathbb{R}^N)$,*

$$F'(\sigma)(\tau) = \sum_{j=1}^{N} F_{y_j}(\sigma)\tau_j, \tag{4.1}$$

where $F_{y_j}(\sigma)$ is the real-valued function defined on Ω by (3.2).

Proof. The key of the proof is Corollary 2.3. Let $f \in C^{m+1}(\bar{\Omega} \times U)$, with U an open subset of $\mathbb{R}^N$ and $m \geq 1$, and let r be an integer such that $r \geq 0$ and $(m + r)p > n$. Note that, as $D_{y_j}f \in C^m(\bar{\Omega} \times U)$, we have $F_{y_j}(\sigma) \in W^{m,p}(\Omega)$ in view of Theorem 3.1; consequently, $\tau_j F_{y_j}(\sigma) \in W^{m,p}(\Omega)$ because of Corollary 2.3. Note also that, since $(m + r)p > n$, $W^{m+r,p}(\Omega)$ can be continuously embedded in $C^1(\bar{\Omega})$.

We first prove that the mapping $\sigma \mapsto F(\sigma)$, of $\{\sigma \in W^{m+r,p}(\Omega, \mathbb{R}^N): \sigma(\bar{\Omega}) \subset U\}$ into $W^{m,p}(\Omega)$ (see Theorem 3.1), is differentiable at any σ and that $F'(\sigma)$ is defined by (4.1), namely, that

$$\lim_{\|\tau\|_{m+r,p} \to 0} \frac{\left\| F(\sigma + \tau) - F(\sigma) - \sum_{j=1}^{N} \tau_j F_{y_j}(\sigma) \right\|_{m,p}}{\|\tau\|_{m+r,p}} = 0. \tag{4.2}$$

Clearly, for any $\tau \in W^{m+r,p}(\bar{\Omega}, \mathbb{R}^N)$ with $\|\tau\|_{m+r,p}$ small enough to have $\sup_{x \in \Omega} |\tau(x)| <$ distance from $\sigma(\bar{\Omega})$ to $\mathbb{R}^N \setminus U$, we can write

$$F(\sigma + \tau) - F(\sigma) - \sum_{j=1}^{N} \tau_j F_{y_j}(\sigma) = \sum_{j=1}^{N} \tau_j \int_0^1 (F_{y_j}(\sigma + t\tau) - F_{y_j}(\sigma))\, dt.$$

Thus, using Corollary 2.3, we obtain

$$\left\| F(\sigma + \tau) - F(\sigma) - \sum_{j=1}^{N} \tau_j F_{y_j}(\sigma) \right\|_{m,p}$$

$$\leq c \|\tau\|_{m+r,p} \sum_{j=1}^{N} \left\| \int_0^1 (F_{y_j}(\sigma + t\tau) - F_{y_j}(\sigma))\, dt \right\|_{m,p}, \quad (4.3)$$

with c a number > 0 independent of τ. By virtue of Theorem 3.1, $\sigma \mapsto F_{y_j}(\sigma)$ is a continuous mapping from $\{\sigma \in W^{m+r,p}(\Omega, \mathbb{R}^N): \sigma(\bar{\Omega}) \subset U\}$ into $W^{m,p}(\Omega)$. Thus (4.2) easily follows from (4.3) because

$$\left\| \int_0^1 (F_{y_j}(\sigma + t\tau) - F(\sigma)\, dt \right\|_{m,p} \leq \int_0^1 \|F_{y_j}(\sigma + t\tau) - F_{y_j}(\sigma)\|_{m,p}\, dt.$$

It remains to prove that the operator $\sigma \mapsto F'(\sigma)$, of $\{\sigma \in W^{m+r,p}(\Omega, \mathbb{R}^n): \sigma(\bar{\Omega}) \subset U\}$ into the space of continuous, linear mappings from $W^{m+r,p}(\Omega, \mathbb{R}^N)$ into $W^{m,p}(\Omega)$, equipped with the topology of bounded convergence, is continuous, namely, that for any fixed $\bar{\sigma} \in W^{m+r,p}(\Omega, \mathbb{R}^N)$ with $\sigma(\bar{\Omega}) \subset U$,

$$\lim_{\|\sigma - \bar{\sigma}\|_{m+r,p} \to 0} \left\{ \frac{\|(F'(\sigma) - F'(\bar{\sigma}))(\tau)\|_{m,p}}{\|\tau\|} : 0 \neq \tau \in W^{m+r,p}(\Omega, \mathbb{R}^N) \right\} = 0.$$

This follows from the continuity of the mapping $\sigma \mapsto F_{y_j}(\sigma)$ from $\{\sigma \in W^{m+r,p}(\Omega, \mathbb{R}^N): \sigma(\bar{\Omega}) \subset U\}$ into $W^{m,p}(\Omega)$ and from the fact that, in view of Corollary 2.3, (4.1) yields

$$\|(F'(\sigma) - F'(\bar{\sigma}))(\tau)\|_{m,p} \leq \sum_{j=1}^{N} \|(F_{y_j}(\sigma) - F_{y_j}(\bar{\sigma}))\tau_j\|_{m,p}$$

$$\leq Nc \|F_{y_j}(\sigma) - F_{y_j}(\bar{\sigma})\|_{m,p} \|\tau_j\|_{m+r,p},$$

with c a number > 0 independent of τ and σ. $\square$

Theorem 4.2. *Let Ω be such that (1.1) holds. If $f \in C^{2+\sup(1,m)}(\bar{\Omega} \times U)$ with U an open subset of $\mathbb{R}^N$ and $m \geq 0$, then, for any $\lambda \in]0,1]$, $\sigma \mapsto F(\sigma)$ is a C^1 mapping from the subset $\{\sigma \in C^{m,\lambda}(\bar{\Omega}, \mathbb{R}^N): \sigma(\bar{\Omega}) \subset U\}$ of $C^{m,\lambda}(\bar{\Omega}, \mathbb{R}^N)$ into $C^{m,\lambda}(\bar{\Omega})$ and its differential at any σ is the mapping $F'(\sigma)$ defined by (4.1).*

Proof. We can deduce Theorem 4.2 from Theorem 3.3 using the fact (pointed out in §1) that $C^{m,\lambda}(\bar{\Omega})$ is a Banach algebra provided (1.1) holds. The deduction of Theorem 4.2 from Theorem 3.3 is quite analogous to that of Theorem 4.1 from Theorem 3.1. $\square$

Theorem 4.3. *Assume that Ω has the cone property and that m, k are integers $\geqq 1$. If $f \in C^{m+k}(\bar{\Omega} \times U)$ with U an open subset of $\mathbb{R}^N$, then, for any integer $r \geqq 0$ such that $(m + r)p > n$, $\sigma \mapsto F(\sigma)$ is a C^1 mapping from the subset $\{\sigma \in W^{m+r,p}(\Omega, \mathbb{R}^N): \sigma(\bar{\Omega}) \subset U\}$ of $W^{m+r,p}(\Omega, \mathbb{R}^N)$ into $W^{m,p}(\Omega)$ and, for $\sigma \in W^{m+r,p}(\Omega, \mathbb{R}^N)$, $\sigma(\bar{\Omega}) \subset U$, and $(\tau^1, \ldots, \tau^k) \in (W^{m+r,p}(\Omega, \mathbb{R}^N))^k$, we have†*

$$F^{(k)}(\sigma)(\tau^1, \ldots, \tau^k) = \sum_{j_1, \ldots, j_k}^N F_{y_{j_1}, \ldots, y_{j_k}}(\sigma)\tau_{j_1}^1 \ldots \tau_{j_k}^k, \tag{4.4}$$

where $F_{y_{j_1}, \ldots, y_{j_k}}(\sigma)$ is the real-valued function defined on Ω by

$$F_{y_{j_1}, \ldots, y_{j_k}}(\sigma)(x) = D_{y_{j_1}} \ldots D_{y_{j_k}} f(x, \sigma(x)).$$

Proof (by induction on k). Theorem 4.1 states that Theorem 4.3 is true when $k = 1$. We now suppose that the statement of Theorem 4.3 is true for a $k \geqq 1$ and we show that, consequently, it holds when replacing k with $k + 1$. Accordingly, let $f \in C^{m+k+1}(\bar{\Omega} \times U)$ with U an open subset of $\mathbb{R}^N$, and let $\sigma \in W^{m+r,p}(\Omega, \mathbb{R}^N)$ such that $\sigma(\bar{\Omega}) \subset U$. By the induction hypothesis, $\sigma \mapsto F(\sigma)$ is a C^k mapping from $\{\sigma \in W^{m+r,p}(\Omega, \mathbb{R}^N): \sigma(\bar{\Omega}) \subset U\}$ into $W^{m,p}(\Omega)$ and (4.4) holds. We denote by $\mathscr{L}_k(W^{m+r,p}(\Omega, \mathbb{R}^N), W^{m,p}(\Omega))$ the (real) Banach space of continuous, k-linear mappings from $(W^{m+r,p}(\Omega, \mathbb{R}^N))$ into $W^{m,p}(\Omega)$ endowed with the norm $\|\cdot\|_k$ defined by (see Appendix I)

$$\|\varphi\|_k = \sup\{\|\varphi(\tau^1, \ldots, \tau^k)\|_{m,p}: \|\tau^1\|_{m+r,p} \leqq 1, \ldots, \|\tau^k\|_{m+r,p} \leqq 1\}.$$

As a first step, we prove that the mapping $\sigma \mapsto F^{(k)}(\sigma)$, of $\{\sigma \in W^{m+r,p}(\Omega, \mathbb{R}^N): \sigma(\bar{\Omega}) \subset U\}$ into $\mathscr{L}_k(W^{m+r,p}(\Omega, \mathbb{R}^N), W^{m,p}(\Omega))$, is differentiable at any σ and its differential at σ is the following (continuous, linar) mapping from $W^{m+r,p}(\Omega, \mathbb{R}^N)$ into $\mathscr{L}_k(W^{m+r,p}(\Omega, \mathbb{R}^N), W^{m,p}(\Omega))$:

$$W^{m+r,p}(\Omega, \mathbb{R}^N) \ni \tau \mapsto \left[(\tau^1, \ldots, \tau^k) \mapsto \sum_{j_{k+1}}^N \left(\sum_{j_1, \ldots, j_k}^N F_{y_{j_1}, \ldots, y_{j_{k+1}}}(\sigma)\tau_{j_1}^1 \ldots \tau_{j_k}^k \right) \tau_{j_{k+1}} \right].$$

This means that

$$\lim_{\|\tau\|_{m+r,p} \to 0} \frac{1}{\|\tau\|_{m+r,p}} \sup \left\{ \left\| \sum_{j_1, \ldots, j_k}^N \left(F_{y_{j_1}, \ldots, y_{j_k}}(\sigma + \tau) - F_{y_{j_1}, \ldots, y_{j_k}}(\sigma) \right. \right. \right.$$

$$\left. \left. - \sum_{j_{k+1}}^N F_{y_{j_1}, \ldots, y_{j_{k+1}}}(\sigma)\tau_{j_{k+1}} \right) \tau_{j_1}^1 \ldots \tau_{j_k}^k \right\|_{m,p} : \|\tau^1\|_{m+r,p}$$

$$\left. \leqq 1, \ldots, \|\tau^k\|_{m+r,p} \leqq 1 \right\} = 0.$$

To justify this it suffices to remark that, in view of Corollary 2.3, we

---

† $F^{(k)}(\sigma)$ denotes the (k-linear mapping associated with the) kth differential of $\sigma \mapsto F(\sigma)$ at σ (see Appendix I), while $D_{y_{j_1}}, \ldots, D_{y_{j_k}}$ denote the partial derivatives with respect to $y_{j_1}, \ldots, y_{j_k}$, respectively.

have

$$\frac{1}{\|\tau\|_{m+r,p}} \sup \left\{ \left\| \sum_{j_1,\ldots,j_k}^{N} \left(F_{y_{j_1},\ldots,y_{j_k}}(\sigma + \tau) - F_{y_{j_1},\ldots,y_{j_k}}(\sigma) \right. \right.\right.$$

$$\left.\left.\left. - \sum_{j_{k+1}}^{N} F_{y_{j_1},\ldots,y_{j_{k+1}}}(\sigma)\tau_{j_{k+1}} \right) \tau_{j_1}^1 \cdots \tau_{j_k}^k \right\|_{m,p} : \|\tau^1\|_{m+r,p} \leqq 1, \ldots, \|\tau^k\|_{m+r,p} \leqq 1 \right\}$$

$$\leqq c \frac{1}{\|\tau\|_{m+r,p}} \sum_{j_1,\ldots,j_k}^{N} \left\| F_{y_{j_1},\ldots,y_{j_k}}(\sigma + \tau) - F_{y_{j_1},\ldots,y_{j_k}}(\sigma) \right.$$

$$\left. - \sum_{j_{k+1}}^{N} F_{y_{j_1},\ldots,y_{j_{k+1}}}(\sigma)\tau_{j_{k+1}} \right\|_{m,p},$$

with c a number > 0 independent of τ and note that, as $D_{y_{j_1}} \ldots$ $D_{y_{j_k}} f \in C^{m+1}(\bar{\Omega} \times U)$, Theorem 4.1 ensures that $\sigma \mapsto F_{y_{j_1},\ldots,y_{j_k}}(\sigma)$ is a differentiable mapping from $\{\sigma \in W^{m+r,p}(\Omega, \mathbb{R}^N): \sigma(\bar{\Omega}) \subset U\}$ into $W^{m,p}(\Omega)$ and its differential at σ is the mapping

$$W^{m+r,p}(\Omega, \mathbb{R}^N) \ni \tau \mapsto \sum_{j_{k+1}}^{N} F_{y_{j_1},\ldots,y_{j_{k+1}}}(\sigma)\tau_{j_{k+1}}.$$

As a second step of our proof, we show that $\sigma \mapsto F^{(k+1)}(\sigma)$ is a continuous mapping from $\{\sigma \in W^{m+r,p}(\Omega, \mathbb{R}^N): \sigma(\bar{\Omega}) \subset U\}$ into

$$\mathscr{L}_{k+1}(W^{m+r,p}(\Omega, \mathbb{R}^N), W^{m,p}(\Omega)).$$

To this end, we observe that if $\sigma, \bar{\sigma} \in W^{m+r,p}(\Omega, \mathbb{R}^N)$, $\sigma(\bar{\Omega}) \subset U$, and $\bar{\sigma}(\bar{\Omega}) \subset U$, then (by Corollary 2.3) we have

$$\sup \left\{ \left\| \sum_{j_1,\ldots,j_{k+1}}^{N} (F_{y_{j_1},\ldots,y_{j_{k+1}}}(\sigma) - F_{y_{j_1},\ldots,y_{j_{k+1}}}(\bar{\sigma}))\tau_{j_1}^1 \cdots \tau_{j_{k+1}}^{k+1} \right\|_{m+r,p} : \right.$$

$$\left. \|\tau^1\|_{m+r,p} \leqq 1, \ldots, \|\tau^{k+1}\|_{m+r,p} \leqq 1 \right\}$$

$$\leqq c \sum_{j_1,\ldots,j_{k+1}}^{N} \left\| F_{y_{j_1},\ldots,y_{j_k}}(\sigma) - F_{y_{j_1},\ldots,y_{j_k}}(\bar{\sigma}) \right\|_{m,p},$$

with c a number > 0 independent of σ and $\bar{\sigma}$, and note that, as $D_{y_{j_1}} \ldots D_{y_{j_{k+1}}} f \in C^m(\bar{\Omega} \times U)$, then $\sigma \mapsto F_{y_{j_1},\ldots,y_{j_{k+1}}}(\sigma)$ is a continuous mapping from $\{\sigma \in W^{m+n,p}(\Omega, \mathbb{R}^N): \sigma(\bar{\Omega}) \subset U\}$ into $W^{m,p}(\Omega)$ in view of Theorem 3.1. □

Analogously, using Theorem 3.3 and the fact that $C^{m,\lambda}(\bar{\Omega})$ is a Banach algebra provided Ω has the property (1.1) (see §1), we can prove

Theorem 4.4. *Let Ω be such that (1.1) holds and let $m \geqq 0$. If $f \in C^{1+k+\sup(1,m)}(\bar{\Omega} \times U)$ with U an open subset of $\mathbb{R}^N$ and $k \geqq 1$, then for any $\lambda \in {]}0, 1]$ $\sigma \mapsto F(\sigma)$ is a C^k mapping from the subset $\{\sigma \in C^{m,\lambda}(\bar{\Omega}, \mathbb{R}^N):$*

$\sigma(\bar{\Omega}) \subset U\}$ of $C^{m,\lambda}(\bar{\Omega}, \mathbb{R}^N)$ into $C^{m,\lambda}(\bar{\Omega})$ and (4.4) holds for any σ, τ^1, ..., $\tau^k \in C^{m,\lambda}(\bar{\Omega}, \mathbb{R}^N)$ such that $\sigma(\bar{\Omega}) \subset U$.

Let us now deal with the differentiability of the mapping $(f, \sigma) \mapsto F(\sigma)$.

Theorem 4.5. *Assume that Ω has the cone property and that $(m + r)p > n$ with m an integer ≥ 1 and r an integer ≥ 0. Let $\mathscr{A}_{m+r,p}$ be an open subset of $W^{m+r,p}(\Omega, \mathbb{R}^N)$ and let K be a convex, bounded, open subset of $\mathbb{R}^N$ such that $\sigma(x) \in K$, $\forall(x, \sigma) \in \bar{\Omega} \times \mathscr{A}_{m+r,p}$. Then $(f, \sigma) \mapsto F(\sigma)$ is a continuously differentiable mapping from $C^{m+1}(\bar{\Omega} \times \bar{K}) \times \mathscr{A}_{m+r,p}$ into $W^{m,p}(\Omega)$ and its differential at any $(\bar{f}, \bar{\sigma}) \in C^{m+1}(\bar{\Omega} \times \bar{K}) \times \mathscr{A}_{m+r,p}$ is the mapping*

$$(f, \sigma) \mapsto \sum_{j=1}^{N} \bar{F}_{y_j}(\bar{\sigma})\sigma_j + F(\bar{\sigma}) \tag{4.5}$$

from $C^{m+1}(\bar{\Omega} \times \bar{K}) \times W^{m+r,p}(\Omega, \mathbb{R}^N)$ into $W^{m,p}(\Omega)$, where $\bar{F}_{y_j}(\bar{\sigma})$ denotes the real-valued function defined on Ω by

$$\bar{F}_{y_j}(\bar{\sigma})(x) = D_{y_j}\bar{f}(x, \bar{\sigma}(x)), \qquad x \in \Omega.$$

Proof. From Theorem 3.1 it follows that $(f, \sigma) \mapsto F(\sigma)$ maps $C^m(\bar{\Omega} \times \bar{K}) \times \mathscr{A}_{m+r,p}$ into $W^{m,p}(\Omega)$. Then, in view of Corollary 2.3, we have $\bar{F}_{y_j}(\bar{\sigma})\sigma_j \in W^{m,p}(\Omega)$ for all $\sigma \in W^{m+r,p}(\Omega, \mathbb{R}^N)$ and there is a number $c > 0$, independent of σ, such that

$$\|\bar{F}_{y_j}(\bar{\sigma})\sigma_j\|_{m,p} \leq c \|\bar{F}_{y_j}(\bar{\sigma})\|_{m,p} \|\sigma_j\|_{m+r,p}.$$

Moreover, it is easy to recognize that if $f \in C^m(\bar{\Omega} \times \bar{K})$ and $\sigma \in W^{m,p}(\Omega, \mathbb{R}^N)$, then

$$\|F(\sigma)\|_{m,p} \leq c_{m,p}(\sigma)|f|_m, \tag{4.6}$$

where

$$|f|_m = \sum_{|\alpha| \leq m} \sup_{x \in \Omega \times K} |D^\alpha f(x)|$$

and $c_{m,p}(\sigma)$ is a number > 0 independent of f. Thus we have proved that the linear mapping (4.5) is continuous from $C^{m+1}(\bar{\Omega} \times \bar{K}) \times W^{m+r,p}(\Omega, \mathbb{R}^N)$ into $W^{m,p}(\Omega)$.

On the other hand, it is easy to see that the derivative of the mapping $(f, \sigma) \mapsto F(\sigma)$ at $(\bar{f}, \bar{\sigma}) \in C^{m+1}(\bar{\Omega} \times \bar{K}) \times \mathscr{A}_{m+r,p}$ with respect to $(f, \sigma) \in C^{m+1}(\bar{\Omega} \times \bar{K}) \times W^{m+r,p}(\Omega, \mathbb{R}^N)$ is $\sum_{j=1}^{N} \bar{F}_{y_j}(\bar{\sigma})\sigma_j + F(\bar{\sigma})$. Therefore (see Appendix I), to prove Theorem 4.5 it suffices to show that the mapping sending $(\bar{f}, \bar{\sigma})$ onto the function (4.5) is continuous from $C^{m+1}(\bar{\Omega} \times \bar{K}) \times \mathscr{A}_{m+r,p}$ into the space of continuous, linear mappings of $C^{m+1}(\bar{\Omega} \times \bar{K}) \times W^{m+r,p}(\Omega, \mathbb{R}^N)$ into $W^{m,p}(\Omega)$, equipped with the topology of bounded convergence. It is not difficult to see that this occurs provided the following two facts are true.

(i) The mappings $(f, \sigma) \mapsto F_{y_j}(\sigma)$ $(j = 1, ..., N)$, are continuous from $C^{m+1}(\bar{\Omega} \times \bar{K}) \times \mathscr{A}_{m+r,p}$ into $W^{m,p}(\Omega)$.

(ii) For any $\bar{\sigma} \in \mathscr{A}_{m+r,p}$ there is a number $c(\bar{\sigma}) > 0$ such that

$$\|F(\sigma) - F(\bar{\sigma})\|_{m,p} \leq c(\bar{\sigma})|f|_{m+1}\|\sigma - \bar{\sigma}\|_{m+r,p},$$

$$\forall (f, \sigma) \in C^{m+1}(\bar{\Omega} \times \bar{K}) \times \mathscr{A}_{m+r,p}.$$

Accordingly, let us show that, under our assumptions, (i) and (ii) hold. To justify (i) we observe that, for any $\sigma \in \mathscr{A}_{m+r,p}$, the linear mapping $f \mapsto F_{y_j}(\sigma)$ is continuous from $C^{m+1}(\bar{\Omega} \times \bar{K})$ into $W^{m,p}(\Omega)$, because (4.5) yields

$$\|F_{y_j}(\sigma)\|_{m,p} \leq c_{m,p}(\sigma)|D_{y_j}f|_m \leq c_{m,p}(\sigma)|f|_{m+1}, \qquad \forall f \in C^{m+1}(\bar{\Omega} \times \bar{K}).$$

We also note that from Lemma 2.1 it follows that, for any $f \in C^{m+1}(\bar{\Omega} \times \bar{K})$, the mappings $\sigma \mapsto F_{y_j}(\sigma)$ $(j = 1, \ldots, N)$, are continuous from $\mathscr{A}_{m+r,p}$ into $W^{m,p}(\Omega)$. Thus (ii) is true in view of a well-known result concerning the continuity of separately continuous mappings (see BOURBAKI [1967a], Chapitre 3, Sect. 6, Théorème 3). Finally, (ii) can be proved by iterated applications of the inequality $|f(x, \sigma(x)) - f(x, \bar{\sigma}(x))| \leq |f|_1|\sigma(x) - \bar{\sigma}(x)|$, $\forall x \in \bar{\Omega}$, which holds (by Taylor's expansion of order 1 of f) for any $f \in C^1(\bar{\Omega} \times \bar{K})$ and any $\sigma, \bar{\sigma} \in C^0(\bar{\Omega})$. Thus the proof is complete. $\square$

Theorem 4.6. *Let* Ω *be such that* (1.1) *holds and let* $\mathscr{A}_{m,\lambda}$ *be an open subset of* $C^{m,\lambda}(\bar{\Omega}, \mathbb{R}^N)$ *($m \geq 0$, $\lambda \in]0, 1]$) and K a convex, bounded, open subset of $\mathbb{R}^N$ such that $\sigma(x) \in K$, $\forall (x, \sigma) \in \bar{\Omega} \times \mathscr{A}_{m,\lambda}$. Then $(f, \sigma) \mapsto F(\sigma)$ is a continuously differentiable mapping of $C^{2+\sup(1,m)}(\bar{\Omega} \times \bar{K}) \times \mathscr{A}_{m,\lambda}$ into $C^{m,\lambda}(\bar{\Omega})$ and its differential at any $(\bar{f}, \bar{\sigma}) \in C^{2+\sup(1,m)}(\bar{\Omega} \times \bar{K}) \times \mathscr{A}_{m,\lambda}$ is the (continuous, linear) mapping* (4.5) *of* $C^{2+\sup(1,m)}(\bar{\Omega} \times \bar{K}) \times C^{m,\lambda}(\bar{\Omega}, \mathbb{R}^N)$ *into* $C^{m,\lambda}(\bar{\Omega})$.

Proof. We can proceed essentially as in the proof of Theorem 4.5, keeping in mind Theorem 3.3 and the fact that (under the assumptions made on Ω and K) $C^{m,\lambda}(\bar{\Omega})$ is a Banach algebra and $C^{m+1}(\bar{\Omega})$ and $C^{m+1}(\bar{\Omega} \times \bar{K})$ can be continuously embedded in $C^{m,\lambda}(\bar{\Omega})$ and $C^{m,\lambda}(\bar{\Omega} \times \bar{K})$, respectively. We only remark that a Hölder version of (4.6) holds. Indeed, we now prove that, if $f \in C^{m+1}(\bar{\Omega} \times \bar{K})$ and $\sigma \in C^{m,\lambda}(\bar{\Omega}, \mathbb{R}^N)$, we have

$$\|F(\sigma)\|_{m,\lambda} \leq c_{m,\lambda}(\sigma)\|f\|_{m+1,\lambda} \tag{4.7}$$

with $c_{m,\lambda}(\sigma)$ a number > 0 independent of f.

In order to show (4.7) we first observe that (4.7) is true for $m = 0$ because, if $x', x'' \in \bar{\Omega}$, $x' \neq x''$, evidently we have $|F(\sigma)|_0 \leq |f|_0$ and

$$\frac{|f(x', \sigma(x')) - f(x'', \sigma(x''))|}{|x' - x''|^\lambda}$$

$$= \frac{|f(x', \sigma(x')) - f(x'', \sigma(x''))|}{|x' - x''| + |\sigma(x') - \sigma(x'')|}\left(|x' - x''|^{1-\lambda} + \frac{|\sigma(x') - \sigma(x'')|}{|x' - x''|^\lambda}\right)$$

$$\leq \|f\|_{0,1}\left(\sup_{x',x'' \in \Omega}|x' - x''|^{1-\lambda} + \|\sigma\|_{0,\lambda}\right).$$

We now suppose that (4.7) holds for an $m \geq 0$ when $f \in C^{m+1}(\bar{\Omega} \times \bar{K})$ and $\sigma \in C^{m,\lambda}(\bar{\Omega}, \mathbb{R}^N)$ and we prove that, consequently, if $f \in C^{m+2}(\bar{\Omega} \times \bar{K})$ and $\sigma \in C^{m+1,\lambda}(\bar{\Omega}, \mathbb{R}^N)$, then we have

$$\|F(\sigma)\|_{m+1,\lambda} \leq c_{m+1,\lambda}(\sigma)\|f\|_{m+2,\lambda}, \qquad (4.8)$$

where $c_{m+1,\lambda}(\sigma)$ is a number > 0 independent of f. Since

$$\|F(\sigma)\|_{m+1,\lambda} = |F(\sigma)|_0 + \sum_{i=1}^{n} \|D_i F(\sigma)\|_{m,\lambda} \qquad \text{and} \qquad |F(\sigma)|_0 \leq |f|_0,$$

in order to prove (4.8) we need only show that

$$\|D_i F(\sigma)\|_{m,\lambda} \leq c_i(\sigma)\|f\|_{m+2,\lambda} \qquad (4.9)$$

with $c_i(\sigma)$ a number > 0 independent of f. Equation (4.9) is true because, in view of the induction hypothesis and the fact that $C^{m,\lambda}(\bar{\Omega})$ is a Banach algebra, we have

$$\|D_i F(\sigma)\|_{m,\lambda} \leq \|F_{x_i}(\sigma)\|_{m,\lambda} + \sum_{j=1}^{n} \|F_{y_j}(\sigma)D_i\sigma_j\|_{m,\lambda}$$

$$\leq c_{m,\lambda}(\sigma)\left(\|D_{x_i}f\|_{m+1,\lambda} + c'_{m,\lambda}\sum_{j=1}^{n} \|D_{y_j}f\|_{m+1,\lambda}\|D_i\sigma\|_{m,\lambda} \right)$$

$$\leq c_{m,\lambda}(\sigma)\|f\|_{m+2,\lambda}(1 + c'_{m,\lambda}\|\sigma\|_{m+1,\lambda}),$$

where $c'_{m,\lambda}$ is a number > 0 independent of f and σ. $\square$

§5. On Analyticity of Composition Operators in Sobolev and Schauder Spaces

Let U be an open subset of $\mathbb{R}^N$ and let $(x, y) \mapsto f(x, y)$ be a real-valued function of class C^∞ on $\bar{\Omega} \times U$. We will say that f is *analytic in* y *at* y_0, *uniformly with respect to* x if for each $x_0 \in \bar{\Omega}$ there is a neighborhood U_0 of (x_0, y_0) in $\bar{\Omega} \times U$ such that

$$f(x, y) = \sum_{k=0}^{+\infty} \frac{1}{k!} f_y^{(k)}(x, y_0)(y - y_0)^k \qquad (5.1)$$

for every $(x, y) \in U_0$, where $f_y^{(k)}(x, y_0)$ is the kth-order differential at y_0 of the function $y \mapsto f(x, y)$ and $(y - y_0)^k$ denotes the element $(y - y_0, \ldots, y - y_0)$ of the diagonal of $(\mathbb{R}^N)^k$. Note that, as $\bar{\Omega}$ is compact, f is analytic in y at y_0 uniformly with respect to x, if and only if there is a neighborhood V_0 of y_0 in U such that (5.1) holds for every $(x, y) \in \bar{\Omega} \times V_0$. We will say that f is *analytic in* y *uniformly with respect to* x if, for every $y_0 \in U$, f is analytic in y at y_0 uniformly with respect to x. Of course, the same definition holds if f takes its values in $\mathbb{R}^M$, where M is any integer ≥ 1.

Obviously, if f is analytic, then f is analytic in y uniformly with respect to x. For $\alpha = (\alpha_1, \ldots, \alpha_n) \in \mathbb{N}^n$ and $\beta = (\beta_1, \ldots, \beta_N) \in \mathbb{N}^N$ let $D_x^\alpha = D_{x_1}^{\alpha_1} \ldots D_{x_n}^{\alpha_n}$ and $D_y^\beta = D_{y_1}^{\beta_1} \ldots D_{y_N}^{\beta_N}$. Standard arguments show that f is analytic in y uniformly with respect to x if and only if, for every compact K of U, there is a number $c_K > 0$ such that

$$|D_y^\beta f(x, y)| \leq c_K^{1+|\beta|} \beta!, \qquad \forall (x, y, \beta) \in \bar{\Omega} \times K \times \mathbb{N}^N,$$

where $\beta! = \beta_1! \ldots \beta_N!$. We realize that the partial derivatives $(x, y) \mapsto D_y^\beta f(x, y)$ are analytic in y uniformly with respect to x provided f is analytic in y uniformly with respect to x.

Theorem 5.1. *Let U be an open subset of $\mathbb{R}^N$. Assume that Ω has the cone property and that $m \geq 1$. If $f \in C^\infty(\bar{\Omega} \times U)$ and the functions $(x, y) \mapsto D_x^\alpha f(x, y)$, $|\alpha| \leq m$, are analytic in y uniformly with respect to x, then $\sigma \mapsto F(\sigma)$ is an analytic mapping from the subset $\{\sigma \in W^{m+r, p}(\Omega, \mathbb{R}^N): \sigma(\bar{\Omega}) \subset U\}$ of $W^{m+r, p}(\Omega, \mathbb{R}^N)$ into $W^{m, p}(\Omega)$, where r is an integer ≥ 0 such that $(m + r)p > n$.*

Proof. To simplify the rather lengthy proof we prove the theorem in the case $U = \mathbb{R}^N$; the slight modifications of the proof, tending to prove the theorem in the general case, are evident when one bears in mind that if B is a closed subset of $\mathbb{R}^N$ and $g \in C^\infty(\bar{\Omega} \times B)$, then g has a C^∞ extension to $\mathbb{R}^n \times \mathbb{R}^N$ (see §1).

Then let $f \in C^\infty(\bar{\Omega} \times \mathbb{R}^N)$. From Theorem 4.3 it follows that $\sigma \mapsto F(\sigma)$ is a C^∞ mapping from $W^{m+r, p}(\Omega, \mathbb{R}^N)$ into $W^{m, p}(\Omega)$ for any integer $r \geq 0$ such that $(m + r)p > n$. We will prove that if the functions $(x, y) \mapsto D_x^\alpha f(x, y)$, $|\alpha| \leq m$, are analytic in y uniformly with respect to x, then for any integer $r \geq 0$ such that $(m + r)p > n$, the following proposition holds:

For any $\bar{\sigma} \in W^{m+r, p}(\Omega, \mathbb{R}^N)$ and any number $\varepsilon > 0$ there are two numbers > 0, $\rho_{m, p, r, \varepsilon, \bar{\sigma}}$ and $c_{m, p, r, \varepsilon, \bar{\sigma}}$, such that

$$\|F^{(k)}(\sigma)(\tau^1, \ldots, \tau^k)\|_{m, p}$$

$$\leq c_{m, p, r, \varepsilon, \bar{\sigma}} \frac{k!}{(\rho_{m, p, r, \varepsilon, \bar{\sigma}})^k} \|\tau^1\|_{m+r, p} \cdots \|\tau^k\|_{m+r, p}$$

for all $k \in \mathbb{N}$, for all $\tau^1, \ldots, \tau^k \in W^{m+r, p}(\Omega, \mathbb{R}^N)$, and for all $\sigma \in W^{m+r, p}(\Omega, \mathbb{R}^N)$ with $\|\sigma - \bar{\sigma}\|_{m+r, p} \leq \varepsilon$. (5.2)

Property (5.2) implies that the C^∞ mapping $\sigma \mapsto F(\sigma)$ of $W^{m+r, p}(\Omega, \mathbb{R}^N)$ into $W^{m, p}(\Omega)$ with $(m + r)p > n$ is analytic. Indeed, from (5.2) it follows that for all $k \in \mathbb{N}$ and all $\sigma \in W^{m+r, p}(\Omega, \mathbb{R}^N)$ with $\|\sigma - \bar{\sigma}\|_{m+r, p} \leq \varepsilon$,

$$\|F^{(k)}(\sigma)\| \leq c_{m, p, r, \varepsilon, \bar{\sigma}} \frac{k!}{(\rho_{m, p, r, \varepsilon, \bar{\sigma}})^k},$$

where

$$\|F^{(k)}(\sigma)\| = \sup\{\|F^{(k)}(\sigma)(\tau^1, \ldots, \tau^k)\|_{m,p} \colon \|\tau^1\|_{m+r,p} \leqq 1, \ldots, \|\tau^k\|_{m+r,p} \leqq 1\};$$

therefore, as the numbers $\rho_{m,p,r,\varepsilon,\bar{\sigma}}$ and $c_{m,p,r,\varepsilon,\bar{\sigma}}$ are independent of k and σ, $\sigma \mapsto F(\sigma)$ is an analytic mapping from $W^{m+r,p}(\Omega, \mathbb{R}^n)$ into $W^{m,p}(\Omega)$, by virtue of a known result concerning analytic mappings (see Appendix I).

In order to prove (5.2) we recall that if $(m+r)p > n$ (the pointwise) multiplication is a continuous mapping from $W^{m+r,p}(\Omega) \times W^{m+r,p}(\Omega)$ into $W^{m+r,p}(\Omega)$ and from $W^{m,p}(\Omega) \times W^{m+r,p}(\Omega)$ into $W^{m,p}(\Omega)$; consequently, from (4.4) it follows that

$$\|F^{(k)}(\sigma)(\tau^1, \ldots, \tau^k)\|_{m,p} \leqq c_{m,p,r}(c_{m+r,p})^k \sum_{|\beta|=k} \|F_{y^\beta}(\sigma)\|_{m,p} \|\tau^1\|_{m+r,p} \cdots \|\tau^k\|_{m+r,p},$$

where $c_{m,p,r}$ and $c_{m+r,p}$ are numbers > 0 independent of σ, $\tau^1, \ldots, \tau^k$ and $F_{y^\beta}(\sigma)$ is the real-valued function defined on Ω by

$$F_{y^\beta}(\sigma)(x) = D_y^\beta f(x, \sigma(x)) \qquad [= D_{y_1}^{\beta_1} \ldots D_{y_N}^{\beta_N} f(x, \sigma(x))]. \tag{5.3}$$

Thus, if $(m+r)p > n$, and in order to prove proposition (5.2), it suffices to prove the following proposition:

For any $\bar{\sigma} \in W^{m+r,p}(\Omega, \mathbb{R}^N)$ and any number $\varepsilon > 0$ there are two numbers > 0, $\rho'_{m,p,r,\varepsilon,\bar{\sigma}}$ and $c'_{m,p,r,\varepsilon,\bar{\sigma}}$, such that

$$\sum_{|\beta|=k} \|F_{y^\beta}(\sigma)\|_{m,p} \leqq c'_{m,p,r,\varepsilon,\bar{\sigma}} \frac{k!}{(\rho'_{m,p,r,\varepsilon,\bar{\sigma}})^k}$$

for all $k \in \mathbb{N}$ and all $\sigma \in W^{m+r,p}(\Omega, \mathbb{R}^N)$ with $\|\sigma - \bar{\sigma}\|_{m+r,p} \leqq \varepsilon$. (5.4)

Let us prove (5.4) by using induction on m. We begin by showing that if the functions $(x, y) \mapsto D_x^\alpha f(x, y)$, $|\alpha| \leqq 1$, are analytic in y uniformly with respect to x and $(1+r)p > n$, then for any $\bar{\sigma} \in W^{1+r,p}(\Omega, \mathbb{R}^N)$ and any number $\varepsilon > 0$ there are numbers $\rho'_{1,p,r,\varepsilon,\bar{\sigma}} > 0$ and $c'_{1,p,r,\varepsilon,\bar{\sigma}} > 0$ such that

$$\sum_{|\beta|=k} \|F_{y^\beta}(\sigma)\|_{1,p} \leqq c'_{1,p,r,\varepsilon,\bar{\sigma}} \frac{k!}{(\rho'_{1,p,r,\varepsilon,\bar{\sigma}})^k} \tag{5.5}$$

for all $k \in \mathbb{N}$ and all $\sigma \in W^{1+r,p}(\Omega, \mathbb{R}^N)$ with $\|\sigma - \bar{\sigma}\|_{1+r,p} \leqq \varepsilon$. Accordingly, we suppose that $(1+r)p > n$ and that the functions $(x, y) \mapsto D_x^\alpha f(x, y)$, $|\alpha| \leqq m$, are analytic in y uniformly with respect to x and we fix $\bar{\sigma} \in W^{1+r,p}(\Omega, \mathbb{R}^N)$ and $\varepsilon > 0$. Let $(e_j)_{j=1,\ldots,N}$ be the canonical base of $\mathbb{R}^N$. Proceeding as in the proof of Lemma 2.1 we arrive at

$$D_i F_{y^\beta}(\sigma) = F_{x_i, y^\beta}(\sigma) + \sum_{j=1}^{N} F_{y^{\beta+e_j}}(\sigma) D_i \sigma_j, \tag{5.6}$$

where $F_{x_i, y^\beta}(\sigma)$ denotes the real-valued function defined on Ω by

$$F_{x_i, y^\beta}(\sigma)(x) = D_{x_i} D_{y^\beta} f(x, \sigma(x)).$$

Then

$$\sum_{|\beta|=k} \|F_{y^\beta}(\sigma)\|_{1,p} \leq \sum_{|\beta|=k} \|F_{y^\beta}(\sigma)\|_{0,p} + \sum_{i=1}^{n} \|F_{x_i,y^\beta}(\sigma)\|_{0,p}$$

$$+ \sum_{\substack{i=1,\ldots,n \\ j=1,\ldots,N}} \|F_{y^{\beta+e_i}}(\sigma)D_i\sigma_j\|_{0,p} \quad . \tag{5.7}$$

Recall that (by the Sobolev embedding theorem) each $\sigma \in W^{1+r,p}(\Omega, \mathbb{R}^N)$ can be identified with a bounded function and

$$\sup_{x\in\Omega} |\sigma(x)| \leq \delta_{1+r,p} \|\sigma\|_{1+r,p},$$

with $\delta_{1,p}$ a number > 0 independent of σ. Hence

$$\sup_{x\in\Omega} |\sigma(x)| \leq \delta_{1+r,p}(\|\sigma - \bar{\sigma}\|_{1+r,p} + \|\bar{\sigma}\|_{1+r,p}) \leq \delta_{1+r,p}(\varepsilon + \|\bar{\sigma}\|_{1+r,p}).$$

Let us set $R_{r,p,\varepsilon,\bar{\sigma}} = \delta_{1+r,p}(\varepsilon + \|\bar{\sigma}\|_{1+r,p})$ and remark that the hypotheses made on f yields, for all multi-index $\beta = (\beta_1, \ldots, \beta_N)$ and all $i = 1, \ldots, n$, $j = 1, \ldots, N$,

$$\begin{cases} \sup\{|D_y^\beta f(x,y)|: x \in \Omega, |y| \leq R_{r,p,\bar{\sigma}}\} \leq (\eta_{r,p,\varepsilon,\bar{\sigma}})^{1+|\beta|}\beta!, \\ \sup\{D_y^{\beta+e_j} f(x,y): x \in \Omega, |y| \leq R_{r,p,\bar{\sigma}}\} \leq (\eta_{r,p,\varepsilon,\bar{\sigma}})^{1+|\beta|}\beta!, \\ \sup\{D_y^\beta D_{x_i} f(x,y): x \in \Omega, |y| \leq R_{r,p,\bar{\sigma}}\} \leq (\eta_{r,p,\varepsilon,\bar{\sigma}})^{1+|\beta|}\beta!, \end{cases}$$

where $\eta_{r,p,\varepsilon,\bar{\sigma}}$ is a number > 0 independent of β, i, and j. Then from (5.7) it follows that

$$\sum_{|\beta|=k} \|F_{y^\beta}(\sigma)\|_{1,p} \leq \sum_{|\beta|=k} [((\eta_{r,p,\varepsilon,\bar{\sigma}})^{1+|\beta|}\beta! + n(\eta_{r,p,\varepsilon,\bar{\sigma}})^{1+|\beta|}\beta!)(\mathrm{vol}(\Omega))^{1/p}$$

$$+ (\eta_{r,p,\varepsilon,\bar{\sigma}})^{1+|\beta|}\beta! \|\sigma\|_{1,p}],$$

namely,

$$\sum_{|\beta|=k} \|F_{y^\beta}(\sigma)\|_{1,p} \leq (\eta_{r,p,\varepsilon,\bar{\sigma}})^{1+k} \sum_{|\beta|=k} \beta![(1+n)(\mathrm{vol}(\Omega))^{1/p} + \|\sigma\|_{1,p}]. \tag{5.8}$$

We now remark that

$$\sum_{|\beta|=k} \beta! \leq k! \, N^k \tag{5.9}$$

because the cardinal of the set $\{\beta \in \mathbb{N}^N: |\beta| = k\}$ is $\leq N^k$ and, if $|\beta| = k$, we have $\beta! \leq k!$. Therefore, since

$$\|\sigma\|_{1,p} \leq \varepsilon + \|\bar{\sigma}\|_{1+r,p},$$

from (5.8) it follows that

$$\sum_{|\beta|=k} \|F_{y^\beta}(\sigma)\|_{1,p} \leq (\eta_{r,p,\varepsilon,\bar{\sigma}})^{1+k} N^k k! \, [(1+n)(\mathrm{vol}(\Omega))^{1/p}(\varepsilon + \|\bar{\sigma}\|_{1+r,p})]$$

$$= (\eta_{r,p,\varepsilon,\bar{\sigma}}N)^k k! \, \eta_{r,p,\varepsilon,\bar{\sigma}}[(1+n)(\mathrm{vol}(\Omega))^{1/p} + \varepsilon + \|\bar{\sigma}\|_{1+r,p}].$$

Thus (5.5) holds with

$$\rho'_{1,p,r,\varepsilon,\bar{\sigma}} = (N\eta_{r,p,\varepsilon,\bar{\sigma}})^{-1} \qquad \text{and}$$

$$c'_{1,p,r,\varepsilon,\bar{\sigma}} = \eta_{r,p,\varepsilon,\bar{\sigma}}[(1+n)(\text{vol}(\Omega))^{1/p} + \varepsilon + \|\bar{\sigma}\|_{1+r,p}].$$

As the next step of the induction procedure we suppose that proposition (5.4) holds for an $m \geq 1$ if $(m+r)p > n$, and the functions $(x,y) \mapsto D_x^\alpha f(x,y)$, $|\alpha| \leq m$, are analytic in y uniformly with respect to x; and we prove that, consequently, (5.4) holds when m is replaced by $m+1$ provided $(m+1+r)p > n$ and the functions $(x,y) \mapsto D_x^\alpha f(x,y)$, $|\alpha| \leq m+1$, are analytic in y uniformly with respect to x. Let us then suppose that $(m+1+r)p > n$ and that the functions $(x,y) \mapsto D_x^\alpha f(x,y)$, $|\alpha| \leq m+1$, are analytic in y uniformly with respect to x. Moreover, let ε be a number > 0 and let $\bar{\sigma} \in W^{m+1+r,p}(\Omega, \mathbb{R}^N)$. Note (in view of Lemma 2.1 and its proof) that if $\sigma \in W^{m+1+r,p}(\Omega, \mathbb{R}^N)$ then $F_{y^\beta}(\sigma)$, $F_{x_i,y^\beta}(\sigma)$, and $F_{y^{\beta+e_j}}(\sigma)$ belong to $W^{m+1+r,p}(\Omega)$ and (5.6) holds. Note also that, by the Sobolev embedding theorem, $W^{m+r+1,p}(\Omega)$ can be continuously embedded in $W^{m,q}(\Omega)$ with

$$q = \frac{np}{n - (r+1)p} \quad \text{if } (r+1)p < n \qquad \text{and}$$

$$q = (m+r+1)p \quad \text{if } (m+1)p \geq n;$$

hence, there is a number $d'_{m,p,r} > 0$ such that

$$\|v\|_{m,q} \leq d'_{m,p,r}\|v\|_{m+r+1,p}, \qquad \forall v \in W^{m+r+1,p}(\Omega).$$

Since $q \geq p$, there is a number $d''_{m,p} > 0$ such that

$$\|v\|_{m,p} \leq d''_{m,p}\|v\|_{m,q}, \qquad \forall v \in W^{m,q}(\Omega).$$

Furthermore, as (by hypothesis) $(m+r+1)p > n$, we have $mq > n$; then, from Lemma 2.1 it follows that (the pointwise) multiplication is a continuous mapping from $W^{m,q}(\Omega) \times W^{m,p}(\Omega)$ into $W^{m,p}(\Omega)$; thus, there is a number $d'''_{m,p} > 0$ such that

$$\|uv\|_{m,p} \leq d'''_{m,p,r}\|u\|_{m,q}\|v\|_{m,p}, \qquad \forall (u,v) \in W^{m,q}(\Omega) \times W^{m,p}(\Omega).$$

Then, setting $d_{m,p,r} = \sup(d'_{m,p,r}, d''_{m,p}, d'''_{m,p})$, we have, for any $\sigma \in W^{m+r+1,p}(\Omega, \mathbb{R}^N)$, any $\beta \in \mathbb{N}^N$, and any $i = 1, \ldots, n, j = 1, \ldots, N$,

$$\begin{cases} \|F_{y^{\beta+e_j}}(\sigma)D_i\sigma_j\|_{m,p} \leq d_{m,p,r}\|F_{y^{\beta+e_j}}(\sigma)\|_{m,q}\|D_i\sigma_j\|_{m,p}, \\ \|F_{y^{\beta+e_j}}(\sigma)\|_{m,q} \leq d_{m,p,r}\|F_{y^{\beta+e_j}}(\sigma)\|_{m+r+1,p}, \\ \|F_{x_i,y^\beta}(\sigma)\|_{m,p} \leq d_{m,p,r}\|F_{x_i,y^\beta}(\sigma)\|_{m,q}. \end{cases}$$

Therefore, from (5.6) we deduce

$$\sum_{|\beta|=k} \|F_{y^\beta}(\sigma)\|_{m+1,p} \leq \sum_{|\beta|=k}\left[\|F_{y^\beta}(\sigma)\|_{0,p} + d_{m,p,r}\left(\sum_{i=1}^n \|F_{x_i,y^\alpha}(\sigma)\|_{m,q}\right.\right.$$

$$\left.\left. + \sum_{\substack{i=1,\ldots,n \\ j=1,\ldots,N}} \|F_{y^{\alpha+e_j}}(\sigma)\|_{m,q}\|D_i\sigma_j\|_{m,p}\right)\right]. \qquad (5.10)$$

Remark that, as $mq > n$, the induction hypothesis applied to the functions $D_{x_i}f$ and $D_{y_j}f$ gives

$$\begin{cases} \sum_{|\beta|=k} \|F_{x_i, y^\beta}(\sigma)\|_{m,q} \leq c''_{m,p,r,\varepsilon,\bar{\sigma}} \dfrac{k!}{(\rho_{m,p,r,\varepsilon,\bar{\sigma}})^k}, \\[3mm] \sum_{|\beta|=k} \|F_{y^{\beta+e_j}}(\sigma)\|_{m,q} \leq c''_{m,p,r,\varepsilon,\bar{\sigma}} \dfrac{k!}{(\rho'_{m,p,r,\varepsilon,\bar{\sigma}})^k} \end{cases} \qquad (5.11)$$

for all $k \in \mathbb{N}$, $i = 1, \ldots, n$, $j = 1, \ldots, N$ and all $\sigma \in W^{m,q}(\Omega, \mathbb{R}^N)$ with $\|\sigma - \bar{\sigma}\|_{m,q} \leq \varepsilon$, where $c''_{m,p,r,\varepsilon,\bar{\sigma}}$ and $\rho'_{m,p,r,\varepsilon,\bar{\sigma}}$ are numbers > 0 independent of k, σ, i, j. Let $\sigma \in W^{m+1+r,p}(\Omega, \mathbb{R}^N)$ with $\|\sigma - \bar{\sigma}\|_{m+r+1,p} \leq \varepsilon$. By the Sobolev embedding theorem we have

$$\sup_{x \in \Omega} |\sigma(x)| \leq \delta_{m+r+1,p} \|\sigma\|_{m+r+1,p}$$

with $\delta_{m+r+1,p}$ being a number > 0 independent of σ. Thus

$$\sup_{x \in \Omega} |\sigma(x)| \leq \delta_{m+r+1,p}(\|\bar{\sigma}\|_{m+r+1,p} + \varepsilon)$$

and hence, in view of the hypotheses made on f, there is a number $\eta_{m,p,r,\varepsilon,\bar{\sigma}} > 0$ such that

$$\sup\{|D_y^\beta f(x, y)|: x \in \Omega, |y| \leq \delta_{m+r+1,p}(\|\bar{\sigma}\|_{m+r+1,p} + \varepsilon)\}$$

$$\leq (\eta_{m,p,r,\varepsilon,\bar{\sigma}})^{1+|\beta|}\beta!, \qquad \forall \beta \in \mathbb{N}^N. \qquad (5.12)$$

Recalling (5.9), and observing that the cardinal of $\{\beta \in \mathbb{N}^N: |\beta| = k\}$ is $\leq N^k$, from (5.10), (5.11), and (5.12) we easily deduce

$$\sum_{|\beta|=k} \|F_{y^\beta}(\sigma)\|_{m+1,p} \leq k! \, N^k \Bigg((\eta_{m,p,r,\varepsilon,\bar{\sigma}})^{1+k} (\mathrm{vol}(\Omega))^{1/p}$$

$$+ n d_{m,p}(1 + \|\sigma\|_{m+1,p}) \frac{c''_{m,p,r,\varepsilon,\bar{\sigma}}}{(\rho''_{m,p,r,\varepsilon,\bar{\sigma}})^k} \Bigg).$$

Therefore

$$\sum_{|\beta|=k} \|F_{y^\beta}(\sigma)\|_{m+1,p}$$

$$\leq k! \, N^k[1 + \varepsilon + \|\bar{\sigma}\|_{m+r+1,p} + \eta_{m,p,r,\varepsilon,\bar{\sigma}}(\mathrm{vol}(\Omega))^{1/p}]$$

$$\times \left(\frac{n d_{m,p,r}c''_{m,p,r,\varepsilon,\bar{\sigma}}}{(\rho''_{m,p,r,\varepsilon,\bar{\sigma}})^k} + (\eta_{m,p,r,\varepsilon,\bar{\sigma}})^k \right)$$

$$\leq k! \left(\frac{N}{\rho''_{m,p,r,\varepsilon,\bar{\sigma}}} \right)^k [1 + \varepsilon + \|\bar{\sigma}\|_{m+r+1,p} + \eta_{m,p,r,\varepsilon,\bar{\sigma}}(\mathrm{vol}(\Omega))^{1/p}]$$

$$\times [n d_{m,p,r}c''_{m,p,r,\varepsilon,\bar{\sigma}} + (\eta_{m,p,r,\varepsilon,\bar{\sigma}}\rho''_{m,p,r,\varepsilon,\bar{\sigma}})^k]$$

$$\leq k! \, (1 + \varepsilon + \|\bar{\sigma}\|_{m+r+1,p} + \eta_{m,p,r,\varepsilon,\bar{\sigma}}(\mathrm{vol}(\Omega))^{1/p}$$

$$\times (1 + n d_{m,p,r}c''_{m,p,r,\varepsilon,\bar{\sigma}}) \left(\frac{N \sup(1, \eta_{m,p,r,\varepsilon,\bar{\sigma}}\rho''_{m,p,r,\varepsilon,\bar{\sigma}})}{\rho''_{m,p,r,\varepsilon,\bar{\sigma}}} \right)^k.$$

Then, setting

$$\begin{cases} c_{m+1,p,r,\varepsilon,\bar{\sigma}}=(1+\varepsilon+\|\bar{\sigma}\|_{m+r+1,p}+(\text{vol}(\Omega))^{1/p}\eta_{m,p,r,\varepsilon,\bar{\sigma}})(1+nd_{m,p,r}c''_{m,p,r,\varepsilon,\bar{\sigma}}), \\ \rho_{m+1,p,r,\varepsilon,\bar{\sigma}}=\dfrac{\rho''_{m,p,r,\varepsilon,\bar{\sigma}}}{N\sup(1,\eta_{m,p,r,\varepsilon,\bar{\sigma}}\rho''_{m,p,r,\varepsilon,\bar{\sigma}})}, \end{cases}$$

we have, for all $k \in \mathbb{N}$ and all $\sigma \in W^{m+r+1,p}(\Omega, \mathbb{R}^N)$ with $\|\sigma - \bar{\sigma}\|_{m+r+1,p} \leq \varepsilon$,

$$\|F_{y^\beta}(\sigma)\|_{m+1,p} \leq c_{m+1,p,r,\varepsilon,\bar{\sigma}}\frac{k!}{(\rho_{m+1,p,r,\varepsilon,\bar{\sigma}})^k}.$$

Since the numbers $c_{m+1,p,r,\varepsilon,\bar{\sigma}}$ and $\rho_{m+1,p,r,\varepsilon,\bar{\sigma}}$ are independent of k and σ, the proof is achieved. □

Theorem 5.2. *Let Ω be such that (1.1) holds and let $m \geq 0$. If $f \in C^\infty(\bar{\Omega} \times U)$ with U an open subset of $\mathbb{R}^N$ and the functions $(x, y) \mapsto D^\alpha f(x, y)$, $|\alpha| \leq \sup(1, m)$, are analytic in y uniformly with respect to x, then $\sigma \mapsto F(\sigma)$ is an analytic mapping from $\{\sigma \in C^{m,\lambda}(\bar{\Omega}, \mathbb{R}^N): \sigma(\bar{\Omega}) \subseteq U\}$ into $C^{m,\lambda}(\bar{\Omega})$, where $\lambda \in]0, 1]$.*

Proof. As we did in the proof of Theorem 5.1, we give the proof of Theorem 5.2 in the simplified case when $U = \mathbb{R}^N$, for the sake of the complexity of the matter. Bearing in mind the well-known extension theorems of a C^∞ function on a closed subset of $\mathbb{R}^M$ (see §1) the reader will have no difficulty in seeing how we proceed when U is any open subset of $\mathbb{R}^N$. The scheme of the proof is substantially analogous to that of the proof of Theorem 5.1.

Let $f \in C^\infty(\bar{\Omega} \times \mathbb{R}^N)$. We first note that, by Theorem 4.4, $\sigma \to F(\sigma)$ is a C^∞ mapping of $C^{m,\lambda}(\bar{\Omega}, \mathbb{R}^N)$ into $C^{m,\lambda}(\bar{\Omega})$ for all $m \geq 0$ and $\lambda \in]0, 1]$. We will prove that if the functions $(x, y) \mapsto D_x^\alpha f(x, y)$, $|\alpha| \leq \sup(1, m)$, are analytic in y uniformly with respect to x, then the following proposition holds:

For any $\bar{\sigma} \in C^{m,\lambda}(\bar{\Omega} \times \mathbb{R}^N)$ and any number $\varepsilon > 0$ there are two numbers $\rho_{m,\lambda,\varepsilon,\bar{\sigma}} > 0$ and $c_{m,\lambda,\varepsilon,\bar{\sigma}} > 0$ such that

$$\|F^{(k)}(\sigma)(\tau^1, \ldots, \tau^k)\|_{m,\lambda} \leq c_{m,\lambda,\varepsilon,\bar{\sigma}}\frac{k!}{(\rho_{m,\lambda,\varepsilon,\bar{\sigma}})^k}\|\tau^1\|_{m,\lambda}\cdots\|\tau^k\|_{m,\lambda}$$

for all $k \in \mathbb{N}$, for all $\tau^1, \ldots, \tau^k \in C^{m,\lambda}(\bar{\Omega}, \mathbb{R}^N)$, and for all $\sigma \in C^{m,\lambda}(\bar{\Omega}, \mathbb{R}^N)$ with $\|\sigma - \bar{\sigma}\|_{m,\lambda} \leq \varepsilon$.

This proposition ensures that $\sigma \mapsto F(\sigma)$ is analytic from $C^{m,\lambda}(\bar{\Omega}, \mathbb{R}^N)$ into $C^{m,\lambda}(\bar{\Omega})$ (see Appendix I).

Recalling that (by (1.1)) $C^{m,\lambda}(\bar{\Omega})$ is a Banach algebra and that (by Theorem 4.4) (4.4) holds, we easily recognize that the, under the hypotheses made on f, aforesaid proposition is satisfied when the following proposition is true:

For any $\bar{\sigma} \in C^{m,\lambda}(\bar{\Omega}, \mathbb{R}^N)$ and any number $\varepsilon > 0$ there are two numbers $\rho'_{m,\lambda,\varepsilon,\bar{\sigma}} > 0$ and $c'_{m,\lambda,\varepsilon,\bar{\sigma}} > 0$ such that

$$\sum_{|\beta|=k} \|F_{y^\beta}(\sigma)\|_{m,\lambda} \leq c'_{m,\lambda,\varepsilon,\bar{\sigma}} \frac{k!}{(\rho'_{m,\lambda,\varepsilon,\bar{\sigma}})^k}$$

for all $k \in \mathbb{N}$ and all $\sigma \in C^{m,\lambda}(\bar{\Omega}, \mathbb{R}^N)$ with $\|\sigma - \bar{\sigma}\|_{m,\lambda} \leq \varepsilon$.   (5.13)

We will prove (5.13) by using induction on m. We start with the case when $m = 0$. Then let the functions $(x, y) \mapsto D_x^\alpha f(x, y)$, $|\alpha| \leq 1$, be analytic in y uniformly with respect to x, let $\bar{\sigma} \in C^{m,\lambda}(\bar{\Omega}, \mathbb{R}^N)$, and let ε be a number > 0. We propose to show that there are two numbers $\rho'_{0,\lambda,\varepsilon,\bar{\sigma}} > 0$ and $c'_{0,\lambda,\varepsilon,\bar{\sigma}} > 0$ such that

$$\sum_{|\beta|=k} \|F_{y^\beta}(\sigma)\|_{0,\lambda} \leq c'_{1,\lambda,\varepsilon,\bar{\sigma}} \frac{k!}{(\rho'_{0,\lambda,\varepsilon,\bar{\sigma}})^k}$$ (5.14)

for all $k \in \mathbb{N}$ and $\sigma \in C^{0,\lambda}(\bar{\Omega}, \mathbb{R}^N)$ with $\|\sigma - \bar{\sigma}\|_{0,\lambda \leq \varepsilon}$. By the hypotheses made on f we have

$$\begin{cases} \sup\{|D_y^\beta f(x, y)|: x \in \Omega, |y| \leq (\varepsilon + \|\bar{\sigma}\|_{0,\lambda})\} \leq (\eta_{\lambda,\varepsilon,\bar{\sigma}})^{1+|\beta|}\beta!, \\ \sup\{|D_y^\beta D_{y_j} f(x, y)|: x \in \Omega, |y| \leq (\varepsilon + \|\bar{\sigma}\|_{0,\lambda})\} \leq (\eta_{\lambda,\varepsilon,\bar{\sigma}})^{1+|\beta|}\beta!, \\ \sup\{|D_y^\beta D_{x_i} f(x, y)|: x \in \Omega, |y| \leq (\varepsilon + \|\bar{\sigma}\|_{0,\lambda})\} \leq (\eta_{\lambda,\varepsilon,\bar{\sigma}})^{1+|\beta|}\beta!, \end{cases}$$ (5.15)

for all multi-index $\beta = (\beta_1, \ldots, \beta_N)$ and all $i = 1, \ldots, n$, $j = 1, \ldots, N$, where $\eta_{\lambda,\varepsilon,\bar{\sigma}}$ is a number > 0 independent of β, i, and j.

Consider, for $(x', y'), (x'', y'') \in \bar{\Omega} \times \mathbb{R}^N$, the obvious inequality

$$|D_y^\beta f(x', y') - D_y^\beta f(x'', y'')| \leq |D_y^\beta f(x', y') - D_y^\beta f(x'', y')|$$
$$+ |D_y^\beta f(x'', y') - D_y^\beta f(x'', y'')|.$$

Note that, by (1.1), for any $y' \in \mathbb{R}^N$, any $\beta \in \mathbb{N}^N$, and any $\lambda \in \,]0, 1]$ we have

$$\sup_{\substack{x',x'' \in \Omega \\ x' \neq x''}} \frac{|D_y^\beta f(x', y') - D_y^\beta f(x'', y')|}{|x' - x''|^\lambda} \leq d \sup_{i=1,\ldots,n} \left(\sup_{x \in \Omega} |D_y^\beta D_{x_i} f(x, y')| \right),$$

with d a number > 0 independent of y', β, and λ. Moreover, a Taylor expansion of order 1 of the function $y \mapsto D_y^\beta f(x'', y)$ yields, for any $x'' \in \Omega$ and $\beta \in \mathbb{N}^N$,

$$\sup\{|D_y^\beta f(x'', y') - D_y^\beta f(x'', y'')|: |y'| \leq \varepsilon + \|\bar{\sigma}\|_{0,\lambda}, |y''| \leq \varepsilon + \|\bar{\sigma}\|_{0,\lambda}, y' \neq y''\}$$

$$\leq \sum_{j=1}^N (\sup\{|D_{y_j} D_y^\beta f(x'', y)|: |y| \leq \varepsilon + \|\bar{\sigma}\|_{0,\lambda}\})|y'_j - y''_j|.$$

Therefore, if $\|\sigma - \bar{\sigma}\|_{0,\lambda} \leqq \varepsilon$, we obtain

$$\|F_{y^\beta}(\sigma)\|_{0,\lambda} \leqq d \sup_{i=1,\ldots,n} \left(\sup\{|D_y^\beta D_{x_i} f(x, y)|: x \in \Omega, |y| \leqq \varepsilon + \|\bar{\sigma}\|_{0,\lambda}\}\right)$$

$$+ \sup_{\substack{x', x'' \in \Omega \\ x' \neq x''}} \left(\sum_{j=1}^{N} \sup|D_{y_j} D_y^\beta f(x, y)|: x \in \Omega, |y| \leqq \varepsilon + \|\bar{\sigma}\|_{0,\lambda}\} \frac{|\sigma_j(x') - \sigma_j(x'')|}{|x' - x''|^\lambda}\right)$$

and hence, recalling (5.9) and (5.15), we obtain

$$\sum_{|\beta|=k} |F_{y^\beta}(\sigma)\|_{0,\lambda} \leqq k!\, N!\, (\eta_{\lambda,\varepsilon,\bar{\sigma}})^{1+k}(d + \|\bar{\sigma}\|_{0,\lambda})$$

$$\leqq k!\, (N\eta_{\lambda,\varepsilon,\bar{\sigma}})^k \eta_{\lambda,\varepsilon,\bar{\sigma}}(d + \varepsilon + \|\bar{\sigma}\|_{0,\lambda}).$$

Thus (5.14) holds with

$$\rho'_{0,\lambda,\varepsilon,\bar{\sigma}} = (N\eta_{\lambda,\varepsilon,\bar{\sigma}})^{-1} \quad \text{and} \quad c'_{1,\lambda,\varepsilon,\bar{\sigma}} = \eta_{\lambda,\varepsilon,\bar{\sigma}}(d + \varepsilon + \|\bar{\sigma}\|_{0,\lambda}).$$

To carry out the induction procedure, we now assume that (5.13) holds for an m when the functions $(x, y) \mapsto D_x^\alpha f(x, y)$, $|\alpha| \leqq \sup(1, m)$, are analytic in y uniformly with respect to x. Moreover, we suppose that the functions $(x, y) \mapsto D_x^\alpha f(x, y)$, $|\alpha| \leqq m + 1$, are analytic in y uniformly with respect to x and we prove that for any $\bar{\sigma} \in C^{m+1,\lambda}(\bar{\Omega}, \mathbb{R}^N)$ and any number $\varepsilon > 0$ there are two numbers $\rho'_{m+1,\lambda,\varepsilon,\bar{\sigma}} > 0$ and $c'_{m+1,\lambda,\varepsilon,\bar{\sigma}} > 0$ such that

$$\sum_{|\beta|=k} \|F_{y^\beta}(\sigma)\|_{m+1,\lambda} \leqq c'_{m+1,\lambda,\varepsilon,\bar{\sigma}} \frac{k!}{(\rho'_{m+1,\lambda,\varepsilon,\bar{\sigma}})^k} \qquad (5.16)$$

for all $k \in \mathbb{N}$ and all $\sigma \in C^{m+1,\lambda}(\bar{\Omega}, \mathbb{R}^N)$ with $\|\sigma - \bar{\sigma}\|_{m+1,\lambda} \leqq \varepsilon$. Then let $\bar{\sigma} \in C^{m+1,\lambda}(\bar{\Omega}, \mathbb{R}^N)$ and let ε be a number > 0. Recalling (5.6) and the fact that $C^{m,\lambda}\bar{\Omega}$ is a Banach algebra (because of (1.1)), we get, for any $\sigma \in C^{m+1,\lambda}(\bar{\Omega}, \mathbb{R}^N)$,

$$\sum_{|\beta|=k} \|F_{y^\beta}(\sigma)\|_{m+1,\lambda} = \sum_{|\beta|=k} \left(\sup_{x \in \Omega} |D_y^\beta f(x, \sigma(x))| + \sum_{i=1}^{n} \|F_{x_i, y^\beta}(\sigma)\|_{m,\lambda}\right.$$

$$\left. + \sum_{\substack{i=1,\ldots,n \\ j=1,\ldots,N}} \|F_{y^{\beta+e_j}}(\sigma)D_i\sigma_j\|_{m,\lambda}\right)$$

$$\leqq \sum_{|\beta|=k} \left(\sup_{x \in \Omega} |D_y^\beta f(x, \sigma(x))| + \sum_{i=1}^{n} \|F_{x_i, y^\beta}(\sigma)\|_{m,\lambda}\right.$$

$$\left. + c_{m,\lambda}\|\sigma\|_{m+1,\lambda} \sup_{j=1,\ldots,N} \|F_{y^{\beta+e_j}}(\sigma)D_i\sigma_j\|_{m,\lambda}\right),$$

with $c_{m,\lambda}$ a number > 0 independent of σ. By the induction hypothesis applied to the functions $D_{x_i} f$ and $D_{y_j} f$, there are two numbers $\rho''_{m,\lambda,\varepsilon,\bar{\sigma}} > 0$

and $c''_{m,\lambda,\varepsilon,\bar{\sigma}} > 0$ such that

$$
\begin{cases}
\sum_{|\beta|=k} \|F_{x_i,y^a}(\sigma)\|_{m,\lambda} \leqq c''_{m,\lambda,\varepsilon,\bar{\sigma}} \dfrac{k!}{(\rho''_{m,\lambda,\varepsilon,\bar{\sigma}})^k}, \\[2ex]
\sum_{|\beta|=k} \|F_{y^{\beta+e_j}}(\sigma)\|_{m,\lambda} \leqq c''_{m,\lambda,\varepsilon,\bar{\sigma}} \dfrac{k!}{(\rho''_{m,\lambda,\varepsilon,\bar{\sigma}})^k},
\end{cases}
$$

for all $\sigma \in C^{m+1,\lambda}(\bar{\Omega}, \mathbb{R}^N)$ with $\|\sigma - \bar{\sigma}\|_{m+1,\lambda} \leqq \varepsilon$, for all $\beta \in \mathbb{N}^N$, and for $i = 1, \ldots, n$, $j = 1, \ldots, N$. Then, in view of (5.9) and (5.15), we obtain, for all $k \in \mathbb{N}$ and all $\sigma \in C^{m+1,\lambda}(\bar{\Omega}, \mathbb{R}^N)$ with $\|\sigma - \bar{\sigma}\|_{m+1,\lambda} \leqq \varepsilon$,

$$
\begin{aligned}
\|F_{y^\beta}(\sigma)\|_{m+1,\lambda} &\leqq k! \, N^k \left((\eta_{\lambda,\varepsilon,\bar{\sigma}})^{1+k} + \frac{nc''_{m,\lambda,\varepsilon,\bar{\sigma}} + (\varepsilon + \|\bar{\sigma}\|_{m+1,\lambda}) c_{m,\lambda} c''_{m,\lambda,\varepsilon,\bar{\sigma}}}{(\rho''_{m,\lambda,\varepsilon,\bar{\sigma}})^k} \right) \\[2ex]
&\leqq k! \, (\eta_{\lambda,\varepsilon,\bar{\sigma}} + c''_{m,\lambda,\varepsilon,\bar{\sigma}}(n + \varepsilon c_{m,\lambda} + c_{m,\lambda}\|\bar{\sigma}\|_{m+1,\lambda})) \\[1ex]
&\quad \times \left(\frac{N \sup(1, \eta_{\lambda,\varepsilon,\bar{\sigma}} c_{m,\lambda,\bar{\sigma}})}{\rho''_{m,\lambda,\varepsilon,\bar{\sigma}}} \right)^k.
\end{aligned}
$$

Thus we have proved that (5.16) holds with

$$
\begin{cases}
c'_{m+1,\lambda,\varepsilon,\bar{\sigma}} = \eta_{\lambda,\varepsilon,\bar{\sigma}} + c''_{m,\lambda,\varepsilon,\bar{\sigma}}(n + \varepsilon c_{m,\lambda} + c_{m,\lambda}\|\bar{\sigma}\|_{m+1,\lambda}), \\[2ex]
\rho'_{m+1,\lambda,\varepsilon,\bar{\sigma}} = \dfrac{\rho''_{m,\lambda,\varepsilon,\bar{\sigma}}}{N\eta_{\lambda,\varepsilon,\bar{\sigma}} \sup((\eta_{\lambda,\varepsilon,\bar{\sigma}})^{-1}, c_{m,\lambda}\rho''_{m,\lambda,\varepsilon,\bar{\sigma}})}.
\end{cases}
$$

This completes the proof. $\square$

§6. A Theorem on Failure of Differentiability for Composition Operators

In this section $(x, Z) \mapsto f(x, Z)$ denotes a real-valued function defined on $\Omega \times \mathbb{M}_{m \times n}$ with m, n integers $\geqq 1$. We recall that Ω is a bounded, open subset of $\mathbb{R}^n$ and that $\mathbb{M}_{m \times n}$ denotes the linear space of real $m \times n$ matrices. For any function $\sigma: \Omega \to \mathbb{M}_{m \times n}$ let $F(\sigma)$ be the real-valued function defined on Ω by (3.1).

Suppose that f satisfies the Carathéodory condition, i.e., that f is continuous in Z for almost all $x \in \Omega$ and measurable in x for each fixed Z. Then we can prove (see, e.g., KRASNOSELSKIJ et al. [1976]) that $F(\sigma)$ is a measurable function whenever σ is measurable. Thus $\sigma \mapsto F(\sigma)$ maps $L^p(\Omega, \mathbb{M}_{m \times n})$ into $L^q(\Omega)$ with p, q real numbers $\geqq 1$, provided f satisfies the growth condition

$$|f(x, Z)| \leqq \varphi(x) + c|Z|^{p/q}, \tag{6.1}$$

where $\varphi \in L^q(\Omega)$ and c is a number > 0 independent of (x, Z). Furthermore, it is known that $\sigma \mapsto F(\sigma)$ maps $L^p(\Omega, \mathbb{M}_{m \times n})$ into $L^q(\Omega)$ (if and) only if (6.1) holds, and that whenever $\sigma \mapsto F(\sigma)$ maps $L^p(\Omega, \mathbb{M}_{m \times n})$ into $L^q(\Omega)$ it is continuous from $L^p(\Omega, \mathbb{M}_{m \times n})$ into $L^q(\Omega)$ (see KRASNOSELSKIJ *et al.* [1976]).

As far as the differentiability of $\sigma \mapsto F(\sigma)$ is concerned, it is not very difficult to show the following (rather surprising) fact: when $p \leq q$ the mapping $F: L^p(\Omega, \mathbb{M}_{m \times n}) \to L^q(\Omega)$ is differentiable at some $\sigma \in L^p(\Omega, \mathbb{M}_{m \times n})$ (if and) only if, for almost all $x \in \Omega$, the function $Z \mapsto f(x, Z)$ is affine (see, again, KRASNOSELSKIJ *et al.* [1976], Theorem 20.1 and PRODI & AMBROSETTI [1973], Teorema 36 for the case $p = q = 2$). In many questions connected with boundary value problems for second-order nonlinear differential operators (in particular, for operators of the finite elastostatic type) it is interesting to establish whether an analogous fact holds for the mapping

$$F \circ D: V \to L^q(\Omega),$$

where V is any subspace of $W^{1,p}(\Omega, \mathbb{R}^m)$ containing $W_0^{1,p}(\Omega, \mathbb{R}^m)$. The answer is positive for $V = W_0^{1,p}(\Omega, \mathbb{R}^m)$ and hence for all subspace V of $W^{1,p}(\Omega, \mathbb{R}^m)$ containing $W_0^{1,p}(\Omega, \mathbb{R}^m)$, as stated in the following theorem.

Theorem 6.1. *Let* $(x, Z) \mapsto f(x, Z)$ *be a real-valued function defined on* $\Omega \times \mathbb{M}_{m \times n}$ *such that for almost all* $x \in \Omega$ *the function* $Z \mapsto f(x, Z)$ *is of class* C^1 *in* $\mathbb{M}_{m \times n}$, *and that for all* $Z \in \mathbb{M}_{m \times n}$ *the functions* $x \mapsto F(x, Z)$ *and* $x \mapsto D_{Z_{ij}} f(x, Z)$ $(i = 1, \dots, m; j = 1, \dots, n)$ *are measurable. Assume further that* $p \leq q$ *and that the functions* f *and* $D_{Z_{ij}} f$ *satisfy a growth condition of the type (6.1) with* $\varphi \in L^q(\Omega)$ *and* $c \in \mathbb{R}$. *Then the (continuous) mapping* $v \mapsto F(Dv)$ *of* $W_0^{1,p}(\Omega, \mathbb{R}^m)$ *into* $L^q(\Omega)$ *is differentiable at some* $\bar{v} \in W_0^{1,p}(\Omega, \mathbb{R}^m)$ *(if and) only if, for almost* $x \in \Omega$, *the function* $Z \mapsto f(x, D\bar{v}(x) + Z) - f(x, D\bar{v}(x))$ *is linear.*

Proof. Without loss of generality, we can suppose that $v = 0$ and $f(x, 0) = 0$, $\forall x \in \Omega$ (otherwise, we replace the function f with the function $(x, Z) \mapsto f(x, D\bar{v}(x) + Z) - f(x, D\bar{v}(x)))$. Then let $v \mapsto F(Dv)$ be differentiable at 0 when it is regarded as a mapping of $W_0^{1,p}(\Omega, \mathbb{M}_{m \times n})$ into $L^q(\Omega)$. We first show that its differential at 0 carries the function $v \in W_0^{1,p}(\Omega, \mathbb{M}_{m \times n})$ to the function

$$x \mapsto \sum_{\substack{i=1,\dots,m \\ j=1,\dots,n}} D_{Z_{ij}} f(x, 0) D_j v_i(x), \tag{6.2}$$

where $v_i(x) = v(x) \cdot e_i$ and $D_{Z_{ij}}$ is the partial, derivative operator with respect to the coordinate Z_{ij} of Z. Accordingly, let us denote by N the mapping $v \mapsto F(Dv)$; so, for any $v \in W_0^{1,p}(\Omega, \mathbb{M}_{m \times n})$, $N(v)$ is the real-valued function defined on Ω by setting $N(v)(x) = f(x, Dv(x))$, $\forall x \in \Omega$. As $f(x, y) = 0$, $\forall x \in \Omega$, we have for any $v \in W_0^{1,p}(\Omega, \mathbb{M}_{m \times n})$

$$\lim_{\mathbb{R} \ni t \to 0} \left\| \frac{N(tv)}{t} - N'(0)(v) \right\|_{0,q} = 0.$$

Then, since any sequence in $L^q(\Omega)$ converging to 0 admits a subsequence which converges to 0 almost everywhere, there is a decreasing sequence $(t_h)_{h \in \mathbb{N}}$ in $\mathbb{R}$ converging to 0 such that for almost all $x \in \Omega$

$$\lim_{h \to 0} \frac{f(x, t_h Dv(x))}{t_h} = N'(0)(v)(x).$$

On the other hand, recalling (by hypothesis) that for almost all $x \in \Omega$ the function $Z \mapsto f(x, Z)$ is differentiable, we have for almost all $x \in \Omega$

$$\lim_{h \to 0} \frac{f(x, t_h Dv(x))}{t_h} = D_{Z_{ij}} f(x, 0) D_j v_i(x).$$

Thus

$$N'(0)(v)(x) = \sum_{\substack{i=1,\ldots,m \\ j=1,\ldots,n}} D_{Z_{ij}} f(x, 0) D_j v_i(x),$$

as we wanted.

We shall prove that, from the fact that the mapping

$$N: W_0^{1,p}(\Omega, \mathbb{M}_{m \times n}) \to L^q(\Omega)$$

is differentiable at 0, it follows that for almost all $x \in \Omega$ and for all $Z \in \mathbb{M}_{m \times n}$,

$$f(x, Z) = \sum_{\substack{i=1,\ldots,m \\ j=1,\ldots,n}} D_{Z_{ij}} f(x, 0) Z_{ij}.$$

Since, for almost all $x \in \Omega$, the function $Z \mapsto f(x, Z)$ is continuous, in order to prove the previous equality it suffices, evidently, to show that for each $m \times n$ matrix $\bar{Z}$ on $\mathbb{Q} \backslash \{0\}$, with $\mathbb{Q}$ the set of rational numbers, we have

$$f(x, \bar{Z}) = \sum_{\substack{i=1,\ldots,m \\ j=1,\ldots,n}} D_{Z_{ij}} f(x, 0) \bar{Z}_{ij} \tag{6.3}$$

for almost all $x \in \Omega$. We then fix an $m \times n$ matrix on $\mathbb{Q} \backslash \{0\}$. We recall that if $(B_k(x))_{k \in \mathbb{N}}$ is a sequence of balls of Ω centered at x, such that the sequence $(\text{vol}(B_k(x))_{k \in \mathbb{N}}$ is decreasing and infinitesimal as $k \to \infty$, then (by a known fact in measure theory) for almost all $\bar{x} \in \Omega$ we have

$$\lim_{k \to \infty} \frac{\int_{B_k(\bar{x})} \left| f(x, \bar{Z}) - \sum_{\substack{i=1,\ldots,m \\ j=1,\ldots,n}} D_{Z_{ij}} f(x, 0) Z_{ij} \right|^q dx}{\text{vol}(B_k(\bar{x}))}$$

$$= \left| f(\bar{x}, \bar{Z}) - \sum_{\substack{i=1,\ldots,m \\ j=1,\ldots,n}} D_{Z_{ij}} f(x, 0) Z_{ij} \right|. \tag{6.4}$$

We will show that for each $\bar{x} \in \Omega$ there exists a sequence $(B_k(\bar{x}))_{k \in \mathbb{N}}$ of balls of Ω centered at $\bar{x}$, such that the sequence $(\text{vol}(B_k(\bar{x}))_{k \in \mathbb{N}}$ is decreasing and infinitesimal as $k \to \infty$ and that

$$\lim_{k \to \infty} \frac{\displaystyle\int_{B_k(\bar{x})} \left| f(x, \bar{Z}) - \sum_{\substack{i=1,\dots,m \\ j=1,\dots,n}} D_{Z_{ij}} f(x, 0) \bar{Z}_{ij} \right|^q dx}{\mathrm{vol}(B_k(\bar{x}))} = 0. \tag{6.5}$$

Thus, in view of (6.4), we have proved that (6.3) holds for almost all $x \in \Omega$. To this end, we fix (arbitrarily) $\bar{x} = (\bar{x}_j)_{j=1,\dots,n} \in \Omega$. For any $i = 1, \dots, m$ and $k \in \mathbb{N}$ we consider the real-valued function u defined on Ω by setting

$$u_i^k(x) = \begin{cases} 0 & \text{if } \dfrac{|x_j - \bar{x}_j|}{c_{ij}^k} \geq 1, \\[3mm] \dfrac{\bar{Z}_{i1}}{k} - \dfrac{|\bar{Z}_{i1}|}{k} \displaystyle\sum_{j=1}^{n} \dfrac{|x_j - \bar{x}_j|}{c_{ij}^k} & \text{if } \dfrac{|x_j - \bar{x}_j|}{c_{ij}^k} \leq 1, \end{cases}$$

where $c_{ij}^k = (1/k)|\bar{Z}_{i1}/Z_{ij}|$. Let, for $i = 1, \dots, m$ and $k \in \mathbb{N}$,

$$U_{i,k} = \left\{ x \in \mathbb{R}^n : x_j \neq \bar{x}_j, \forall j = 1, \dots, n, \sum_{j=1}^{n} \frac{|x_j - \bar{x}_j|}{c_{ij}^k} < 1, \frac{x_j - \bar{x}_j}{|x_j - \bar{x}_j|} = -\frac{\bar{Z}_{ij}}{|\bar{Z}_{ij}|} \right\}$$

and, for $i = 1, \dots, m$, choose $r^i \in \mathbb{R}^n$ such that $\bar{x} + r^i \in U_{i,1}$, so that

$$\bar{x} + \frac{r^i}{k} \in U_{i,k}, \qquad \forall k \in \mathbb{N}.$$

Now we define, for $i = 1, \dots, m$ and $k \in \mathbb{N}$, a function $v_i^k : \Omega \to \mathbb{R}$ by putting

$$v_i^k(x) = u_i^k\left(x + \frac{r^i}{k} \right)$$

and we put $v^k = (v_i^k)_{i=1,\dots,m}$. Setting

$$V_k = \bigcap_{i=1}^{m} \left(U_{i,k} - \frac{r^i}{k} \right),$$

we have, clearly, $\bar{x} \in V_k$ and

$$Dv^k(x) = \bar{Z}, \qquad \forall x \in V_k.$$

It is easy to see that $v^k \in W^{1,p}(\Omega, \mathbb{R}^m)$ and that, at least for k sufficiently large, the support of v^k is compact in Ω. Hence, at least for k large enough, $v^k \in W_0^{1,p}(\Omega, \mathbb{R}^m)$. Moreover, from the definition of v^k it evidently follows that

$$\lim_{k \to \infty} \|Dv^k\|_{0,p} = 0 \tag{6.6}$$

and that there is a number $c > 0$ such that, for all $i = 1, \dots, m$ and $k \in \mathbb{N}$,

$$\mathrm{vol}(\mathrm{supp}\, v_i^k) \leq c \frac{1}{k^n}. \tag{6.7}$$

We now prove that if $B_k(\bar{x})$ is a ball of Ω centered at $\bar{x}$ and contained in the open subset V^1 of Ω and if ε_1 is its radius, then for all $k \in \mathbb{N}$ the ball $B_k(\bar{x})$ centered at $\bar{x}$ and of radius ε_1/k is contained in V^k. We observe that if $x \in B_k(\bar{x})$, namely, if $\sum_{j=1}^{n}(x_j - \bar{x}_j)^2 < (\varepsilon_1/k)^2$, then $\bar{x} + k(x - \bar{x}) \in B_1(\bar{x}) \subseteq \bigcap_{i=1}^{m}(U_{i,1} - r^i)$; therefore, for any $i = 1, \ldots, m$ and any $j = 1, \ldots, n$, we have

$$
\begin{cases}
\dfrac{k(x_j - \bar{x}) + r_j^i}{|k(x_j - \bar{x}_j) + r_j^i|} = -\dfrac{\bar{Z}_{ij}}{|\bar{Z}_{ij}|}, \\[2ex]
\displaystyle\sum_{j=1}^{n} \dfrac{|k(x_j - \bar{x}_j) + r_j^i|}{c_{ij}^1} < 1, \\[2ex]
k(x_j - \bar{x}_j) + r_j^i \neq 0,
\end{cases}
$$

and, consequently,

$$
\begin{cases}
\dfrac{x_j - \bar{x}_j + (r_j^i/k)}{|x_j - \bar{x}_j + r_j^i/k|} = -\dfrac{\bar{Z}_{ij}}{|\bar{Z}_{ij}|}, \\[2ex]
\displaystyle\sum_{j=1}^{n} \dfrac{|x_j - \bar{x}_j + (r_j^i/k)|}{c_{ij}^k} < 1, \\[2ex]
x_j - \bar{x}_j + (r_j^i/k) \neq 0.
\end{cases}
$$

This is the set of conditions on x defining $U_{i,k} - (r^i/k)$; hence, $x \in \bigcap_{i=1}^{m}(U_{i,k} - (r^i/k))$. Therefore $B_k(\bar{x}) \subseteq V_k$.

In this way we have constructed a sequence $(B_k(\bar{x}))_{k \in \mathbb{N}}$ of balls centered at $\bar{x}$ such that

$$\text{vol}(B_k(\bar{x})) = \frac{c_1}{k^n} \tag{6.8}$$

with c_1 a number > 0 independent of k, and that

$$Dv^k(x) = \bar{Z}, \qquad \forall x \in B_k(\bar{x}). \tag{6.9}$$

Note that, by Poincaré's inequality (see §2), the mapping $v \mapsto \|Dv\|_{0,p}$ is a norm on $W_0^{1,p}(\Omega, \mathbb{R}^m)$ equivalent to the norm $v \mapsto \|v\|_{1,p}$. Hence (6.6) implies

$$\lim_{k \to \infty} \|v^k\|_{1,p} = 0.$$

Then, since the differential at 0 of the mapping $N: W_0^{1,p}(\Omega, \mathbb{R}^m) \to L^q(\Omega)$ maps $v \in W_0^{1,p}(\Omega, \mathbb{R}^m)$ onto the function (6.2), we have

$$\lim_{k \to \infty} \frac{\left(\displaystyle\int_\Omega \left| f(x, Dv^k(x)) - \sum_{\substack{i=1,\ldots,m \\ j=1,\ldots,n}} D_{Z_{ij}} f(x,0) D_j v_i^k(x) \right|^q dx \right)^{1/q}}{\|Dv^k\|_{0,p}} = 0. \tag{6.10}$$

In view of (6.7) and (6.9), from (6.10) it follows that

$$\lim_{k\to\infty} \frac{\left(\int_{B_k(\bar{x})} \left| f(x,\bar{Z}) - \sum_{\substack{i=1,\ldots,m \\ j=1,\ldots,n}} D_{Z_{ij}} f(x,0)\bar{Z}_{ij}\right|^q dx\right)^{1/q}}{\left(\sum_{\substack{i=1,\ldots,m \\ j=1,\ldots,n}} |\bar{Z}_{ij}|^p\right)^{1/p} \left(c\dfrac{1}{k^n}\right)^{1/p}} = 0. \qquad (6.11)$$

Remark that, as $p \leq q$, in view of (6.8) we have

$$\left(c\frac{1}{k^n}\right)^{1/p} \left(c\frac{1}{k^n}\right)^{1/q} = \left(\frac{c}{c_1}(\operatorname{vol}(B_k(\bar{x}))\right)^{1/q}.$$

Then (6.11) yields

$$\left(\sum_{\substack{i=1,\ldots,m \\ j=1,\ldots,n}} |\bar{Z}_{ij}|^p\right)^{1/p} \lim_{k\to\infty} \frac{\left(\int_{B_k(\bar{x})} \left| f(x,\bar{Z}) - \sum_{\substack{i=1,\ldots,m \\ j=1,\ldots,n}} D_{Z_{ij}} f(x,0)\bar{Z}_{ij}\right|^q dx\right)}{\operatorname{vol}(B_k(\bar{x}))} = 0;$$

therefore (6.5) holds and thus the proof is complete. □

This theorem generalizes a theorem of VALENT [1978a] and a theorem of VALENT & ZAMPIERI [1977]; the proof presented here differs only by a slight modification from the one given in VALENT & ZAMPIERI [1977].

CHAPTER III

Dirichlet and Neumann Boundary Problems in Linearized Elastostatics. Existence, Uniqueness, and Regularity

Our approach to boundary problems of finite elastostatics is of local type; hence, it requires the existence of deformations at which the linearized operator is a homeomorphism between suitable Banach spaces. The aim of this chapter is to provide appropriate results of existence, uniqueness, and regularity for the principal boundary problems of linearized elastostatics: namely, for the Dirichlet and Neumann problems.

A generalization of a theorem of LAX & MILGRAM proved in §2 (see Theorems 2.3 and 2.4) gives the general basis of the $W^{1,p}$ treatment, while Theorem 3.3 (with the aid of Dirichlet and Korn inequalities) furnishes a means of connecting some special situations to the general theory.

The regularity theorems are obtained using a method of "continuity," but they could be proved by a different technique. We remark that our regularity results require rather little smoothness of some coefficients: actually, as such results are proved with a view to their use in studying a nonlinear problem, the degree of smoothness of the coefficients cannot be chosen arbitrarily but is imposed by the nonlinear problem.

§1. Korn Inequalities

For $u \in C^1(\Omega, \mathbb{R}^n)$, we denote by $D^S u$ the symmetric part of Du, i.e., we set

$$D^S u = \tfrac{1}{2}(D_j u_i + D_i u_j)_{i,j=1,\ldots,n}.$$

Remark 1.1. *For any open subset Ω of $\mathbb{R}^n$ the following (first Korn inequality) holds*:

$$\int_\Omega |Du|^2 \, dx \le 2 \int_\Omega |D^S u|^2 \, dx, \qquad \forall u \in W_0^{1,2}(\Omega, \mathbb{R}^n). \tag{1.1}$$

Proof. It suffices to prove that (1.1) holds for all $u \in \mathscr{D}(\Omega, \mathbb{R}^n)$, because $\mathscr{D}(\Omega)$ is dense in $W_0^{1,p}(\Omega)$. Then let $u \in \mathscr{D}(\Omega, \mathbb{R}^n)$. We denote by $D^W u$ the skew part of Du, namely, let $D^W u = \frac{1}{2}(D_j u_i - D_i u_j)_{i,j=1,\dots,n}$ and observe that

$$|D^S u|^2 - |D^W u|^2 = (Du) \cdot (Du)^{\mathsf{T}}.$$

It is easily seen that

$$(Du) \cdot (Du)^{\mathsf{T}} = \operatorname{div}((Du)u - (\operatorname{div} u)u) + (\operatorname{div} u)^2.$$

Then, recalling that u vanishes on $\partial\Omega$, we immediately obtain by the use of the divergence theorem

$$\int_\Omega (|D^S u|^2 - |D^W u|^2) \, dx = \int_\Omega (Du) \cdot (Du)^{\mathsf{T}} \, dx = \int_\Omega (\operatorname{div} u)^2 \, dx \ge 0.$$

Therefore

$$\int_\Omega |D^W u|^2 \, dx \le \int_\Omega |D^S u|^2 \, dx,$$

which implies (1.1), because $|Du|^2 = |D^S u|^2 + |D^W u|^2$. $\square$

Let us denote by $\mathscr{R}$ the set of all *rigid deformations* of $\mathbb{R}^n$ (see Chapter I, §2), namely, the set of functions $\rho : \mathbb{R}^n \to \mathbb{R}^n$ of the type

$$\rho(x) = c + Rx, \qquad x \in \mathbb{R}^n,$$

with $c \in \mathbb{R}^n$ and $R \in \mathbb{O}_n^+$. We will regard $\mathscr{R}$ as a submanifold of the (real) topological linear space of all affine functions of $\mathbb{R}^n$ into itself (equipped with the topology of bounded convergence), so that it can be identified with the submanifold $\mathbb{R}^n \times \mathbb{O}_n^+$ of $\mathbb{R}^n \times \mathbb{M}_n$.

Recalling that the function $W \mapsto \exp W$ is a C^∞ diffeomorphism of $U \cap \operatorname{Skew}_n$ onto $V \cap \mathbb{O}_n^+$, with U and V suitable open neighborhoods of O and I, respectively, in $\mathbb{M}_n$; and that its differential at O is the identity function of $\mathbb{M}_n$ into itself (see Appendix II), it is easy to recognize that the tangent space $T_{I_{\mathbb{R}^n}}(\mathscr{R})$ to the manifold $\mathscr{R}$ at the identity function $I_{\mathbb{R}^n} : \mathbb{R}^n \to \mathbb{R}^n$ is the set of functions $r : \mathbb{R}^n \to \mathbb{R}^n$ of the type

$$r(x) = c + Wx, \qquad x \in \mathbb{R}^n,$$

with $c \in \mathbb{R}^n$ and $W \in \operatorname{Skew}_n$. Moreover, as $\mathscr{R}$ is a topological group for the composition mapping, we easily see that if $\rho_0 \in \mathscr{R}$ and $T_{\rho_0}(\mathscr{R})$ denote the tangent space at ρ_0 to $\mathscr{R}$, then

$$T_{\rho_0}(\mathscr{R}) = \{r \circ \rho_0 : r \in T_{\iota_{\mathbb{R}^n}}(\mathscr{R})\}.$$

The set of rigid deformations of $\bar{\Omega}$, i.e., the set of restrictions to $\bar{\Omega}$ of the elements of $\mathscr{R}$, will be denoted by $\mathscr{R}_{\bar{\Omega}}$. Then the tangent space $T_{\iota_{\bar{\Omega}}}(\mathscr{R}_{\bar{\Omega}})$ to the manifold $\mathscr{R}_{\bar{\Omega}}$ at the identity function $\iota_{\bar{\Omega}}: \bar{\Omega} \to \mathbb{R}^n$ is the set of restrictions to $\bar{\Omega}$ of the elements of $T_{\iota_{\mathbb{R}^n}}(\mathscr{R})$; furthermore, for any $\rho_0 \in \mathscr{R}_{\bar{\Omega}}$, we have $T_{\rho_0}(\mathscr{R}_{\bar{\Omega}}) = \{r \circ \rho_0 : r \in T_{\iota_{\mathbb{R}^n}}(\mathscr{R})\}$. The elements of $T_{\iota_{\mathbb{R}^n}}(\mathscr{R})$ [respectively $T_{\iota_{\bar{\Omega}}}(\mathscr{R}_{\bar{\Omega}})$] shall be called *infinitesimal rigid displacements* of $\mathbb{R}^n$ [respectively $\bar{\Omega}$].

Remark 1.2. $T_{\iota_{\mathbb{R}^n}}(\mathscr{R})$ *is the kernel of the mapping* $D^S: \mathscr{D}'(\mathbb{R}^n, \mathbb{R}^n) \to \mathscr{D}'(\mathbb{R}^n, \mathsf{M}_n)$.

Proof. Obviously, $D^S u = 0$, $\forall u \in T_{\iota_{\mathbb{R}^n}}(\mathscr{R})$. Now let $u \in \mathscr{D}'(\mathbb{R}^n, \mathbb{R}^n)$ satisfy $D^S u = 0$. Since the operator D^S is elliptic, we have $u \in C^\infty(\mathbb{R}^n, \mathbb{R}^n)$ (cf., e.g., FICHERA [1972]). Then, as $D_j u_i + D_i u_j = 0$ for $i, j = 1, \ldots, n$, we easily see that

$$D_h D_j u_i = 0 \quad \text{for} \quad i, j, h = 1, \ldots, n,$$

which implies $u(x) = c + Wx$ with $c \in \mathbb{R}^n$ and $W \in \mathsf{M}_n$. Finally, the conditions $D_j u_i + D_i u_j = 0$ for $i, j = 1, \ldots, n$ yield $W + W^{\mathsf{T}} = 0$. Thus $u \in T_{\iota_{\mathbb{R}^n}}(\mathscr{R})$. □

Remark 1.3. *We have*

$$W^{1,p}(\Omega) = \{r|_\Omega : r \in T_{\iota_{\mathbb{R}^n}}(\mathscr{R})\}$$

$$\oplus \left\{ v \in W^{1,p}(\Omega, \mathbb{R}^n) : \int_\Omega v \, dx = 0, \int_\Omega Dv \, dx \in \mathrm{Sym}_n \right\}.$$

Proof. Let

$$\mathscr{V}^{1,p} = \left\{ v \in W^{1,p}(\Omega, \mathbb{R}^n) : \int_\Omega v \, dx = 0, \int_\Omega Dv \, dx \in \mathrm{Sym}_n \right\}. \tag{1.2}$$

Clearly, $\{r|_\Omega : r \in T_{\iota_{\mathbb{R}^n}}(\mathscr{R})\} \cap \mathscr{V}^{1,p} = 0$. Moreover, $W^{1,p}(\Omega, \mathbb{R}^n) = \{r|_\Omega : r \in T_{\iota_{\mathbb{R}^n}}(\mathscr{R})\} + \mathscr{V}^{1,p}$, namely, for any $u \in W^{1,p}(\Omega, \mathbb{R}^n)$ there are $c \in \mathbb{R}^n$ and $W \in \mathrm{Skew}_n$ such that the function $\Omega \ni x \mapsto u(x) - c - Wx$ belongs to $\mathscr{V}^{1,p}$. Indeed, this function belongs to $\mathscr{V}^{1,p}$ if and only if

$$\begin{cases} \displaystyle\int_\Omega u \, dx = \int_\Omega c \, dx + W \int_\Omega x \, dx, \\[2mm] \displaystyle\int_\Omega (Du - (Du)^{\mathsf{T}}) \, dx = 2 \int_\Omega W \, dx, \end{cases}$$

and from these equations we can obtain W and c for any fixed u. □

Theorem 1.4. *If Ω has the cone property, then the following (second Korn inequality) holds:*

$$\|u\|_{1,2}^2 \leqq c \left(\int_\Omega |u|^2 \, dx + \int_\Omega |D^S u|^2 \, dx \right), \qquad \forall u \in W^{1,2}(\Omega, \mathbb{R}^n), \quad (1.3)$$

where c is a number > 0 independent of u.

For the proof we refer to GOBERT [1962] and FICHERA [1972a], while here we point out the following consequence of Theorem 1.4.

Corollary 1.5. *If Ω has the cone property, then*

$$\|v\|_{1,2}^2 \leqq c_1 \int_\Omega |D^S v|^2 \, dx, \qquad \forall v \in \mathscr{V}^{1,2}, \quad (1.4)$$

where c is a number > 0 independent of v and $\mathscr{V}^{1,2}$ is defined by (1.2).

Proof. In view of Theorem 1.4 it suffices to verify that

$$\int_\Omega |v|^2 \, dx \leqq c_2 \int_\Omega |D^S v|^2 \, dx, \qquad \forall v \in \mathscr{V}^{1,2}, \quad (1.5)$$

with c_2 a number > 0 independent of v. Accordingly, suppose that (1.5) fails. Then there is a sequence $(v^k)_{k \in \mathbb{N}}$ in $\mathscr{V}^{1,2}$ such that

$$\int_\Omega |v^k|^2 \, dx = 1, \qquad \lim_{k \to \infty} \int_\Omega |D^S v^k|^2 \, dx = 0. \quad (1.6)$$

Thus, by (1.3), the sequence $(v^k)_{k \in \mathbb{N}}$ is bounded in $W^{1,2}(\Omega, \mathbb{R}^n)$, and hence (as $W^{1,2}(\Omega, \mathbb{R}^n)$ is reflexive) a subsequence of $(v^k)_{k \in \mathbb{N}}$, which we still denote by $(v^k)_{k \in \mathbb{N}}$, converges weakly in $W^{1,2}(\Omega, \mathbb{R}^n)$. If v is its limit, then we have that, evidently, $v \in \mathscr{V}^{1,2}$ and $(D^S v^k)_{k \in \mathbb{N}}$ converge to $D^S v$ weakly in $L^2(\Omega, \mathbb{M}_n)$. Hence, $D^S v = 0$ because of $(1.6)_2$; thus, by Remark 1.2, v is an infinitesimal rigid displacement belonging to $\mathscr{V}^{1,2}$, and so, by Remark 1.3, $v = 0$. Then $(v^k)_{k \in \mathbb{N}}$ converges to 0 weakly in $W^{1,2}(\Omega, \mathbb{R}^n)$ and hence some subsequence of $(v^k)_{k \in \mathbb{N}}$ converges to 0 in $L^2(\Omega, \mathbb{R}^n)$, by virtue of the compactness of the identity function of $W^{1,2}(\Omega)$ into $L^2(\Omega)$ (see Chapter 2, §1). This conflicts with $(1.6)_1$. Thus we can conclude that (1.5) holds. □

§2. A Generalization of a Theorem of Lax and Milgram

Let X, Y be topological linear spaces and X', Y' their duals. The transpose of a continuous, linear mapping $f: X \to Y$ is the (continuous, linear) mapping $f^T: Y' \to X'$ defined by putting, for any $y' \in Y'$,

$$f^T(y') = y' \circ f \quad (2.1)$$

Remark 2.1. *Let X, Y be locally convex topological linear spaces and let $f: X \to Y$ be a continuous, linear mapping. The set $f(X)$ is dense in Y if and only if $f^T: Y' \to X'$ is one-to-one.*

Proof. From (2.1) it follows that $f^T(y') = 0$ if and only if $y'|_{f(X)} = 0$. Hence f^T is one-to-one if and only if

$$y' \in Y', \qquad y'|_{f(X)} = 0 \Rightarrow y' = 0. \tag{2.2}$$

Thus, to conclude the proof it suffices to recall that (by the Hahn–Banach theorem) implication (2.2) is necessary and sufficient for $f(X)$ to be dense in Y. □

Remark 2.2. *Let $(X, \|\cdot\|_X)$, $(Y, \|\cdot\|_Y)$ be normed (linear) spaces and f a linear homeomorphism of X onto Y (for the topological linear structures). Then f^T is a (linear) homeomorphism of the strong dual X' of X onto the strong dual Y' of Y.*

Proof. Recall that the strong topologies on X' and Y' are defined by the norms $\|\cdot\|_{X'}$ and $\|\cdot\|_{Y'}$, respectively, defined by

$$\|x'\|_{X'} = \sup_{0 \neq x \in X} \frac{|x'(x)|}{\|x\|_X}, \qquad \|y'\|_{Y'} = \sup_{0 \neq y \in Y} \frac{|y'(y)|}{\|y\|_Y}. \tag{2.3}$$

It is well known that $(X', \|\cdot\|_{X'})$ and $(Y', \|\cdot\|_{Y'})$ are Banach spaces. Therefore, by the open mapping theorem, to prove that f^T is a homeomorphism of Y' onto X' it suffices to show that $f^T: Y' \to X'$ is a continuous bijection. Now, it is easily seen that $f^T: Y' \to X'$ is continuous. Moreover $f^T: Y' \to X'$ is one-to-one by Remark 2.1, because $f: X \to Y$ is surjective. Finally, $f^T: Y' \to X'$ is onto, i.e., $\forall x' \in X'$, $\exists y' \in Y'$ such that $y' \circ f = x'$; indeed, setting $y' = x' \circ f^{-1}$ for every $x' \in X'$, we have $y' \in Y'$ because (by hypothesis) $f^{-1}: Y \to X$ is continuous. □

Theorem 2.3. *Let $(X, \|\cdot\|_X)$, $(Y, \|\cdot\|_Y)$ be reflexive real Banach spaces and let b be a continuous bilinear form on $X \times Y$. The following three propositions are equivalent:*

(i) *There is a number $c > 0$ such that*

$$\begin{cases} \displaystyle\sup_{0 \neq x \in X} \frac{|b(x, y)|}{\|x\|_X} \geq c\|y\|_Y, & \forall y \in Y, \\[3mm] \displaystyle\sup_{0 \neq y \in Y} \frac{|b(x, y)|}{\|y\|_Y} \geq c\|x\|_X, & \forall x \in X. \end{cases} \tag{2.4}$$

(ii) *For each $y' \in Y'$ there is a unique $x \in X$ such that*

$$y'(y) = b(x, y), \qquad \forall y \in Y.$$

(iii) *For each $x' \in X'$ there is a unique $y \in Y$ such that*

$$x'(x) = b(x, y), \qquad \forall x \in X.$$

Proof. As b is continuous, there is a number $c_1 > 0$ such that

$$|b(x, y)| \leqq c_1 \|x\|_X \|y\|_Y, \qquad \forall(x, y) \in X \times Y. \tag{2.5}$$

We consider on the duals X' and Y' the strong topologies (defined by the norms (2.3)). For any $y \in Y$ let $\tilde{y}$ be the (continuous) linear form on Y' defined by setting

$$\tilde{y}(y') = y'(y), \qquad \forall y' \in Y'. \tag{2.6}$$

By the reflexivity of Y, the mapping $y \mapsto \tilde{y}$ is a linear isometry of Y onto the dual Y'' of Y' equipped with the norm $y'' \mapsto \|y''\|_{Y''}$ with

$$\|y''\|_{Y''} = \sup_{0 \neq y' \in Y'} \frac{|y''(y')|}{\|y'\|_{Y'}}.$$

To each $x \in X$ we associate the linear form $f(x)$ on Y defined by setting

$$f(x)(y) = b(x, y), \qquad \forall y \in Y. \tag{2.7}$$

By (2.4) $f(x)$ is continuous, i.e., $f(x) \in Y'$ and the linear mapping $x \mapsto f(x)$ is continuous from X into Y'. Consider the transpose $f^T \colon Y'' \to X'$ of f. In view of (2.6) and (2.7) we have, for each $y \in Y$,

$$\|f^T(\tilde{y})\|_{X'} = \|\tilde{y} \circ f\|_{X'} = \sup_{0 \neq x \in X} \frac{|\tilde{y}(f(x))|}{\|x\|_X} = \sup_{0 \neq x \in X} \frac{|f(x)(y)|}{\|x\|_X}$$

$$= \sup_{0 \neq x \in X} \frac{|b(x, y)|}{\|x\|_X}.$$

Then, recalling that $\|y\|_Y = \|\tilde{y}\|_{Y''}$, (2.4) can be written in the form

$$\begin{cases} \|f(x)\|_{Y'} \geqq c \|x\|_X, & \forall x \in X, \\ \|f^T(\tilde{y})\|_{X'} \geqq c \|\tilde{y}\|_{Y''}, & \forall y \in Y. \end{cases} \tag{2.8}$$

We are now in a position to prove the equivalence of (i), (ii), (iii).

We first note that (ii) is equivalent to (iii) because of Remark 2.2 and the fact that, under our hypotheses, $(X'', \|\cdot\|_{X''})$ can be identified with $(X, \|\cdot\|_X)$, $(Y'', \|\cdot\|_{Y''})$ can be identified with $(Y, \|\cdot\|_Y)$, and so the transpose of f^T can be identified with f.

To prove that (ii) implies (i) it suffices to recall Remark 2.2 and observe that, by (2.7), (ii) expresses the fact that the (continuous, linear) mapping $f \colon X \to Y'$ is a bijection and hence (by the open mapping theorem) a homeomorphism when Y' is endowed with the strong topology.

Finally, let us prove that (i) implies (ii). We have already remarked that (2.4) is equivalent to (2.8) and that (ii) is equivalent to the fact that f is a bijection of X onto Y'. Then we must show that from (2.8), with

c a number > 0 independent of (x, y), it follows that f is a bijection of X onto Y'. Accordingly, note that $(2.8)_1$ implies that the (continuous) mapping $f: X \to Y'$ is one-to-one and open, so that $f(X)$ is a complete subspace of Y' (because X is complete). Hence $(2.8)_1$ implies that f is one-to-one and that $f(X)$ is a closed subspace of Y', while $(2.8)_2$ clearly implies that $f^T: Y'' \to X'$ is one-to-one, namely, by Remark 2.1, that $f(X)$ is dense in Y'. Thus we have show that from (2.8) it follows that $f: X \to Y'$ is a bijection. $\square$

The following theorem gives a different formulation of the statement of Theorem 2.3, but actually does not differ from it.

Theorem 2.4. *Let* $(X, \|\cdot\|_X)$, $(Y, \|\cdot\|_Y)$ *be reflexive real Banach spaces and let* X', Y' *be their duals endowed with the norms* $\|\cdot\|_{X'}$ *and* $\|\cdot\|_{Y'}$ *defined by* (2.3). *Moreover, let* $f: X \to Y'$ *be a continuous, linear mapping and let* $f^T: Y \to X'$ *be the mapping defined by setting*

$$\langle x, f^T(y) \rangle = \langle y, f(x) \rangle, \qquad \forall (x, y) \in X \times Y,$$

where $\langle \cdot, \cdot \rangle$ *denotes both the natural pairing of* X *and* X' *and the natural pairing of* Y *and* Y'. *Then the following three propositions are equivalent:*

(i) *There is a number* $c > 0$ *such that*

$$\begin{cases} \|f(x)\|_{Y'} \geq c\|x\|_X, & \forall x \in X, \\ \|f^T(y)\|_{X'} \geq c\|y\|_Y, & \forall y \in Y. \end{cases} \qquad (2.9)$$

(ii) f *is a homeomorphism of* X *onto* Y'.
(iii) f *is a homeomorphism of* Y *onto* X'.

Proof. Let b be the (continuous) bilinear form on $X \times Y$ related to f by (2.7). As has been shown in proving Theorem 2.3, propositions (ii) and (iii) of Theorem 2.3 are equivalent to the corresponding propositions (ii) and (iii) of Theorem 2.4. Moreover, (2.4) is evidently equivalent to (2.9). Then Theorem 2.4 is a consequence of Theorem 2.3. $\square$

An immediate application of Theorem 2.3 gives the following result, known as the *Lax–Milgram lemma* in the particular case when X is a Hilbert space.

Corollary 2.5. *Let* $(X, \|\cdot\|)$ *be a reflexive real Banach space and let* b *be a continuous bilinear form on* $X \times X$ *such that*

$$|b(x, x)| \geq c\|x\|^2, \qquad \forall x \in X,$$

with c *a number* > 0 *independent of* x. *Then for each* $x' \in X'$ *there are two uniquely determined elements* x_1, x_2 *of* X *such that*

$$x'(x) = b(x_1, x) = b(x, x_2), \qquad \forall x \in X.$$

§3. Linearized Elastostatics

In view of (4.9) and (5.3) of Chapter I, the equilibrium deformations $\phi: \bar{\Omega} \to \mathbb{R}^n$ for an elastic body are solutions of the equation

$$\text{div } A(D\phi) + \mu b_\phi \circ \phi = 0,$$

where $A(D\phi)$ is the function of Ω to M_n defined by putting

$$A(D\phi)(x) = a(x, D\phi(x)), \qquad \forall x \in \Omega,$$

and b_ϕ is a $\mathbb{R}^n$-valued function somehow assigned on $\phi(\Omega)$.

We recall that the function $(x, Z) \mapsto a(x, Z)$ of $\Omega \times \mathsf{M}_n^+$ into M_n^+ defines the elastic response. In many concrete cases there is a function $(x, y) \mapsto f(x, y)$ of $\Omega \times \mathbb{R}^n$ into $\mathbb{R}^n$ such that

$$(b_\phi \circ \phi)(x) = f(x, \phi(x)), \qquad \forall x \in \Omega.$$

In the case when f does not depend on y the body forces are said to be *dead*; otherwise, the body forces are often said to be *live*. We will refer to

$$\text{div } A(D\phi) \tag{3.1}$$

as to the *elastostatics operator*.

Moreover, for any deformation ϕ_0 of $\bar{\Omega}$, by the (formally) *linearized operator* of (3.1) *at* ϕ_0, we mean the linear operator

$$u \mapsto \text{div } L_{D\phi_0}(Du), \tag{3.2}$$

where u is a $\mathbb{R}^n$-valued (smooth) function defined on Ω and $L_{D\phi_0}(Du)$ denotes the M_n-valued function defined on Ω by

$$L_{D\phi_0}(Du)(x) = l_{D\phi_0}(x, Du(x)), \tag{3.3}$$

with $l_{D\phi_0}$ the M_n-valued function defined on $\Omega \times \mathsf{M}_n^+$ by setting

$$l_{D\phi_0}(x, Z) = \sum_{h,k=1}^{n} D_{Z_{hk}} a(x, D\phi_0(x)) Z_{hk}, \tag{3.4}$$

where $D_{Z_{hk}}$ is the partial derivative operator with respect to the co-ordinate Z_{hk} of Z. The definition is justified by Theorems 4.1 and 4.2 of Chapter II. The (formally) linearized operator at $\iota_{\bar{\Omega}}$ of (3.1) is often called the *operator of linear elastostatics*.

We recall that the balance of angular momentum and the independence of the observer of the elastic response lead us to the following restrictions on the choice of the function a (see Chapter I, (5.4) and (6.10)):

$$a(x, Z)Z^T \in \text{Sym}_n, \qquad \forall(x, Z) \in \Omega \times \mathsf{M}_n^+, \tag{3.5}$$

$$a(x, RZ) = Ra(x, Z), \qquad \forall(x, Z, R) \in \Omega \times \mathsf{M}_n^+ \times \mathbb{O}_n^+. \tag{3.6}$$

The *reference configuration* Ω is said to be *unstressed* if the Cauchy stress at the identity ι_Ω (cf. Chapter I, §4) vanishes, that is, if

$$a(x, I) = 0, \qquad \forall x \in \Omega. \tag{3.7}$$

Remark 3.1. *Let* (3.6) *be satisfied. Then* (3.7) *implies*

$$a(x, R) = 0, \qquad \forall(x,) \in \Omega \times \mathbb{O}_n^+.$$

Indeed, combining (3.6) and (3.7), we obtain $a(x, R) = a(x, RI) = Ra(x, I) = 0, \forall x \in \Omega$.

Remark 3.2. *Let* (3.7) *be satisfied. Then* (3.5) *and* (3.6) *yield, respectively,*

$$D_{Z_{hk}}a(x, R)R^{\mathrm{T}} \in \mathrm{Sym}_n, \qquad \forall h, k = 1, \ldots, n, \tag{3.8}$$

and

$$\sum_{s=1}^{n} D_{Z_{hs}}a(x, R)R_{ks} = \sum_{s=1}^{n} D_{Z_{ks}}a(x, R)R_{hs} \tag{3.9}$$

for every $(x, R) \in \Omega \times \mathbb{O}_n^+$ *such that the function* $Z \mapsto a(x, Z)$ *is differentiable at* R. *In particular,* (3.5) *and* (3.6) *give*

$$D_{Z_{hk}}a(x, I) = D_{Z_{hk}}(a(x, I))^{\mathrm{T}}, \qquad \forall h, k = 1, \ldots, n, \tag{3.10}$$

and

$$D_{Z_{hk}}a(x, I) = D_{Z_{kh}}a(x, I), \qquad \forall h, k = 1, \ldots, n, \tag{3.11}$$

provided the function $Z \mapsto a(x, Z)$ *is differentiable at* I.

Proof. Suppose that $Z \mapsto a(x, Z)$ is differentiable at $R \in \mathbb{O}_n^+$. From (3.5) it follows

$$D_{Z_{hk}}(a(x, R)R^{\mathrm{T}} + (a(x, R)e_k) \otimes e_h \in \mathrm{Sym}_n, \qquad \forall h, k = 1, \ldots, n,$$

where $(e_1, \ldots, e_n)$ is the canonical base of $\mathbb{R}^n$. Thus, by Remark 3.1, (3.5) implies (3.8).

We now prove that (3.6) implies (3.9). We observe that (3.6) yields (see Appendix II)

$$a(x, (\exp(Wt))R) = (\exp(Wt))a(x, R),$$

$$\forall(t, x, W, R) \in \mathbb{R} \times \Omega \times \mathrm{Skew}_n \times \mathbb{O}_n^+.$$

This gives, after differentiation with respect to t at $t = 0$,

$$\sum_{h,k,s=1}^{n} D_{Z_{hs}}a(x, R)W_{hk}R_{ks} = Wa(x, R), \qquad \forall(x, W, R) \in \Omega \times \mathrm{Skew}_n \times \mathbb{O}_n^+,$$

which easily yields (3.9) since, by Remark 3.1, $a(x, R) = 0$. $\square$

When the reference configuration is unstressed, it is customary to assume that

$$\left(\sum_{h,k=1}^{n} D_{Z_{hk}} a(x, I) S_{hk} \right) \cdot S > 0, \qquad \forall (x, S) \in \bar{\Omega} \times (\text{Sym}_n \backslash \{0\}). \qquad (3.12)$$

Note that, for hyperelastic bodies (cf. Chapter I, §5) condition (3.12) becomes

$$\sum_{i,j,h,k=1}^{n} D_{Z_{ij}} D_{Z_{hk}} w(x, I) S_{ij} S_{hk} > 0, \qquad \forall (x, S) \in \bar{\Omega} \times (\text{Sym}_n \backslash \{0\}),$$

where w is the stored-energy function. Thus (if w is sufficiently smooth) condition (3.12), associated with (3.7), implies that for each $x \in \bar{\Omega}$, the function

$$\text{Sym}_n \ni S \mapsto w(x, S)$$

has a strict, local minimum at I.

Theorem 3.3. *Suppose that $a \in C^1(\bar{\Omega} \times \mathbb{M}_n^+, \mathbb{M}_n)$ and that (3.5), (3.6), (3.7), and (3.12) are satisfied. Then for each $R \in \mathbb{O}_n^+$ there is a real number $m_R > 0$ such that*

$$\int_{\Omega} \left(\sum_{h,k=1}^{n} D_{Z_{hk}} a(x, R) D_k v_h(x) \right) \cdot Dv(x) \, dx$$

$$\geqq m_R \int_{\Omega} |D(R^T v) + (D(R^T v))^T|^2 \, dx, \qquad \forall v \in W^{1,2}(\Omega, \mathbb{R}^n). \qquad (3.13)$$

In particular, there is a real number $m > 0$ such that

$$\int_{\Omega} \left(\sum_{h,k=1}^{n} D_{Z_{hk}} a(x, I) D_k v_h(x) \right) \cdot Dv(x) \, dx$$

$$\geqq m \int_{\Omega} \sum_{i,j=1}^{n} |D_j v_i + D_i v_j| \, dx, \qquad \forall v \in W^{1,2}(\Omega, \mathbb{R}^n). \qquad (3.14)$$

Proof. We first prove that

$$\sum_{h,k=1}^{n} (D_{Z_{hk}} a(x, R) \cdot (SR))(SR)_{hk} > 0, \qquad \forall (x, S, R) \in \bar{\Omega} \times (\text{Sym}_n \backslash \{0\}) \times \mathbb{O}_n^+.$$

$$(3.15)$$

Note that (3.12) and (3.15) can be written, respectively, in the forms

$$S \cdot a_z'(x, I)(S) > 0, \qquad \forall (x, S) \in \bar{\Omega} \times (\text{Sym}_n \backslash \{0\}), \qquad (3.16)$$

and

$$(SR) \cdot a_z'(x, R)(SR) > 0, \qquad \forall (x, S, R) \in \bar{\Omega} \times (\text{Sym}_n \backslash \{0\}) \times \mathbb{O}_n^+, \qquad (3.17)$$

where a_z' denotes the partial differential of a with respect to Z.
Let $S \in \text{Sym}_n \backslash \{0\}$ and let $R \in \mathbb{O}_n^+$. As, by (3.6),

$$a(x, ZR) = Ra(x, R^T Z R), \qquad \forall (x, Z) \in \bar{\Omega} \times \mathbb{M}_n^+,$$

we have, for every $t \in \mathbb{R}$ with $|t|$ small enough,

$$a(x, R - tSR) = a(x, (I + tS)R) = Ra(x, R^T(I + tS)R)$$
$$= Ra(x, I + tR^TSR).$$

Hence, by (3.6), (3.7), and Remark 3.1, we obtain

$$a'_z(x, R)(SR) = Ra'_z(x, I)(R^TSR), \qquad \forall x \in \bar{\Omega}.$$

Then

$$(SR) \cdot a'_z(x, R)(SR) = (SR) \cdot (Ra'_z(x, I)(R^TSR)) = (R^TSR) \cdot a'_z(x, I)(R^TSR),$$

which, combined with (3.16), gives (3.17) because $R^TSR \in \mathrm{Sym}_n \backslash \{0\}$ whenever $S \in \mathrm{Sym}_n \backslash \{0\}$. Thus we have proved (3.15).

Now fix $R \in \mathbb{O}_n^+$ and denote by m_R the minimum of the continuous function

$$(x, S) \mapsto \frac{1}{4} \sum_{h,k=1}^n (D_{Z_{hk}} a(x, R) \cdot (SR))(SR)_{hk}$$

on the compact subset $\bar{\Omega} \times \{S \in \mathrm{Sym}_n : |S| = 1\}$ of $\mathbb{R}^n \times \mathbb{M}_n$. By (3.15) we have $m_R > 0$ and hence

$$\frac{1}{4} \sum_{h,k=1}^n (D_{Z_{hk}} a(x, R) \cdot (SR))(SR)_{hk} \geqq m_R > 0,$$

$$\forall (x, S) \in \bar{\Omega} \times \mathrm{Sym}_n \quad \text{with } |S| = 1.$$

Then

$$\frac{1}{4} \sum_{h,k=1}^n (D_{Z_{hk}} a(x, R) \cdot (SR))(SR)_{hk}$$

$$= \frac{1}{4} \sum_{h,k=1}^n \left(D_{Z_{hk}} a(x, R) \cdot \left(\frac{S}{|S|} R \right) \right) \left(\frac{S}{|S|} R \right)_{hk} |S|^2$$

$$\geqq m_R |S|^2, \qquad \forall (x, S) \in \bar{\Omega} \times (\mathrm{Sym}_n \backslash \{0\}). \quad (3.18)$$

Remark that in view of (3.8) and (3.9) we have

$$\sum_{h,k=1}^n (D_{Z_{hk}} a(x, R) \cdot (ZR))(ZR)_{hk}$$

$$= \frac{1}{4} \sum_{h,k=1}^n (D_{Z_{hk}} a(x, R) \cdot ((Z + Z^T)Z))((Z + Z^T)R)_{hk} \quad (3.19)$$

for every $(x, R, Z) \in \bar{\Omega} \times \mathbb{O}_n^+ \times \mathbb{M}_n$; thus (3.18) yields

$$\int_\Omega \sum_{h,k=1}^n (D_{Z_{hk}} a(x, R) \cdot ((Du(x))R))((Du(x))R)_{hk} \, dx$$

$$\geqq m_R \int_\Omega \sum_{i,j=1}^n (D_j u_i(x) + D_i u_j(x))(D_j u_i(x) + D_i u_j(x)) \, dx,$$

$$\forall u \in W^{1,2}(\Omega, \mathbb{R}^n).$$

From this we easily deduce that (3.13) holds. Thus the proof is complete.

□

Corollary 3.4. *Suppose that Ω has the cone property, that*

$$a \in C^1(\overline{\Omega} \times \mathsf{M}_n^+, \mathsf{M}_n),$$

and that (3.5), (3.6), (3.7), *and* (3.12) *are satisfied. Then for each $R \in \mathbb{O}_n^+$ there is a number $c_R > 0$ such that*

$$\int_\Omega \sum_{h,k=1}^n (D_{Z_{hk}} a(x, R) D_k v_h(x)) \cdot Dv(x) \, dx \geq c_R \|v\|_{1,2}^2 \qquad (3.20)$$

for every $v \in W^{1,2}(\Omega, \mathbb{R}^n)$ such that

$$\int_\Omega v \, dx = 0, \qquad R^\mathsf{T} \int_\Omega Dv \, dx \in \mathrm{Sym}_n \qquad (3.21)$$

for every $v \in W_0^{1,2}(\Omega, \mathbb{R}^n)$.

Proof. In view of Corollary 1.5, Remark 1.1, and the Poincaré inequality (see Chapter II, §1) there is a number $c_R' > 0$ such that

$$\|R^\mathsf{T} v\|_{1,2}^2 \leq c_R' \int_\Omega |D(R^\mathsf{T} v) + (D(R^\mathsf{T} v))|^2 \, dx$$

for every $v \in W^{1,2}(\Omega, \mathbb{R}^n)$ satisfying (3.21) and for every $v \in W_0^{1,2}(\Omega, \mathbb{R}^n)$. Moreover, evidently,

$$\|v\|_{1,2} \leq c_R'' \|R^\mathsf{T} v\|_{1,2}, \qquad \forall v \in W^{1,2}(\Omega, \mathbb{R}^n)$$

with c_R'' a number > 0 independent of v. Therefore (3.13) implies (3.20) for every $v \in W_0^{1,2}(\Omega, \mathbb{R}^n)$ satisfying (3.21) and for every $v \in W_0^{1,2}(\Omega, \mathbb{R}^n)$.

□

§4. The Dirichlet Problem in Linearized Elastostatics. Existence and Uniqueness in $W^{1,p}(\Omega, \mathbb{R}^n)$

Let l_{ijhk}, with $i, j, h, k = 1, \ldots, n$, be real-valued functions defined on Ω and let $l: \Omega \times \mathsf{M}_n \to \mathsf{M}_n$ be the function defined by

$$l(x, Z) = \left(\sum_{h,k=1}^n l_{ijhk}(x) Z_{hk} \right)_{i,j=1,\ldots,n}.$$

To any function $\sigma: \Omega \to \mathsf{M}_n$ we can associate the function $L(\sigma): \Omega \to \mathsf{M}_n$ defined by

$$L(\sigma)(x) = l(x, \sigma(x)).$$

This section concerns the (Dirichlet) problem of finding $u: \bar{\Omega} \to \mathbb{R}^n$ such that

$$\begin{cases} -\operatorname{div} L(Du) = f & \text{in } \Omega, \\ \qquad\qquad u = g & \text{on } \partial\Omega, \end{cases} \tag{4.1}$$

where $f: \Omega \to \mathbb{R}^n$ and $g: \partial\Omega \to \mathbb{R}^n$ are given functions. Of course, the differential operators D and div may be understood in a weak sense. Suppose

$$l_{ijhk} \in L^\infty(\Omega).$$

Then the operator

$$u \mapsto \operatorname{div} L(Du) \tag{4.2}$$

maps $W^{1,p}(\Omega, \mathbb{R}^n)$ into $W^{-1,p}(\Omega, \mathbb{R}^n)$ with $1 < p \in \mathbb{R}$. Note that the (formally) linearized operator (3.2) of the elastostatic operator (3.1) at any ϕ_0 is of the type (4.2).

After identifying $W^{-1,p}(\Omega)$ with the dual $(W_0^{1,p'}(\Omega))'$ of $W_0^{1,p'}(\Omega)$ (cf. Chapter II, §1), we define a pairing $(v, f) \mapsto \langle v, f \rangle$ of $W_0^{1,p'}(\Omega, \mathbb{R}^n)$ and $W^{-1,p}(\Omega, \mathbb{R}^n)$ by setting

$$\langle v, f \rangle = \sum_{i=1}^n f_i(v_i), \tag{4.3}$$

where f_i [respectively v_i] $(i = 1, \ldots, n)$, are the scalar components of f_i [respectively v_i]. Thus $W^{-1,p}(\Omega, \mathbb{R}^n)$ can be view as the dual of $W_0^{1,p'}(\Omega, \mathbb{R}^n)$ and we can consider the strong topology on $W^{-1,p}(\Omega, \mathbb{R}^n)$: it is defined by the norm $f \mapsto \|f\|_{-1,p}$ where

$$\|f\|_{-1,p} = \sup \left\{ \frac{|\langle v, f \rangle|}{\|v\|_{1,p'}} : 0 \neq v \in W_0^{1,p'}(\Omega, \mathbb{R}^n) \right\}. \tag{4.4}$$

In view of (4.3) we have, for $u \in W^{1,p}(\Omega, \mathbb{R}^n)$ and $v \in \mathscr{D}(\Omega, \mathbb{R}^n)$,

$$\langle v, -\operatorname{div} L(Du) \rangle = -\int_\Omega \sum_{i,j,h,k=1}^n D_j(l_{ijhk} D_k u_h) v_i \, dx = \int_\Omega l(x, Du(x)) \cdot Dv \, dx.$$

Hence, recalling that $\mathscr{D}(\Omega)$ is dense in $W_0^{1,p}(\Omega)$, we obtain

$$\langle v, -\operatorname{div} L(Du) \rangle = b(v, u), \qquad \forall (v, u) \in W_0^{1,p'}(\Omega, \mathbb{R}^n) \times W^{1,p}(\Omega, \mathbb{R}^n), \tag{4.5}$$

with

$$b(v, u) = \int_\Omega l(x, Du(x)) \cdot Dv \, dx. \tag{4.6}$$

Note that, as $l_{ijhk} \in L^\infty(\Omega)$, the Hölder inequality yields

$$|b(v, u)| \leq c_p \|v\|_{1,p'} \|u\|_{1,p}, \qquad \forall (v, u) \in W^{1,p'}(\Omega, \mathbb{R}^n) \times W^{1,p}(\Omega, \mathbb{R}^n),$$

with c_p a number > 0 independent of u and v; this means that the bilinear form $(v, u) \mapsto b(v, u)$ on $W^{1,p'}(\Omega, \mathbb{R}^n) \times W^{1,p}(\Omega, \mathbb{R}^n)$ is continuous. Thus, from (4.5) and definition (4.4) it follows that

$$\|\operatorname{div} L(Du)\|_{-1,p} \leq c_p \|u\|_{1,p}, \qquad \forall u \in W^{1,p}(\Omega, \mathbb{R}^n),$$

which means that (4.2) is a continuous mapping from $W^{1,p}(\Omega, \mathbb{R}^n)$ into $W^{-1,p}(\Omega, \mathbb{R}^n)$. Note also that, in view of (4.5), if $u \in W^{1,p}(\Omega, \mathbb{R}^n)$ and $f \in W^{-1,p}(\Omega, \mathbb{R}^n)$, then the equality

$$-\operatorname{div} L(Du) = f$$

is equivalent to the condition

$$b(v, u) = \langle v, f \rangle, \qquad \forall v \in W_0^{1,p'}(\Omega, \mathbb{R}^n).$$

Therefore, if there is a number $c > 0$ such that

$$|b(v, v)| \geq c \|v\|_{1,2}^2, \qquad \forall v \in W_0^{1,2}(\Omega, \mathbb{R}^n), \qquad (4.7)$$

then, by Corollary 2.3, for each $f \in W^{-1,2}(\Omega, \mathbb{R}^n)$ there is one and only one $u \in W_0^{1,2}(\Omega, \mathbb{R}^n)$ such that $-\operatorname{div} L(Du) = f$.

More generally, Theorem 2.4, applied to the case when $X = W_0^{1,p}(\Omega, \mathbb{R}^n)$ and $Y = W_0^{1,p'}(\Omega, \mathbb{R}^n)$, asserts that (4.2) is a (linear) homeomorphism of $W_0^{1,p}(\Omega, \mathbb{R}^n)$ onto $W^{-1,p}(\Omega, \mathbb{R}^n)$ with $1 < p \in \mathbb{R}$, if and only if there is a number $c > 0$ such that

$$\begin{cases} \sup \left\{ \dfrac{|b(v, u)|}{\|v\|_{1,p'}} : 0 \neq v \in W_0^{1,p'}(\Omega, \mathbb{R}^n) \right\} \geq c \|u\|_{1,p}, \quad \forall u \in W_0^{1,p}(\Omega, \mathbb{R}^n), \\[3mm] \sup \left\{ \dfrac{|b(v, u)|}{\|u\|_{1,p'}} : 0 \neq u \in W_0^{1,p}(\Omega, \mathbb{R}^n) \right\} \geq c \|v\|_{1,p'}, \quad \forall v \in W_0^{1,p'}(\Omega, \mathbb{R}^n). \end{cases} \qquad (4.8)$$

Summarizing, we can state the following theorem.

Theorem 4.1. *Let $l_{ijhk} \in L^\infty(\Omega)$ $(i, j, h, k = i, \ldots, n)$ and let p be a number > 1. Then (4.2) is a (linear) homeomorphism of $W_0^{1,p}(\Omega, \mathbb{R}^n)$ onto $W^{-1,p}(\Omega, \mathbb{R}^n)$ if and only if (4.8) is satisfied.*

Obviously, (4.7) implies (4.8) in the case $p = 2$. It is possible to prove, using Simader's methods (see SIMADER [1972], Theorem 6.1), that if Ω is of class C^1 and $l_{ijhk} \in C^0(\bar{\Omega})$, then (4.7) implies (4.8) for every real number $p > 1$. Therefore from Theorem 4.1 follows

Theorem 4.2. *Let $l_{ijhk} \in L^\infty(\Omega)$ $(i, j, h, k = 1, \ldots, n)$ and let (4.7) be satisfied. Then (4.2) is a (linear) homeomorphism of $W_0^{1,2}(\Omega, \mathbb{R}^n)$ onto $W^{-1,2}(\Omega, \mathbb{R}^n)$. Moreover, (4.2) is a homeomorphism of $W_0^{1,p}(\Omega, \mathbb{R}^n)$ onto $W^{-1,p}(\Omega, \mathbb{R}^n)$ with p any real number > 1, provided the functions l_{ijhk} are continuous on $\bar{\Omega}$ and Ω is of class C^1.*

Recall that a rigid deformation of $\bar{\Omega}$ is a function $\rho: \bar{\Omega} \to \mathbb{R}^n$ of the type

$$\rho(x) = c + Rx, \qquad x \in \bar{\Omega},$$

with $c \in \mathbb{R}^n$ and $R \in \mathbb{O}_n^+$ (see Chapter I, §2). Note that for each $x \in \bar{\Omega}$ we have $D\rho(x) = R$. Then, in accordance with the notations used in §3 (cf. (3.3) and (3.4)) we set, for each $(x, Z) \in \bar{\Omega} \times \mathbb{M}_n^+$,

$$l_R(x, Z) = \sum_{h,k=1}^{n} D_{Z_{hk}} a(x, R) Z_{hk} \qquad (4.9)$$

and, for each function $\sigma: \bar{\Omega} \to \mathbb{M}_n$, we consider the $\mathbb{M}_n$-valued function $L_R(\sigma)$ defined on $\bar{\Omega}$ by putting

$$L_R(\sigma)(x) = l_R(x, \sigma(x)). \qquad (4.10)$$

Theorem (4.2), combined with Corollary 3.4, gives immediately.

Corollary 4.3. *Let the hypotheses (on Ω and a) of Corollary 3.4 be satisfied. Then for every $R \in \mathbb{O}_n^+$ the mapping*

$$u \mapsto \operatorname{div} L_R(Du) \qquad (4.11)$$

is a homeomorphism of $W_0^{1,2}(\Omega, \mathbb{R}^n)$ onto $W^{-1,2}(\Omega, \mathbb{R}^n)$; furthermore, for every $R \in \mathbb{O}_n^+$ and every real number $p > 1$, the mapping (4.11) is a homeomorphism of $W_0^{1,p}(\Omega, \mathbb{R}^n)$ onto $W^{-1,p}(\Omega, \mathbb{R}^n)$ provided Ω is of class C^1.

The following two corollaries concern the nonhomogeneous Dirichlet problem.

Corollary 4.4. *Let $l_{ijhk} \in L^\infty(\Omega)$ and let Ω be of class C^1. If (4.8) is satisfied, then for each $(f, g) \in W^{-1,p}(\Omega, \mathbb{R}^n) \times W^{1-1/p,p}(\partial\Omega, \mathbb{R}^n)$ with $1 < p \in \mathbb{R}$ there is one and only one solution u of problem (4.1) belonging to $W^{1,p}(\Omega, \mathbb{R}^n)$. In particular, for each $(f, g) \in W^{-1,2}(\Omega, \mathbb{R}^n) \times W^{1/2,2}(\partial\Omega, \mathbb{R}^n)$ there is one and only one solution u of problem (4.1) belonging to $W^{1,2}(\Omega, \mathbb{R}^n)$ provided (4.7) holds.*

Proof. Let $(f, g) \in W^{-1,p}(\Omega, \mathbb{R}^n) \times W^{1-1/p,p}(\partial\Omega, \mathbb{R}^n)$. In view of the trace theorem (see Chapter II, §1) there is a $\bar{u} \in W^{1,p}(\Omega, \mathbb{R}^n)$ such that

$$\bar{u}|_{\partial\Omega} = g.$$

If (4.7) holds, then by Theorem 4.1 there is one and only one $v \in W_0^{1,p}(\Omega, \mathbb{R}^n)$ such that

$$-\operatorname{div} L(Dv) = f - L(D\bar{u}) \quad \text{in } \Omega;$$

hence, $\bar{u} + v$ is a solution of problem (4.1) that belongs to $W^{1,p}(\Omega, \mathbb{R}^n)$ and it is the only solution of (4.1) belonging to do so. $\qquad \square$

Likewise, from Corollary 4.3 we can deduce.

Corollary 4.5. *Assume that Ω is of class C^1, that $a \in C^1(\bar{\Omega} \times \mathsf{M}_n^+, \mathsf{M}_n)$, and that (3.5), (3.6), (3.7), and (3.12) hold. Then for each $(f, g) \in W^{-1,p}(\Omega, \mathbb{R}^n) \times W^{1-1/p,p}(\partial\Omega, \mathbb{R}^n)$ with $1 < p \in \mathbb{R}$ there is one and only one solution $u \in W^{1,p}(\Omega, \mathbb{R}^n)$ of problem*

$$\begin{cases} -\operatorname{div} L_R(Du) = f & in\ \Omega, \\ \qquad\qquad u = g & on\ \partial\Omega, \end{cases}$$

where R is any given element of $\mathbb{O}_n^+$ and L_R is defined by ((4.9), (4.10)).

§5. The Neumann Problem in Linearized Elastostatics. Existence and Uniqueness in $W^{1,p}(\Omega, \mathbb{R}^n)$

In this section we deal with the Neumann problem for operator (4.2), i.e., with the problem of finding $u: \bar{\Omega} \to \mathbb{R}^n$ such that

$$\begin{cases} -\operatorname{div} L(Du) = f & in\ \Omega, \\ \quad\ \ L(Du)v = g & on\ \partial\Omega, \end{cases} \tag{5.1}$$

where $f: \Omega \to \mathbb{R}^n$ and $g: \partial\Omega \to \mathbb{R}^n$ are given functions. As always, v denotes the outward unit normal to $\partial\Omega$.

Remark 5.1. *If Ω, u, l_{ijhk}, f, g are smooth enough (e.g., Ω of class C^1, $u \in C^2(\bar{\Omega}, \mathbb{R}^n)$, $l_{ijhk} \in C^1(\bar{\Omega})$, and f, g continuous), then (5.1) is equivalent to*

$$b(v, u) = \varphi(v), \qquad \forall v \in C^1(\bar{\Omega}, \mathbb{R}^n), \tag{5.2}$$

where $b(v, u)$ is the real number defined by (4.6) and

$$\varphi(v) = \int_\Omega f \cdot v \, dx + \int_{\partial\Omega} g \cdot v \, d\sigma. \tag{5.3}$$

Proof. Suppose that Ω, u, l_{ijhk}, f, and g have sufficient smoothness to make sense of what follows. By the divergence theorem we have, for each $v \in C^1(\bar{\Omega}, \mathbb{R}^n)$,

$$\int_\Omega (-\operatorname{div} L(Du)) \cdot v \, dx + \int_{\partial\Omega} (L(Du)v) \cdot v \, d\sigma = b(v, u);$$

therefore, (5.2) is equivalent to

$$\int_\Omega (-\operatorname{div} L(Du)) \cdot v \, dx + \int_{\partial\Omega} (L(Du)v) \cdot v \, d\sigma$$

$$= \int_\Omega f \cdot v \, dx + \int_{\partial\Omega} g \cdot v \, d\sigma, \qquad \forall v \in C^1(\bar{\Omega}, \mathbb{R}^n). \tag{5.4}$$

Thus we have reduced our problem to verifying that (5.1) is equivalent to (5.4). Now, trivially (5.1) implies (5.4). We then show that (5.1) follows from (5.4). We first note that (5.4) yields, evidently,

$$\int_{\Omega} (\text{div } L(Du) + f) \cdot v \, dx = 0, \qquad \forall v \in \mathscr{D}(\Omega, \mathbb{R}^n),$$

which implies $(5.1)_1$ because $\mathscr{D}(\Omega)$ is dense in $L^2(\Omega)$. Consequently, (5.4) gives

$$\int_{\partial\Omega} (L(Du)v - g) \cdot v \, d\sigma = 0, \qquad \forall v \in C^1(\bar{\Omega}, \mathbb{R}^n)$$

and this implies (5.1) because (by a well-known theorem of Weierstrass) the restrictions to $\partial\Omega$ of the polynomial functions defined on $\mathbb{R}^n$ constitute a dense set in the space of continuous functions defined on $\partial\Omega$ endowed with the topology of uniform convergence.

   Problem (5.2) is said to be the *variational formulation* of the differential problem (5.1).   □

Remark 5.2. *Let Ω be of class C^1 and let $(f, g) \in L^2(\Omega, \mathbb{R}^n) \times L^2(\partial\Omega, \mathbb{R}^n)$. Then φ is a continuous (linear) form on $W^{1,2}(\Omega, \mathbb{R}^n)$.*

Proof. To prove that the linear mapping $\varphi\colon W^{1,2}(\Omega, \mathbb{R}^n) \to \mathbb{R}$, defined by (5.3), is continuous it suffices to remark that

$$|\varphi(x)| \leqq \left(\int_{\Omega} |f|^2 \, dx \right)^{1/2} \left(\int_{\Omega} |v|^2 \, dx \right)^{1/2} + \left(\int_{\partial\Omega} |g|^2 \, d\sigma \right)^{1/2} \left(\int_{\partial\Omega} |v|^2 \, d\sigma \right)^{1/2}$$

and recall that the trace operator $v \mapsto v|_{\partial\Omega}$ is continuous from $W^{1,2}(\Omega, \mathbb{R}^n)$ into $W^{1/2,2}(\partial\Omega, \mathbb{R}^n)$ and that $W^{1/2,2}(\partial\Omega)$ can be continuously embedded in $L^2(\partial\Omega)$.   □

   Let $l_{ijhk} \in L^\infty(\Omega)$, so that $(v, u) \mapsto b(v, u)$ is a continuous bilinear form on $W^{1,p'}(\Omega, \mathbb{R}^n) \times W^{1,p}(\Omega, \mathbb{R}^n)$ (see §4). Let us set, for any real number $p > 1$,

$$\mathscr{N}^{1,p} = \{u \in W^{1,p}(\Omega, \mathbb{R}^n) \colon b(v, u) = 0, \forall v \in W^{1,p'}(\Omega, \mathbb{R}^n)\}.$$

Theorem 5.3. *Let $l_{ijhk} \in L^\infty(\Omega)$, $1 < p \in \mathbb{R}$, $1/p + 1/p' = 1$, and suppose that there are two closed subspaces $\mathscr{V}^{1,p}$ and $\mathscr{V}^{1,p'}$ of $W^{1,p}(\Omega, \mathbb{R}^n)$ and $W^{1,p'}(\Omega, \mathbb{R}^n)$, respectively, such that*

$$W^{1,p}(\Omega, \mathbb{R}^n) = \mathscr{N}^{1,p} \oplus \mathscr{V}^{1,p}, \qquad W^{1,p'}(\Omega, \mathbb{R}^n) = \mathscr{N}^{1,p'} \oplus \mathscr{V}^{1,p'},$$

$$1 < p \in \mathbb{R}. \qquad (5.5)$$

Then the following propositions are equivalent.

(i) *There is a number $c > 0$ such that*

$$\begin{cases} \sup \left\{ \dfrac{|b(v, u)|}{\|v\|_{1, p'}} : 0 \neq v \in W^{1, p'}(\Omega, \mathbb{R}^n) \right\} \geq c \|u\|_{1, p}, & \forall u \in \mathcal{V}^{1, p}, \\[4mm] \sup \left\{ \dfrac{|b(v, u)|}{\|u\|_{1, p}} : 0 \neq u \in W^{1, p}(\Omega, \mathbb{R}^n) \right\} \geq c \|v\|_{1, p'}, & \forall v \in \mathcal{V}^{1, p'}. \end{cases} \tag{5.6}$$

(ii) *For each $\varphi \in (W^{1, p'}(\Omega, \mathbb{R}^n))'$ satisfying*

$$\varphi(v) = 0, \qquad \forall v \in \mathcal{N}^{1, p'},$$

there is one and only one $u \in \mathcal{V}^{1, p}$ such that

$$b(v, u) = \varphi(v), \qquad \forall v \in W^{1, p'}(\Omega, \mathbb{R}^n).$$

(iii) *For each $\varphi \in (W^{1, p}(\Omega, \mathbb{R}^n))'$ satisfying*

$$\varphi(u) = 0, \qquad \forall u \in \mathcal{N}^{1, p},$$

there is one and only one $v \in \mathcal{V}^{1, p'}$ such that

$$b(v, u) = \varphi(u), \qquad \forall u \in W^{1, p}(\Omega, \mathbb{R}^n).$$

Proof. From (5.5) it follows without difficulty that the restriction mappings

$$(W^{1, p}(\Omega, \mathbb{R}^n))' \ni \varphi \mapsto \varphi|_{\mathcal{V}^{1, p}} \qquad \text{and} \qquad (W^{1, p'}(\Omega, \mathbb{R}^n))' \ni \varphi \mapsto \varphi|_{\mathcal{V}^{1, p'}}$$

induce, respectively, a bijection of $\{\varphi \in (W^{1, p}(\Omega, \mathbb{R}^n))' : \varphi(v) = 0, \forall v \in \mathcal{N}^{1, p'}\}$ onto $(\mathcal{V}^{1, p})'$ and a bijection of $\{\varphi \in (W^{1, p'}(\Omega, \mathbb{R}^n))' : \varphi(u) = 0, \forall u \in \mathcal{N}^{1, p}\}$ onto $(\mathcal{V}^{1, p'})'$. Note that $\mathcal{V}^{1, p}$ and $\mathcal{V}^{1, p'}$ are reflexive Banach spaces, because a closed subspace of a reflexive Banach space is reflexive. Then by taking $X = \mathcal{V}^{1, p'}$ and $Y = \mathcal{V}^{1, p}$ we see that Theorem 5.3 is a consequence of Theorem 2.3. □

We now consider problem (5.1) in the case when the linear operator L is obtained by a (formal) linearization of the composition operator A. Precisely, we consider the problem

$$\begin{cases} -\operatorname{div} L_R(Du) = f & \text{in } \Omega, \\ L_R(Du)v = g & \text{on } \partial\Omega, \end{cases} \tag{5.7}$$

where $R \in \mathbb{O}_n^+$ and L_R is the operator defined by ((4.9), (4.10)), i.e., $L_R(Du)$ is the $\mathbb{R}^n$-valued function defined on Ω by

$$L_R(Du)(x) = \sum_{h, k=1}^n D_{Z_{hk}} a(x, R) D_k u_h(x).$$

When the functions $x \mapsto D_{Z_{hk}} a(x, R)$ $(h, k = 1, \ldots, n)$ belong to $L^\infty(\Omega)$, let us denote by b_R the (continuous) bilinear form on $W^{1, p'}(\Omega, \mathbb{R}^n) \times W^{1, p}(\Omega, \mathbb{R}^n)$

defined by

$$b_R(v, u) = \int_\Omega \left(\sum_{h,k=1}^{n} D_{Z_{hk}} a(x, R) D_k u_h(x) \right) \cdot Dv(x)\, dx.$$

Remark 5.4. *Suppose that* $a \in C^1(\bar\Omega \times \mathbb{M}_n^+, \mathbb{M}_n)$ *and that* (3.7) *holds. Then* (3.5) *and* (3.6) *yield, respectively,*

$$b_R(Rv, u) = 0, \qquad \forall(v, u) \in T_{i_{\bar\Omega}}(\mathscr{R}_{\bar\Omega}) \times W^{1,p}(\Omega, \mathbb{R}^n), \qquad (5.8)$$

and

$$b_R(v, Ru) = 0, \qquad \forall(v, u) \in W^{1,p'}(\Omega, \mathbb{R}^n) \times T_{i_{\bar\Omega}}(\mathscr{R}_{\bar\Omega}). \qquad (5.9)$$

Proof. Clearly, putting $a_{ij}(x, R) = a(x, R) \cdot (e_i \otimes e_j)$, with $(e_1, \ldots, e_n)$ the canonical base of $\mathbb{R}^n$, we have

$$b_R(Rv, u) = \int_\Omega \sum_{i,j,h,k,s=1}^{n} D_{Z_{hk}} a_{ij}(x, R) R_{is} D_j v_s(x) D_k u_h(x)\, dx,$$

and

$$b_R(v, Ru) = \int_\Omega \sum_{i,j,h,k,s=1}^{n} D_{Z_{hk}} a_{ij}(x, R) D_j v_i(x) R_{hs} D_k u_s(x)\, dx.$$

By Remark 3.2, (3.5) implies that for every $h, k = 1, \ldots, n$ the matrix

$$\left(\sum_{i=1}^{n} D_{Z_{hk}} a_{ij}(x, R) R_{is} \right)_{j,s=1,\ldots,n}$$

is symmetric, while (3.6) implies that for every $i, j = 1, \ldots, n$ the matrix

$$\left(\sum_{h=1}^{n} D_{Z_{hk}} a_{ij}(x, R) R_{hs} \right)_{k,s=1,\ldots,n}$$

is symmetric. Then (3.5) implies (5.8) and (3.6) implies (5.9), because $Dv(x) \in \text{Skew}_n, \forall v \in T_{i_{\bar\Omega}}(\mathscr{R}_{\bar\Omega})$. $\square$

Let $a \in C^1(\bar\Omega \times \mathbb{M}_n^+, \mathbb{M}_n)$ and let Ω, u, f, g be sufficiently smooth. Then, by Remark 5.1, (5.7) is equivalent to

$$b_R(v, u) = \varphi(v), \qquad \forall v \in C^\infty(\bar\Omega, \mathbb{R}^n);$$

thus, from Remark 5.4 it follows that if a satisfies (3.5) and (3.7), then a necessary condition on the data f, g for problem (5.7) to have a solution is that

$$\varphi(Rv) = 0, \qquad \forall v \in T_{i_{\bar\Omega}}(\mathscr{R}_{\bar\Omega}). \qquad (5.10)$$

Recalling that $T_{i_{\bar\Omega}}(\mathscr{R}_{\bar\Omega})$ is the set of functions from $\bar\Omega$ to $\mathbb{R}^n$ of the type $x \mapsto c + Wx$, with $c \in \mathbb{R}^n$ and $W \in \text{Skew}_n$, it is easy to verify that condition

(5.10) is equivalent to

$$
\begin{cases}
\displaystyle\int_\Omega f\,dx + \int_{\partial\Omega} g\,d\sigma = 0, \\[2mm]
\displaystyle\int_\Omega \iota_\Omega \wedge R^{\mathrm T} f\,dx + \int_{\partial\Omega} \iota_{\partial\Omega} \wedge R^{\mathrm T} g\,d\sigma = 0.
\end{cases}
\tag{5.11}
$$

We will say that a pair $(f, g) \in L^1(\Omega, \mathbb{R}^n) \times L^1(\partial\Omega, \mathbb{R}^n)$ is *equilibrated with respect to a deformation ϕ of $\bar\Omega$* if

$$
\begin{cases}
\displaystyle\int_\Omega f\,dx + \int_{\partial\Omega} g\,d\sigma = 0, \\[2mm]
\displaystyle\int_\Omega \phi \wedge f\,dx + \int_{\partial\Omega} \phi \wedge g\,d\sigma = 0.
\end{cases}
$$

When (f, g) is equilibrated with respect to the identity function $\iota_{\bar\Omega}\colon \bar\Omega \to \mathbb{R}^n$, namely, when

$$
\begin{cases}
\displaystyle\int_\Omega f\,dx + \int_{\partial\Omega} g\,d\sigma = 0, \\[2mm]
\displaystyle\int_\Omega \iota_\Omega \wedge f\,dx + \int_{\partial\Omega} \iota_{\partial\Omega} \wedge g\,d\sigma = 0,
\end{cases}
$$

we will say, more simply, that (f, g) is *equilibrated*.

Note that *conditions (5.11) express the fact that (f, g) is equilibrated with respect to any rigid deformation of $\bar\Omega$ having R as the gradient at any point $x \in \bar\Omega$.* To realize that it suffices to remark that, for any $a, b \in \mathbb{R}^n$ and any $R \in \mathbb{O}_n^+$, we have

$$
R^{\mathrm T}(Ra \wedge b)R = a \wedge R^{\mathrm T} b.
$$

The last equality follows from the fact that

$$
\begin{cases}
Ra \wedge b = R(a \otimes b) - (b \otimes a)R^{\mathrm T}, \\
a \wedge R^{\mathrm T} b = (a \otimes b)R - R^{\mathrm T}(b \otimes a).
\end{cases}
$$

Then, *when the function a satisfies (3.5) and (3.7), a necessary condition for problem (5.7) to have a solution is that the pair $(R^{\mathrm T} f, R^{\mathrm T} g)$ be equilibrated.*

Theorem 5.5. *Suppose that $a \in C^1(\bar\Omega \times \mathbb{M}_n^+, \mathbb{M}_n)$ and that (3.5), (3.6), (3.7) are satisfied. Then, for every real number $p > 1$ and every $R \in \mathbb{O}_n^+$, the following three propositions are equivalent.*

(i) *There is a number $c > 0$ such that*

$$
\begin{cases}
\displaystyle\sup\left\{\frac{|b_R(v, u)|}{\|v\|_{1,p'}} : 0 \neq v \in W^{1,p'}(\Omega, \mathbb{R}^n)\right\} \geq c\|u\|_{1,p}, \qquad \forall u \in \mathscr{V}_R^{1,p}, \\[4mm]
\displaystyle\sup\left\{\frac{|b_R(v, u)|}{\|u\|_{1,p}} : 0 \neq u \in W^{1,p}(\Omega, \mathbb{R}^n)\right\} \geq c\|v\|_{1,p'}, \qquad \forall v \in \mathscr{V}_R^{1,p'},
\end{cases}
\tag{5.12}
$$

where

$$\begin{cases} \mathcal{V}_R^{1,p} = \left\{ u \in W^{1,p}(\Omega, \mathbb{R}^n) : \int_\Omega u \, dx = 0, R^\mathrm{T} \int_\Omega Du \, dx \in \mathrm{Sym}_n \right\}, \\ \mathcal{V}_R^{1,p'} = \left\{ v \in W^{1,p'}(\Omega, \mathbb{R}^n) : \int_\Omega v \, dx = 0, R^\mathrm{T} \int_\Omega Dv \, dx \in \mathrm{Sym}_n \right\}. \end{cases}$$

(ii) *For each* $\varphi \in (W^{1,p'}(\Omega, \mathbb{R}^n))'$ *satisfying*

$$\varphi(Rv) = 0, \qquad \forall v \in T_{i_{\bar\Omega}}(\mathcal{R}_{\bar\Omega})$$

there is one and only one $u \in \mathcal{V}_R^{1,p}$ *such that*

$$b_R(v, u) = \varphi(v), \qquad \forall v \in W^{1,p'}(\Omega, \mathbb{R}^n).$$

(iii) *For each* $\varphi \in (W^{1,p}(\Omega, \mathbb{R}^n))'$ *satisfying*

$$\varphi(Ru) = 0, \qquad \forall u \in T_{i_{\bar\Omega}}(\mathcal{R}_{\bar\Omega})$$

there is one and only one $v \in \mathcal{V}_R^{1,p'}$ *such that*

$$b_R(v, u) = \varphi(u), \qquad \forall u \in W^{1,p}(\Omega, \mathbb{R}^n).$$

Proof. Let

$$\mathcal{N}_R^{1,p} = \{ u \in W^{1,p}(\Omega, \mathbb{R}^n) : b_R(v, u) = 0, \forall v \in W^{1,p'}(\Omega, \mathbb{R}^n) \}. \qquad (5.13)$$

We will prove that

$$\mathcal{N}_R^{1,p} = \mathcal{N}_R^{1,p'} = \{ Rv : v \in T_{i_{\bar\Omega}}(\mathcal{R}_{\bar\Omega}) \}, \qquad (5.14)$$

and

$$W^{1,p}(\Omega, \mathbb{R}^n) = \mathcal{V}_R^{1,p} \oplus \mathcal{N}_R^{1,p}, \qquad W^{1,p'}(\Omega, \mathbb{R}^n) = \mathcal{V}_R^{1,p'} \oplus \mathcal{N}_R^{1,p'}, \qquad (5.15)$$

so that, evidently, Theorem 5.5 follows from Theorem 5.3.

Accordingly, note that, in view of Remark 1.3,

$$\begin{cases} W^{1,p}(\Omega, \mathbb{R}^n) = \mathcal{V}_R^{1,p} \oplus \{ Rv : v \in T_{i_{\bar\Omega}}(\mathcal{R}_{\bar\Omega}) \}, \\ W^{1,p'}(\Omega, \mathbb{R}^n) = \mathcal{V}_R^{1,p'} \oplus \{ Rv : v \in T_{i_{\bar\Omega}}(\mathcal{R}_{\bar\Omega}) \}, \end{cases} \qquad (5.16)$$

and that (5.12) implies

$$\mathcal{V}_R^{1,p} \cap \mathcal{N}_R^{1,p} = \{0\}, \qquad \mathcal{V}_R^{1,p'} \cap \mathcal{N}_R^{1,p'} = \{0\}. \qquad (5.17)$$

Note also that, by Remark 5.4,

$$\{ Rv : v \in T_{i_{\bar\Omega}}(\mathcal{R}_{\bar\Omega}) \} \subseteq \mathcal{N}_R^{1,p} \cap \mathcal{N}_R^{1,p'}. \qquad (5.18)$$

Now, if $w \in \mathcal{N}_R^{1,p}$ we have (by (5.16)) $w = u + Rv$ with $u \in \mathcal{V}_R^{1,p}$ and $v \in T_{i_{\bar\Omega}}(\mathcal{R}_{\bar\Omega})$; then, since (by (5.18)) $Rv \in \mathcal{N}_R^{1,p}$ we have $u \, (= w - Rv) \in \mathcal{N}_R^{1,p} \cap \mathcal{V}_R^{1,p}$, so that $u = 0$ in view of (5.17) and hence $w \in \{ Rv : v \in T_{i_{\bar\Omega}}(\mathcal{R}_{\bar\Omega}) \}$. Therefore $\mathcal{N}_R^{1,p} = \{ Rv : v \in T_{i_{\bar\Omega}}(\mathcal{R}_{\bar\Omega}) \}$. Analogously, we see that $\mathcal{N}_R^{1,p'} = \{ Rv : v \in T_{i_{\bar\Omega}}(\mathcal{R}_{\bar\Omega}) \}$. Thus (5.16) is true and, consequently, (5.16) yields (5.15). $\square$

In view of Corollary 3.4, from Theorem 5.5 clearly follows

Corollary 5.6. *Suppose that* Ω *has the cone property, that*

$$a \in C^1(\bar{\Omega} \times \mathsf{M}_n^+, \mathsf{M}_n),$$

and that (3.5), (3.6), (3.7), *and* (3.12) *hold. Moreover, let* $R \in \mathbb{O}_n^+$. *Then, for each* $\varphi \in (W^{1,2}(\Omega, \mathbb{R}^n))'$ *satisfying the condition* $\varphi(Rv) = 0$, $\forall v \in T_{\bar{\Omega}}(\mathscr{R}_{\bar{\Omega}})$, *there is one and only one* $u \in W^{1,2}(\Omega, \mathbb{R}^n)$ *such that*

$$\int_\Omega u \, dx = 0, \qquad R^T \int_\Omega Du \, dx \in \mathrm{Sym}_n, \tag{5.19}$$

and

$$b_R(v, u) = \varphi(v), \qquad \forall v \in W^{1,2}(\Omega, \mathbb{R}^n).$$

Upon recalling that, if $(f, g) \in L^2(\Omega, \mathbb{R}^n) \times L^2(\partial\Omega, \mathbb{R}^n)$, condition $\varphi(Rv) = 0$, $\forall v \in T_{\bar{\Omega}}(\mathscr{R}_{\bar{\Omega}})$, expresses the fact that the pair $(R^T f, R^T g)$ is equilibrated, we see that Corollary 5.6 gives immediately

Corollary 5.7. *Let the assumptions of Corollary 5.6 be satisfied. Then, for each* $(f, g) \in L^2(\Omega, \mathbb{R}^n) \times L^2(\partial\Omega, \mathbb{R}^n)$ *such that* $(R^T f, R^T g)$ *is equilibrated, there is one and only one solution* $u \in W^{1,2}(\Omega, \mathbb{R}^n)$ *of problem* (5.7) (*in a generalized sense*) *which satisfies* (5.19).

§6. Some Basic Inequalities for Elliptic Operators

Operator (4.2) is *elliptic* at a point $x \in \Omega$ if

$$\det l(x, \xi \otimes \xi) \neq 0, \qquad \forall \xi \in \mathbb{R}^n \backslash \{0\}.$$

If

$$(\eta \otimes \eta) \cdot l(x, \xi \otimes \xi) \neq 0, \qquad \forall \xi, \eta \in \mathbb{R}^n \backslash \{0\},$$

then operator (4.2) is *strongly elliptic* at x. Clearly, strong ellipticity implies ellipticity. It is possible to prove (see FICHERA [1972a], Section 5) the following theorem.

Theorem 6.1. *Let the function* l_{ijhk} *be continuous in* $\bar{\Omega}$. *A necessary and sufficient condition for*

$$(\eta \otimes \eta) \cdot l(x, \xi \otimes \xi) > 0, \qquad (x, \xi, \eta) \in \bar{\Omega} \times (\mathbb{R}^n \backslash \{0\}) \times (\mathbb{R}^n \backslash \{0\}) \tag{6.1}$$

is that there be two numbers c_0 *and* λ_0, *with* $c_0 > 0$ *and* $\lambda_0 > 0$, *such that*

$$\sum_{i,j,h,k=1}^n \int_\Omega l_{ijhk} D_j v_i D_k v_h \, dx + \lambda_0 \|v\|_{0,2}^2 \geq c_0 \|v\|_{1,2}^2, \qquad \forall v \in W_0^{1,2}(\Omega, \mathbb{R}^n).$$

Note that (6.1) implies

$$\det l(x, \xi \otimes \xi) > 0, \qquad \forall(x, \xi) \in \bar{\Omega} \times (\mathbb{R}^n \backslash \{0\})$$

and that, if $l_{ijhk} \in C^0(\bar{\Omega})$, the latter inequality gives

$$\det l(x, \xi \otimes \xi) \geqq c_1 |\xi|^{2n}, \qquad \forall(x, \xi) \in \Omega \times \mathbb{R}^n, \tag{6.2}$$

c_1 being a number > 0 independent of (x, ξ); indeed, (6.2) holds with c_1 the minimum on $\bar{\Omega} \times \{\xi \in \mathbb{R}^n : |\xi| = 1\}$ of the continuous function $(x, \xi) \mapsto \det l(x, \xi \otimes \xi)$.

Now we consider operator (4.2). Let m be an integer $\geqq 0$. We observe that

$$p(m + 1) > n, \qquad u \in W^{m+2, p}(\Omega, \mathbb{R}^n) \Rightarrow \operatorname{div} L(Du) \in W^{m, p}(\Omega, \mathbb{R}^n)$$

provided Ω has the cone property and

$$l_{ijhk} \in W^{m+1, p}(\Omega).$$

Indeed, if Ω has the cone property and $p(m + 1) > n$ then $W^{m+1, p}(\Omega)$ is a Banach algebra (see Chapter II, Corollary 2.3), so that $L(Du) \in W^{m+1, p}(\Omega, \mathbb{R}^n)$ if $l_{ijhk} \in W^{m+1, p}(\Omega)$ and $u \in W^{m+2, p}(\Omega, \mathbb{R}^n)$.

The following estimate (6.3) under hypothesis (6.1), is well known when $l_{ijhk} \in C^{m+1}(\bar{\Omega})$ (see AGMON, DOUGLIS & NIRENBERG [1964], Theorem 10.5). We now prove that this estimate still holds when $l_{ijhk} \in W^{m+1, p}(\Omega)$. We emphasize the fact that the hypothesis $l_{ijhk} \in W^{m+1, p}(\Omega)$, besides being the natural one in order that operator (4.2) map $W^{m+2, p}(\Omega, \mathbb{R}^n)$ into $W^{m, p}(\Omega)$, is just the one required when the linear operator (4.2) is obtained by linearization of the elastostatics operator (3.1), thought of as acting from $W^{m+2, p}(\Omega, \mathbb{R}^n)$ into $W^{m, p}(\Omega, \mathbb{R}^n)$.

Theorem 6.2. *Let* $l_{ijhk} \in W^{m+1, p}(\Omega)$ *with* $p > n$ *and let* Ω *be of class* C^{m+2}. *Then* (6.1) *implies*

$$\|\operatorname{div} L(Du)\|_{m, p} + \|u\|_{0, p} \geqq c \|u\|_{m+2, p}, \qquad \forall u \in W_0^{1, p}(\Omega, \mathbb{R}^n) \cap W^{m+2, p}(\Omega, \mathbb{R}^n), \tag{6.3}$$

where c *is a number* > 0 *independent of* u *which depends on the functions* l_{ijhk} *only through an ellipticity constant* c_1 *(appearing in* (6.2)), *and an upper bound for the numbers* $\|l_{ijhk}\|_{m+1, p}$ $(i, j, h, k = 1, \ldots, n)$, *and for* $1/\delta$ *where* δ *is the strictly positive real functional that in* AGMON, DOUGLIS, & NIRENBERG [1964] *is called the "minor constant" (of the Dirichlet problem for operator* (4.2)).

Proof. Recall that, by the Sobolev imbedding theorem, $W^{m+1, p}(\Omega)$ can be embedded in $C^m(\bar{\Omega})$. Note that, for $v \in W^{m+2, p}(\Omega, \mathbb{R}^n)$, we have

$$\operatorname{div} L(Du) = T_1(u) + T_2(u), \tag{6.4}$$

with

$$T_1(u) = \left(\sum_{j,h,k=1}^{n} l_{ijhk} D_j D_k u_h \right)_{i=1,\ldots,n}, \qquad T_2(u) = \left(\sum_{j,h,k=1}^{n} l_{ijhk} D_k u_h \right)_{i=1,\ldots,n}$$

Suppose that (6.1) is satisfied. Then, as the coefficients of T_1 belong to $C^m(\overline{\Omega})$, the following estimate for operator T holds (see AGMON, DOUGLIS, & NIRENBERG [1964], Theorem 10.5, and recall that the Dirichlet problem for a strongly elliptic operator satisfies the "complementing condition"):

$$\|T_1(u)\|_{m,p} + \|u\|_{0,p} \geq c_2 \|u\|_{m+2,p}, \qquad \forall u \in W_0^{1,p}(\Omega, \mathbb{R}^n) \cap W^{m+2,p}(\Omega, \mathbb{R}^n)$$
(6.5)

with c_2 a number > 0 independent of u, which depends on the functions l_{ijhk} only through an ellipticity constant c_1, a function from $\mathbb{R}^+ \backslash \{0\}$ into $\mathbb{R}$ dominating a modulus of continuity of the functions l_{ijhk}, and an upper·bound for the numbers $\sup\{|D^\alpha l_{ijhk}(x): x \in \Omega\}$, with $|\alpha| \leq m$ and i, j, h, $k = 1, \ldots, n$, and for the "minor constant," say δ_1, of Dirichlet's problem for T_1. Recalling that, by the hypotheses made on p and Ω, $W^{m+1,p}(\Omega)$ can be continuously embedded in $C^{m,\lambda}(\overline{\Omega})$ with $0 < \lambda < 1 - n/p$ (see Chapter II, §1), we realize without difficulty that, in (6.5), c_2 depends on the functions l_{ijhk} only through c_1 and an upper bound for the numbers $\|l_{ijhk}\|_{m+1,p}$ $(i, j, h, k = 1, \ldots, n)$, and $1/\delta$.

On the other hand, from the fact that $D_j l_{ijhk} \in W^{m,p}(\Omega)$ and that $W^{m,p}(\Omega)$ is a Banach algebra (see Chapter II, Corollary 2.3), it evidently follows that

$$\|T_2 u\|_{m,p} \leq c_3 \|u\|_{m+1,p}, \qquad \forall u \in W^{m+2,p}(\Omega, \mathbb{R}^n),$$

where c_3 is a number > 0 depending on the functions l_{ijhk} only through an upper bound for the numbers $\|l_{ijhk}\|_{m+1,p}$ $(i, j, h, k = 1, \ldots, n)$. Moreover, since the embeddings of $W^{m+2,p}(\Omega)$ in $W^{m+1,p}(\Omega)$ and of $W^{m+1,p}(\Omega)$ in $L^p(\Omega)$ are compact (see Chapter II, §1), Lions's lemma (see Chapter II, §1) ensures that for each number $\varepsilon > 0$ there is a number $\alpha(\varepsilon) > 0$ such that

$$\|u\|_{m+1,p} \leq \varepsilon \|u\|_{m+2,p} + \alpha(\varepsilon) \|u\|_{0,p}, \qquad \forall u \in W^{m+2,p}(\Omega, \mathbb{R}^n);$$

thus

$$\|T_2(u)\|_{m,p} \leq \varepsilon c_3 \|u\|_{m+2,p} + c_3 \alpha(\varepsilon) \|u\|_{0,p}, \qquad \forall u \in W^{m+2,p}(\Omega, \mathbb{R}^n). \quad (6.6)$$

Combining (6.4) with (6.5) and (6.6) we obtain, for every $u \in W_0^{1,p}(\Omega, \mathbb{R}^n) \cap W^{m+2,p}(\Omega, \mathbb{R}^n)$,

$$\|\operatorname{div} L(Du)\|_{m,p} + \|u\|_{0,p} \geq (c_2 - \varepsilon c_3) \|u\|_{m+2,p} - \alpha(\varepsilon) c_3 \|u\|_{0,p},$$

which clearly implies

$$\|\operatorname{div} L(Du)\|_{m,p} + \|u\|_{0,p} \geq \frac{c_2 - \varepsilon c_3}{1 + \alpha(\varepsilon) c_3} \|u\|_{m+2;p}.$$

Therefore (6.3) holds with

$$c = \frac{c_2 - \varepsilon c_3}{1 + \alpha(\varepsilon)c_3} \quad \text{and} \quad \varepsilon < \frac{c_2}{c_3}. \quad \square$$

We remark that we could obtain estimate (6.3) directly from the estimates for boundary problems in a half-space in the case when the coefficients are constant, that have been proved by AGMON, DOUGLIS, & NIRENBERG [1964] (cf. Theorem 10.2).

Remark 6.3. *In Lemma* 6.2, *inequality* (6.3) *can be replaced with*

$$\|\text{div } L(Du)\|_{m,p} + \|u\|_{0,2} \geq c' \|u\|_{m+2,p}, \quad \forall u \in W_0^{1,p}(\Omega, \mathbb{R}^n) \cap W^{m+2,p}(\Omega, \mathbb{R}^n)$$

with c' *a number* > 0 *independent of* u *and depending on the functions* l_{ijhk} *in the same way as* c.

Proof. The proof is trivial if $p \leq 2$: in this case we can take $c' = c$. If $p > 2$, it suffices to observe that, because of the compactness of the embedding of $W^{m+2,p}(\Omega)$ in $L^p(\Omega)$ and he continuity of the embedding of $L^p(\Omega)$ in $L^2(\Omega)$, from Lions's lemma (see Chapter II, §1) it follows that for each number $\vartheta > 0$ there is a number $\beta(\vartheta) > 0$ such that

$$\|u\|_{0,p} \leq \vartheta \|u\|_{m+2,p} + \beta(\vartheta)\|u\|_{0,2}, \quad \forall u \in W^{m+2,p}(\Omega). \quad \square$$

Inequality (6.3) has a Schauder version, as specified in the following theorem, which derives from a general theorem proved by AGMON, DOUGLIS, & NIRENBERG [1964] (see Theorem 9.3 and recall that the Dirichlet problem for a strongly elliptic operator satisfies the "complementing condition"). Before stating the theorem we note that, by Remark 1.1 of Chapter II, $u \mapsto \text{div } L(Du)$ is a continuous mapping from $C^{m+2,\lambda}(\bar{\Omega}, \mathbb{R}^n)$ into $C^{m,\lambda}(\bar{\Omega}, \mathbb{R}^n)$ provided $l_{ijhk} \in C^{m+1,\lambda}(\bar{\Omega})$ and $\bar{\Omega}$ is of class C^1.

Theorem 6.4. *Let* $l_{ijhk} \in C^{m+1,\lambda}(\bar{\Omega})$ *with* $0 < \lambda < 1$ *and let* Ω *be of class* $C^{m+2,\lambda}$. *Then* (6.1) *implies*

$$\|\text{div } L(Du)\|_{m,\lambda} + \|u\|_{0,1} \geq c\|u\|_{m+2,\lambda}, \quad \forall u \in C_0^0(\bar{\Omega}, \mathbb{R}^n) \cap C^{m+2,\lambda}(\bar{\Omega}, \mathbb{R}^n),$$

where $C_0^0(\bar{\Omega}, \mathbb{R}^n) = \{u \in C^0(\bar{\Omega}, \mathbb{R}^n): u|_{\partial\Omega} = 0\}$ *and* c *is a number* > 0 *independent of* u, *which depends on the functions* l_{ijhk} *only through an ellipticity constant* c_1 (*cf.* (6.2)) *and an upper bound for the numbers* $\|l_{ijhk}\|_{m+1,\lambda}$ ($i, j, h, k = 1, \ldots, n$), *and for* $1/\delta$ *with* δ *the "minor constant" of the Dirichlet problem for operator* (4.2). *Here* $\|\cdot\|_{0,1}$ *denotes the norm of* $L^1(\Omega, \mathbb{R}^n)$.

* * * * *

Now let us deal with estimates for solutions of the Neumann problem (5.1).

Remark 6.5. *If Ω is of class C^{m+2} [respectively $C^{m+2,\lambda}$], then the linear operators*

$$f \mapsto f|_{\partial\Omega} v_j, \qquad j = 1, \ldots, n, \tag{6.7}$$

map $W^{m+1,p}(\Omega)$ into $W^{m+1-1/p,p}(\partial\Omega)$ [respectively $C^{m+1,\lambda}(\bar{\Omega})$ into $C^{m+1,\lambda}(\partial\Omega)$] and are continuous.

Proof. We first consider the case of Sobolev spaces. Accordingly, suppose that Ω is of class C^{m+2}. Then, for each $j = 1, \ldots, n$, there is a C^{m+1} function $g_j \colon \mathbb{R}^n \to \mathbb{R}$ extending v_j. Evidently, if $f \in W^{m+1,p}(\Omega)$ then $fg \in W^{m+1,p}(\Omega)$ and there is a positive number c_j independent of f such that $\|fg_j\|_{m+1,p} \leq c\|f\|_{m+1,p}$. Since $fg \in W^{m+1,p}(\Omega)$, then, by a trace theorem for Sobolev spaces (see Chapter II, §1), $(fg_j)|_{\partial\Omega} \in W^{m+1-1/p,p}(\partial\Omega)$ and there is a number $c_j' > 0$, independent of f, such that $\|(fg_j)|_{\partial\Omega}\|_{m+1-1/p,p,\partial\Omega} \leq c_j'\|fg_j\|_{m+1,p}$, where $\|\cdot\|_{m+1-1/p,p,\partial\Omega}$ denotes a norm on $W^{m+1-1/p,p}(\partial\Omega)$ defining its topology. Hence, as $(fg_j)|_{\partial\Omega} = f|_{\partial\Omega} v_j$, we have

$$\|f|_{\partial\Omega} v_j\|_{m+1-1/p,p,\partial\Omega} \leq c_j' c_j \|u\|_{m+1,p}.$$

Thus (6.7) is a continuous mapping from $W^{m+1,p}(\Omega)$ into $W^{m+1-1/p,p}(\partial\Omega)$.

We now prove that (6.7) is a continuous mapping from $C^{m+1,\lambda}(\bar{\Omega})$ into $C^{m+1,\lambda}(\partial\Omega)$ provided Ω is of class $C^{m+2,\lambda}$. Then let Ω be of class $C^{m+2,\lambda}$ and for every $x \in \partial\Omega$ let U_x and t_x be, respectively, an open neighborhood of x in $\mathbb{R}$ and a $C^{m+1,\lambda}$ diffeomorphism of U_x onto the ball $\{\xi \in \mathbb{R}^n : |\xi| < 1\}$ such that $t_x(\bar{\Omega} \cap U) = \{\xi \in \mathbb{R}^n : |\xi| < 1, \xi_n \geq 0\}$. Moreover, let $f \in C^{m+1,\lambda}(\bar{\Omega})$. Obviously, $f \circ t_x^{-1} \in C^{m+1,\lambda}(B)$ with $B = \{\xi \in \mathbb{R}^n : |\xi| < 1, \xi_n = 0\}$. Now, $v_j \in C^{m+1,\lambda}(\partial\Omega)$ and thus $v_j \circ t_x^{-1}$ is an element of $C^{m+1,\lambda}(B)$. Since $C^{m+1,\lambda}(B)$ is a Banach algebra (see Remark 1.1 of Chapter II), it follows that $(f|_{\partial\Omega} v_j) \circ t_x^{-1} \in C^{m+1,\lambda}(B)$ and $\|(f|_{\partial\Omega} v_j) \circ t_x^{-1}\|_{m+1,\lambda,B} \leq c_{m+1,\lambda}\|f \circ t_x^{-1}\|_{m+1,\lambda,B}$, where $c_{m+1,\lambda}$ is a number > 0 independent of f and $\|\cdot\|_{m+1,\lambda,B}$ denotes a norm on $C^{m+1,\lambda}(B)$ defining its topology. Then

$$\|f|_{\partial\Omega} v_j\|_{m+1,\lambda,B} = \sup_{x \in \partial\Omega} \|f|_{\partial\Omega} v_j) \circ t_x^{-1}\|_{m+1,\lambda,B}$$

$$\leq \sup_{x \in \partial\Omega} \|f \circ t_x^{-1}\|_{m+1,\lambda,B} \sup_{x \in \partial\Omega} \|v_j \circ t_x^{-1}\|_{m+1,\lambda,B}$$

$$= \|f|_{\partial\Omega}\|_{m+1,\lambda,\partial\Omega} \|v_j\|_{m+1,\lambda,\partial\Omega}.$$

Thus we have verified that (6.7) is a continuous mapping from $C^{m+1,\lambda}(\bar{\Omega})$ into $C^{m+1,\lambda}(\partial\Omega)$. □

We have already remarked that if Ω has the cone property [respectively if Ω is of class C^1] and $l_{ijhk} \in W^{m+1,p}(\Omega)$ with $p(m + 1) > n$ [respectively $l_{ijhk} \in C^{m+1,\lambda}(\bar{\Omega})$], then $u \mapsto \operatorname{div} L(Du)$ is a continuous mapping from $W^{m+2,p}(\Omega, \mathbb{R}^n)$ into $W^{m,p}(\Omega, \mathbb{R}^n)$ [respectively from $C^{m+2,\lambda}(\bar{\Omega}, \mathbb{R}^n)$ to $C^{m,\lambda}(\bar{\Omega}, \mathbb{R}^n)$]. Then, taking into account Remark 6.5, we deduce that if Ω is of class C^{m+2} [respectively $C^{m+2,\lambda}$] and $l_{ijhk} \in W^{m+1,p}(\Omega)$ with $p(m + 1) > n$ [respectively $l_{ijhk} \in C^{m+1,\lambda}(\bar{\Omega})$], then $u \mapsto (\operatorname{div} L(Du), L(Du)|_{\partial\Omega} v)$

is a continuous mapping from $W^{m+2,p}(\Omega, \mathbb{R}^n)$ *into* $W^{m,p}(\Omega, \mathbb{R}^n) \times$ $W^{m+1-1/p,p}(\partial\Omega, \mathbb{R}^n)$ [*respectively from* $C^{m+2,\lambda}(\bar{\Omega}, \mathbb{R}^n)$ *into* $C^{m,\lambda}(\bar{\Omega}, \mathbb{R}^n) \times$ $C^{m+1,\lambda}(\partial\Omega, \mathbb{R}^n)$].

Theorem 6.6. *Assume that* Ω *is of class* C^{m+2}, *that* $p(m+1) > n$, *that* $l_{ijhk} \in C^{m+1}(\Omega)$, *and that*

$$\int_\Omega \sum_{i,j,h,k=1}^n l_{ijhk} D_j u_i D_k u_h \, dx + \|u\|_{0,2}^2 \geqq c \|u\|_{1,2}^2, \qquad \forall u \in W^{1,2}(\Omega, \mathbb{R}^n), \quad (6.8)$$

with c *a number* > 0 *independent of* u. *Then there is a number* $c' > 0$ *such that*

$$\|\text{div } L(Du)\|_{m,p} + \|L(Du)|_{\partial\Omega}\nu\|_{m+1-1/p,p,\partial\Omega} + \|u\|_{0,2}$$
$$\geqq c' \|u\|_{m+2,p}, \qquad \forall u \in W^{m+2,p}(\Omega, \mathbb{R}^n).$$

The number c' *depends on the functions* l_{ijhk} *only through an ellipticity constant*† c_1 (*see* (6.2)) *and an upper bound for the numbers* $|l_{ijhk}|_{m+1}$ $(i, j, h, k = 1, \ldots, n)$, *and for* $1/\delta$ *where* δ *denotes the "minor constant" of problem* (5.1) (*see* AGMON, DOUGLIS, & NIRENBERG [1964], p. 43).

Theorem 6.7. *Assume that* Ω *is of class* $C^{m+2,\lambda}$ *with* $0 < \lambda < 1$, *that* $l_{ijhk} \in C^{m+1,\lambda}(\bar{\Omega})$, *and that* (6.8) *holds with* c *independent of* u. *Then there is a number* $c'' > 0$ *such that*

$$\|\text{div } L(Du)\|_{m,\lambda} + \|L(Du)|_{\partial\Omega}\nu\|_{m+1,\lambda,\partial\Omega} + \|u\|_{0,1}$$
$$\geqq c'' \|u\|_{m+2,\lambda}, \qquad \forall u \in C^{m+2,\lambda}(\bar{\Omega}, \mathbb{R}^n).$$

The number c'' *depends on the functions* l_{ijhk} *only through an ellipticity constant* c_1 *and an upper bound for the numbers* $\|l_{ijhk}\|_{m+1,\lambda}$ $(i, j, h, k = 1, \ldots, n)$, *and for* $1/\delta$ *with* δ *the "minor constant" of problem* (5.1) (*see* AGMON, DOUGLIS, & NIRENBERG [1964], p. 43).

Theorems 6.6 and 6.7 are a consequence of Theorems 10.5 and 9.3, respectively, of AGMON, DOUGLIS, & NIRENBERG [1964], and the fact that condition (6.8) implies that the operator $u \mapsto \text{div } L(Du)$ is strongly elliptic (by Theorem 6.1) and that problem (5.1) satisfies the "complementing condition" in view of Theorem 12 of THOMPSON [1968].

§7. Regularity Theorems for Dirichlet and Neumann Problems in Linearized Elastostatics

We will prove some regularity theorems for the Dirichlet and Neumann boundary problems relative to operator (4.2), using a method of continuity: essentially, we will show that it is possible to reduce to the Laplace

---

† Recall that, in view of Theorem 6.1, condition (6.8) implies (6.1) and hence (6.2).

operator. However, the conclusions could be obtained by a different procedure. We begin with the Dirichlet problem.

Theorem 7.1. *Let* $l_{ijhk} \in W^{m+1,p}(\Omega)$ *(i, j, h, k = 1, ..., n) with* $m \geq 0$, $p > n$ *and with* Ω *of class* C^{m+2}. *If (4.7) holds with c a number* > 0 *independent of v, then (4.2) is a (linear) homeomorphism of the subspace* $W_0^{1,p}(\Omega, \mathbb{R}^n) \cap W^{m+2,p}(\Omega, \mathbb{R}^n)$ *of* $W^{m+2,p}(\Omega, \mathbb{R}^n)$ *onto* $W^{m,p}(\Omega, \mathbb{R}^n)$.

Proof. Suppose that (4.7) is satisfied. For every $(x, t) \in \Omega \times [0, 1]$ we set

$$m_{ijhk}(x, t) = t l_{ijhk}(x) + (1 - t) I_{ih} I_{jk},$$

where I_{ij} is the (i, j)th coordinate of I. Moreover, for each function $\sigma : \Omega \to \mathbb{M}_n$ let $M_t(\sigma)$ be the $\mathbb{M}_n$-valued function defined on Ω by setting

$$M_t(\sigma)(x) = \left(\sum_{h,k=1}^{n} m_{ijhk}(x, t)\sigma_{hk}(x) \right)_{i,j=1,...,n}.$$

Proceeding as in §4 we easily see that there is a number $c_2 > 0$ such that

$$\|\operatorname{div} M_t(Du)\|_{-1,2} \geq c_2 \|u\|_{1,2}, \qquad \forall (u, t) \in W_0^{1,2}(\Omega, \mathbb{R}^n) \times [0, 1]. \quad (7.1)$$

We recall that, by the Sobolev embedding theorem, $W^{m+1,p}(\Omega)$ can be embedded in $C^m(\bar{\Omega})$. Then, in view of Theorem 6.1, (6.1) holds and thus

$$\sum_{i,j,h,k=1}^{n} m_{ijhk}(x, t)\xi_i \xi_k \eta_i \eta_h > 0,$$

$$\forall (t, x, \xi, \eta) \in [0, 1] \times \bar{\Omega} \times (\mathbb{R}^n \setminus \{0\}) \times (\mathbb{R}^n \setminus \{0\}). \quad (7.2)$$

This implies (see AGMON, DOUGLIS, & NIRENBERG [1964], p. 43) that the Dirichlet problem for the operator $u \mapsto \operatorname{div} M_t(Du)$ satisfies the "complementing condition" $\forall t \in [0, 1]$ and, consequently, the "minor constant," say δ_t, of such a problem is strictly positive $\forall t \in [0, 1]$. Therefore, because the function $t \mapsto \delta_t$ is continuous, we have $\inf\{\delta_t : t \in [0, 1]\} > 0$.

Note that, since the l_{ijhk} are (identifiable with) continuous functions, from (7.2) it follows that there is a number $c_1 > 0$ such that

$$\det\left(\sum_{j,k=1}^{n} m_{ijhk}(x, t)\xi_j \xi_k \right)_{i,h=1,...,n} \geq c_1 |\xi|^{2n}, \qquad \forall (t, x, \xi) \in [0, 1] \times \bar{\Omega} \times \mathbb{R}^n;$$

we can take as c_1 the minimum of the strictly positive, continuous function

$$(t, x, \xi) \mapsto \det\left(\sum_{j,k=1}^{n} m_{ijhk}(x, t)\xi_j \xi_k \right)_{i,h=1,...,n}$$

on the compact subset $[0, 1] \times \bar{\Omega} \times \{\xi \in \mathbb{R}^n : |\xi| = 1\}$ of $\mathbb{R}^{2n+1}$.

Now, recalling Theorem 6.2, Remark 6.3, and the discussion following (7.2), we can claim that there is a number $c_3 > 0$, independent of u and t, such that

$$\|\mathrm{div}\, M_t(Du)\|_{m,p} + \|u\|_{0,2} \geq c_3 \|u\|_{m+2,p},$$

$$\forall (t, u) \in [0, 1] \times (W_0^{1,p}(\Omega, \mathbb{R}^n) \cap W^{m+2,p}(\Omega, \mathbb{R}^n)). \tag{7.3}$$

Combining (7.1) and (7.3) and putting $\alpha = c_3^{-1}$ and $\beta = (c_2 c_3^{-1})$, we have

$$\|u\|_{m+2,p} \leq \alpha \|\mathrm{div}\, M_t(Du)\|_{m,p} + \beta \|\mathrm{div}\, M_t(Du)\|_{-1,2},$$

$$\forall (t, u) \in [0, 1] \times (W_0^{1,p}(\Omega, \mathbb{R}^n) \cap W^{m+2,p}(\Omega, \mathbb{R}^n)). \tag{7.4}$$

Observe that α and β are independent of u and t. Now remark that, for each $x \in \Omega$,

$$(\mathrm{div}\, M_0(Du))(x) = \Delta u(x) I,$$

where Δ is the Laplace operator $\sum_{j=1}^n D_j D_j$ and recall that, if Ω is of class C^{m+2}, Δ is a homeomorphism of $W_0^{1,p}(\Omega, \mathbb{R}^n) \cap W^{m+2,p}(\Omega, \mathbb{R}^n)$ onto $W^{m,p}(\Omega, \mathbb{R}^n)$ (see, e.g., SIMADER [1972]). Then the set, say J, of those $t \in [0, 1]$ such that the operator

$$u \mapsto \mathrm{div}\, M_t(Du)$$

is a (linear) homeomorphism of $W_0^{1,p}(\Omega, \mathbb{R}^n) \cap W^{m+2,p}(\Omega, \mathbb{R}^n)$ onto $W^{m,p}(\Omega, \mathbb{R}^n)$, is not empty. We also remark that J is an open subset of $[0, 1]$, because the mapping

$$t \mapsto \mathrm{div}(M_t \circ D)$$

from $[0, 1]$ into the space of continuous, linear mappings of $W_0^{1,p}(\Omega, \mathbb{R}^n) \cap W^{m+2,p}(\Omega, \mathbb{R}^n)$ onto $W^{m,p}(\Omega, \mathbb{R}^n)$ equipped with the topology of bounded convergence is continuous, and because of the well-known fact that the set of linear homeomorphisms of a Banach space X onto a Banach space Y is an open subset of the space of all continuous, linear mappings from X into Y equipped with the topology of bounded convergence.

We now prove, using (7.4), that J is closed, so that $J = [0, 1]$ and thus, in particular, the mapping $u \mapsto \mathrm{div}\, M_1(Du)$, namely the mapping (4.2), is a homeomorphism of $W_0^{1,p}(\Omega, \mathbb{R}^n) \cap W^{m+2,p}(\Omega, \mathbb{R}^n)$ onto $W^{m,p}(\Omega, \mathbb{R}^n)$. To prove that J is closed we consider any sequence $(t_r)_{r \in \mathbb{N}}$ in J which converges to a point $t \in [0, 1]$ and we show that $t \in J$.

Let $f \in W^{m,p}(\Omega, \mathbb{R}^n)$. Since $t_r \in J$, $\forall r \in \mathbb{N}$, for each $r \in \mathbb{N}$ there is an $u(r) \in W_0^{1,p}(\Omega, \mathbb{R}^n) \cap W^{m+2,p}(\Omega, \mathbb{R}^n)$ such that $\mathrm{div}\, M_{t_r}(Du(r)) = f$. By (7.4) the sequence $(u(r))_{r \in \mathbb{N}}$ is bounded in $W^{m+2,p}(\Omega, \mathbb{R}^n)$. Since the subspace $W_0^{1,p}(\Omega, \mathbb{R}^n) \cap W^{m+2,p}(\Omega, \mathbb{R}^n)$ of $W^{m+2,p}(\Omega, \mathbb{R}^n)$ is reflexive, there is a subsequence of $(u(r))_{r \in \mathbb{N}}$, which we still denote by $(u(r))_{r \in \mathbb{N}}$, converging weakly in $W^{m+2,p}(\Omega, \mathbb{R}^n)$ to an element $u \in W_0^{1,p}(\Omega, \mathbb{R}^n) \cap W^{m+2,p}(\Omega, \mathbb{R}^n)$. Therefore, the sequence $(\mathrm{div}\, M_{t_r}(Du(r)))_{r \in \mathbb{N}}$ converges in $W^{m,p}(\Omega, \mathbb{R}^n)$ to $\mathrm{div}\, M_t(Du)$; indeed, the mapping $u \mapsto \mathrm{div}\, M_t(Du)$ is continuous from $W_0^{1,p}(\Omega, \mathbb{R}^n) \cap W^{m+2,p}(\Omega, \mathbb{R}^n)$ into $W^{m,p}(\Omega, \mathbb{R}^n)$, and, hence it is continuous for the weak topologies of $W^{m+2,p}(\Omega, \mathbb{R}^n)$ and $W^{m,p}(\Omega, \mathbb{R}^n)$. Consequently, recalling that div $M_{t_r}(Du(r)) = f$, we conclude that div $M_t(Du) = f$. Thus $u \mapsto$ div $M_t(Du)$ is a (continuous, linear) mapping of $W_0^{1,p}(\Omega, \mathbb{R}^n) \cap W^{m+2,p}(\Omega, \mathbb{R}^n)$ into $W^{m,p}(\Omega, \mathbb{R}^n)$. On the other hand, this mapping is one-to-one by (7.4). Hence, in view of the open mapping theorem, $t \in J$. $\square$

Theorem 7.2. *Let* $l_{ijhk} \in C^{m+1,\lambda}(\bar{\Omega})$ *$(i, j, h, k = 1, \ldots, n)$ with $m \geqq 0$, $0 < \lambda < 1$ and Ω be of class C^{m+2}. If (4.7) holds with c a number > 0 independent of v, then (4.2) is a (linear) homeomorphism of the subspace $C_0^0(\bar{\Omega}, \mathbb{R}^n) \cap C^{m+2,\lambda}(\bar{\Omega}, \mathbb{R}^n)$ of $C^{m+2,\lambda}(\bar{\Omega}, \mathbb{R}^n)$ onto $C^{m,\lambda}(\bar{\Omega}, \mathbb{R}^n)$.*

Proof. We can proceed as in the proof of Theorem 7.1, after remarking that from Theorem 6.4 it follows that

$$\|\text{div } M_t(Du)\|_{m,\lambda} + \|u\|_{0,1} \geqq c\|u\|_{m+2,\lambda},$$

$$\forall(t, u) \in [0, 1] \times C_0^0(\bar{\Omega}, \mathbb{R}^n) \cap C^{m+2,\lambda}(\bar{\Omega}, \mathbb{R}^n)$$

with c a number > 0 independent of (t, u). $\square$

We now consider the Neumann problem (5.7). Combining Remark 6.5, Theorems 3.1 and 3.3 with Remark 1.1 of Chapter II we easily obtain

Lemma 7.3. *Let* $R \in \mathbb{O}_n^+$. *If* Ω *is of class* C^{m+2} *[respectively* $C^{m+2,\lambda}$*] and* $a \in C^{m+1}(\bar{\Omega} \times \mathbb{M}_n^+, \mathbb{M}_n)$ *[respectively* $a \in C^{m+2}(\bar{\Omega} \times \mathbb{M}_n^+, \mathbb{M}_n)$*], then*

$$u \mapsto (-\text{div } L_R(Du), L_R(Du)|_{\partial\Omega}v), \tag{7.5}$$

with $L_R(Du)$ *defined by* $((4.9), (4.10))$, *is a continuous mapping of* $W^{m+2,p}(\Omega, \mathbb{R}^n)$ *into* $W^{m,p}(\Omega, \mathbb{R}^n) \times W^{m+1-1/p,p}(\partial\Omega, \mathbb{R}^n)$ *[respectively of* $C^{m+2,\lambda}(\bar{\Omega}, \mathbb{R}^n)$ *into* $C^{m,\lambda}(\bar{\Omega}, \mathbb{R}^n) \times C^{m+1,\lambda}(\partial\Omega, \mathbb{R}^n)$*].*

Remark 7.4. *Let* Ω *be regular (in the sense made precise in Chapter I, §2) and let* $(u, f, g) \in W^{2,p}(\Omega, \mathbb{R}^n) \times L^p(\Omega, \mathbb{R}^n) \times L^p(\partial\Omega, \mathbb{R}^n)$. *If*

$$\begin{cases} -\text{div } Du = f & \text{in } \Omega, \\ (Du)v = g & \text{on } \partial\Omega, \end{cases} \tag{7.6}$$

then

$$\begin{cases} \displaystyle\int_\Omega f \, dx + \int_{\partial\Omega} g \, d\sigma = 0, \\[2ex] \displaystyle\int_\Omega \iota_\Omega \wedge f \, dx + \int_{\partial\Omega} \iota_{\partial\Omega} \wedge g \, d\sigma = \int_\Omega ((Du)^{\mathsf{T}} - Du) \, dx. \end{cases} \tag{7.7}$$

Proof. We note that $(7.7)_1$ follows immediately from (7.6) in view of the divergence theorem. To deduce $(7.7)_2$ from (7.6) we observe that (7.6) yields

$$\begin{cases} -\iota_\Omega \otimes (\operatorname{div} Du) = \iota_\Omega \otimes f & \text{in } \Omega, \\ \iota_{\partial\Omega} \otimes ((Du)v) = \iota_{\partial\Omega} \otimes g & \text{on } \partial\Omega, \end{cases}$$

whence, denoting by u_i, f_i, and g_i the ith component of u, f, and g, respectively, we conclude that

$$\begin{cases} -\operatorname{div}(\iota_\Omega \otimes Du_i) + Du_i = f_i\iota_\Omega & \text{in } \Omega, \\ (\iota_\Omega \otimes Du_i)v = g_i\iota_{\partial\Omega} & \text{on } \partial\Omega, \end{cases}$$

which by use of the divergence theorem easily implies $(7.7)_2$. $\square$

Lemma 7.5. *If Ω is of class C^{m+2} with $m \geq 0$, then for any real number $p > 1$*

$$u \mapsto (-\operatorname{div} Du, (Du)|_{\partial\Omega}v) \tag{7.8}$$

is a (linear) homeomorphism of the subspace $\{u \in W^{m+2,p}(\Omega, \mathbb{R}^n): \int_\Omega u \, dx = 0\}$ of $W^{m+2,p}(\Omega, \mathbb{R}^n)$ onto the subspace

$$\left\{(f, g) \in W^{m,p}(\Omega, \mathbb{R}^n) \times W^{m+1-1/p,p}(\partial\Omega, \mathbb{R}^n): \int_\Omega f \, dx + \int_{\partial\Omega} g \, d\sigma = 0\right\}$$

of $W^{m,p}(\Omega, \mathbb{R}^n) \times W^{m+1-1/p,p}(\partial\Omega, \mathbb{R}^n)$.

Lemma 7.5 is a well-known result (see, e.g., BROWDER [1959]). For any $R \in \mathbb{O}_n^+$, k integer ≥ 0, p real > 1, and $\lambda \in]0, 1]$, we set

$$\begin{cases} \mathcal{V}_R^{k,p} = \left\{u \in W^{k,p}(\Omega, \mathbb{R}^n): \int_\Omega u \, dx = 0, R^T \int_\Omega Du \, dx \in \operatorname{Sym}_n\right\}, \\ \mathcal{V}_R^{k,\lambda} = \left\{u \in C^{k,\lambda}(\bar\Omega, \mathbb{R}^n): \int_\Omega u \, dx = 0, R^T \int_\Omega Du \, dx \in \operatorname{Sym}_n\right\}, \end{cases}$$

and

$$\begin{cases} \mathscr{E}_R^{k,p} = \{(f, g) \in W^{k,p}(\Omega, \mathbb{R}^n) \times W^{k+1-1/p,p}(\partial\Omega, \mathbb{R}^n): \\ \qquad\qquad (R^T f, R^T g) \text{ is equilibrated}\}, \\ \mathscr{E}_R^{k,\lambda} = \{(f, g) \in C^{k,\lambda}(\bar\Omega, \mathbb{R}^n) \times C^{k+1,\lambda}(\partial\Omega, \mathbb{R}^n): \\ \qquad\qquad (R^T f, R^T g) \text{ is equilibrated}\}, \end{cases}$$

Theorem 7.6. *Assume that Ω is of class C^{m+2}, that $(m + 1)p > n$, that $a \in C^{m+2}(\bar\Omega \times \mathbb{M}_n^+, \mathbb{M}_n)$, and that (3.5), (3.6), (3.7), and (3.12) hold. Then for each $R \in \mathbb{O}_n^+$, mapping (7.8) is a (linear) homeomorphism of $\mathcal{V}_R^{m+2,p}$ onto $\mathscr{E}_R^{m,p}$.*

Proof. Let, for every $(t, R, x) \in [0, 1] \times \mathbb{O}_n^+ \times \Omega$,

$$m_{ijhk}(x, t, R) = tD_{z_{hk}}a_{ij}(x, R) + (1 - t)I_{ih}I_{jk}, \tag{7.9}$$

where $a_{ij}(x, R) = a(x, R) \cdot (e_i \otimes e_j)$ and $I_{ij} = I \cdot (e_i \otimes e_j)$ with $e_1, \ldots, e_n$ the canonical base of $\mathbb{R}^n$. Moreover, for every function $\sigma: \Omega \to M_n$, let $M_{t, R}(\sigma)$ be the M_n-valued function defined on Ω by setting

$$M_{t, R}(\sigma)(x) = \left(\sum_{h, k=1}^{n} m_{ijhk}(x, t, R)\sigma_{hk}(x) \right)_{i, j=1, \ldots, n}$$

and let $u \in W^{m+2, p}(\Omega, \mathbb{R}^n)$. Note that

$$M_{t, R}(Du) = tL_R(Du) + (1 - t)Du.$$

Combining Lemma 7.3 with Remark 5.4 and the discussion following the latter, we conclude that (7.5) maps $W^{m+2, p}(\Omega, \mathbb{R}^n)$ into $\mathscr{E}_R^{m, p}$. Moreover, in view of Remark 7.4,

$$u \mapsto (-\operatorname{div} Du, (Du)|_{\partial\Omega} v)$$

maps $\mathscr{V}_R^{m+2, p}$ into $\mathscr{E}_R^{m, p}$. Therefore

$$u \mapsto (-\operatorname{div} M_{t, R}(Du), M_{t, R}(Du)|_{\partial\Omega} v) \tag{7.10}$$

is a continuous (linear) mapping of $\mathscr{V}_R^{m+2, p}$ into $\mathscr{E}_R^{m, p}$. We remark that

$$\sum_{i, j, h, k=1}^{n} I_{ih} I_{jk} Z_{ij} Z_{hk} = |Z|^2, \qquad \forall Z \in M_n$$

and hence, by (7.9),

$$\int_\Omega \sum_{i, j, h, k=1}^{n} m_{ijhk}(x, t, R)D_j u_i(x)D_k u_h(x) \, dx$$

$$= t \int_\Omega \sum_{i, j, h, k=1}^{n} D_{Z_{hk}} a_{ij}(x, R)D_j u_i(x)D_k u_h(x) \, dx + (1 + t) \int_\Omega |Du|^2 \, dx. \tag{7.11}$$

Then, in view of Corollaries 1.5 and 3.4, we obtain

$$\int_\Omega \sum_{i, j, h, k=1}^{n} m_{ijhk}(x, t, R)D_j u_i(x)D_k u_h(x) \, dx$$

$$\geq d_R \|u\|_{1, 2}^2, \qquad \forall (t, u) \in [0, 1] \times \mathscr{V}_R^{1, 2}, \tag{7.12}$$

with d_R a number > 0 independent of (t, u).

On the other hand, in view of Theorems 1.4 and 3.3, from (7.11) it easily follows that

$$\int_\Omega \sum_{i, j, h, k=1}^{n} m_{ijhk}(x, t, R)D_j u_i(x)D_k u_h(x) \, dx + \|u\|_{0, 2}$$

$$\geq d_R' \|u\|_{1, 2}^2, \qquad \forall u \in W^{1, 2}(\Omega, \mathbb{R}^n), \tag{7.13}$$

with d_R' a number > 0 independent of (t, u). This implies, by Theorem 6.6, that

$$\|\operatorname{div} M_{t, R}(Du)\|_{m, p} + \|M_{t, R}(Du)\|_{m+1-1/p, p, \partial\Omega} + \|u\|_{0, 2}$$

$$\geq c_R \|u\|_{m+2, p}, \qquad \forall (t, u) \in [0, 1] \times W^{m+2, p}(\Omega, \mathbb{R}^n), \tag{7.14}$$

in which c_R is a number > 0 independent of u and dependent upon the functions m_{ijhk} only through an ellipticity constant and an upper bound for the $C^{m+1}(\bar{\Omega})$-norms of the functions $x \mapsto m_{ijhk}(x, t, R)$ and for $1/\delta_{t,R}$, where $\delta_{t,R}$ denotes the "minor constant" of the problem

$$\begin{cases} -\operatorname{div} M_{t,R}(Du) = f & \text{in } \Omega, \\ \quad M_{t,R}(Du)v = g & \text{on } \partial\Omega. \end{cases} \tag{7.15}$$

Now, the number c_R in (7.14) can be chosen independently of t, in view of the following remarks (i) and (ii).

(i) Since (7.13) implies that problem (7.15) satisfies the "complementing condition" $\forall t \in [0, 1]$ (see THOMPSON [1968], Theorem 12) we have $\delta_{t,R} > 0$, $\forall t \in [0, 1]$; thus, by the continuity of the function $t \mapsto \delta_{t,R}$,

$$\inf_{t \in [0, 1]} \delta_{t,R} > 0.$$

(ii) In view of Theorem 6.1, from (7.12) it follows that

$$\det \left(\sum_{h,k=1}^{n} m_{ijhk}(x, t, R)\xi_j\xi_k \right)_{i,h=1,\dots,n} > 0,$$

$$\forall (t, x, \xi) \in [0, 1] \times \bar{\Omega} \times (\mathbb{R}^n \setminus \{0\}),$$

which (observing that, by the hypotheses made on a, the functions $x \mapsto m_{ijhk}(x, t, R)$ are continuous on $\bar{\Omega}$) allows us to prove that

$$\det \left(\sum_{h,k=1}^{n} m_{ijhk}(x, t, R)\xi_j\xi_k \right)_{i,h=1,\dots,n} > c_R|\xi|^2,$$

$$\forall (t, x, \xi) \in [0, 1] \times \bar{\Omega} \times \mathbb{R}^n,$$

with c_R a number > 0 independent of x, t, ξ.

We are now in a position to prove that, for all $t \in [0, 1]$, mapping (7.10) is a (linear) homeomorphism of $\mathscr{V}_R^{m+2,p}$ onto $\mathscr{E}_R^{m,p}$; thus, in particular, (7.5) is a homeomorphism of $\mathscr{V}_R^{m+2,p}$ onto $\mathscr{E}_R^{m,p}$. To prove this we show that the set, say J_R, of those $t \in [0, 1]$ such that (7.10) is a homeomorphism of $\mathscr{V}_R^{m+2,p}$ onto $\mathscr{E}_R^{m,p}$, is a nonempty, both open and closed subset of $[0, 1]$, so that $J_R = [0, 1]$. As was done in the proof of Theorem 7.1, we recall that the set of all linear homeomorphisms of a Banach space X onto a Banach space Y is an open subset of the space of all continuous, linear mappings from X into Y with respect to the topology of bounded convergence; furthermore, we note that the mapping which carries $t \in [0, 1]$ to mapping (7.10) is evidently continuous, regarded as a mapping from $[0, 1]$ to the space of continuous, linear operators of $\mathscr{V}_R^{m+2,p}$ into $\mathscr{E}_R^{m,p}$. Therefore J_R is an open subset of $[0, 1]$. Moreover, in view of Remark 7.4 and Lemma 7.5, we deduce that the operator

$$u \mapsto (-\operatorname{div} M_{0,R}(Du), M_{0,R}(Du)|_{\partial\Omega}v),$$

namely operator (7.8), is a (linear) homeomorphism of $\mathscr{V}_R^{m+2,p}$ into $\mathscr{E}_R^{m,p}$; thus J_R is not empty.

It remains to prove that J_R is closed. To do this we will show, using inequalities (7.12) and (7.14), that if $(t_r)_{r \in \mathbb{N}}$ is a sequence in J_R converging to an element $\bar{t} \in [0, 1]$, then $\bar{t} \in J_R$. Now we let $(t_r)_{r \in \mathbb{N}}$ be a sequence in J_R converging to $\bar{t} \in [0, 1]$ and let $(f, g) \in \mathscr{E}_R^{m,p}$. As $t_r \in J_R$, $\forall r \in \mathbb{N}$, so for each $r \in \mathbb{N}$ there is a $u(r) \in \mathscr{V}_R^{m+2,p}$ such that

$$\begin{cases} -\operatorname{div} M_{t_r, R}(Du(r)) = f & \text{in } \Omega, \\ M_{t_r, R}(Du(r))\nu = g & \text{on } \partial\Omega. \end{cases} \tag{7.16}$$

By Remark 5.1, (7.16) implies

$$\int_\Omega \sum_{i,j,h,k=1}^n m_{ijhk}(x, t, R) D_j v_i(x) D_k u(r)_h(x) \, dx$$

$$= \int_\Omega f \cdot v \, dx + \int_{\partial\Omega} g \cdot v \, d\sigma, \qquad \forall v \in C^1(\bar{\Omega}, \mathbb{R}^n). \tag{7.17}$$

Since, by the Sobolev embedding theorem, $W^{m+2,p}(\Omega)$ can be embedded in $C^1(\bar{\Omega})$, combining (7.17) with (7.12) we deduce that, for all $r \in \mathbb{N}$,

$$\|u(r)\|_{1,2}^2 \leq d_R^{-1} \|\varphi\|_{-1,2}, \tag{7.18}$$

where φ is the continuous (see Remark 5.2), linear form on $W^{1,2}(\Omega, \mathbb{R}^n)$ defined by (5.3).

Because of (7.14) and (7.18) the sequence $(u(r))_{r \in \mathbb{N}}$ is bounded in $W^{m+2,p}(\Omega, \mathbb{R}^n)$; thus, as $\mathscr{V}_R^{m+2,p}$ is a reflexive Banach space, a subsequence of $(u(r))_{r \in \mathbb{N}}$—which we still denote by $(u(r))_{r \in \mathbb{N}}$—converges weakly in $W^{m+2,p}(\Omega, \mathbb{R}^n)$ to an element $u \in \mathscr{V}_R^{m+2,p}$. Hence $(\operatorname{div} M_{t_r, R}(Du(r)))_{r \in \mathbb{N}}$ converges weakly in $W^{m,p}(\Omega, \mathbb{R}^n)$ to $\operatorname{div} M_{\bar{t}, R}(Du)$ and $(M_{t_r, R}(Du(r))|_{\partial\Omega}\nu)_{r \in \mathbb{N}}$ converges weakly in $W^{m+1-1/p,p}(\partial\Omega, \mathbb{R}^n)$ to $M_{\bar{t}, R}(Du)|_{\partial\Omega}\nu$; then, by (7.16), we have $\operatorname{div} M_{\bar{t}, R}(Du) = f$ and $M_{\bar{t}, R}(Du)|_{\partial\Omega}\nu = g$. This proves that the continuous linear operator $u \mapsto (-\operatorname{div} M_{\bar{t}, R}(Du), M_{\bar{t}, R}(Du)|_{\partial\Omega}\nu)$ of $\mathscr{V}_R^{m+2,p}$ into $\mathscr{E}_R^{m,p}$ is surjective. Since, in view of ((7.14), (7.18)) this operator is also one-to-one, we can conclude that $\bar{t} \in J_R$. $\square$

Proceeding essentially as in the proof of the previous theorem and exploiting Theorem 6.7 we can prove

Theorem 7.7. *Assume that Ω is of class $C^{m+2,\lambda}$ with $0 < \lambda < 1$, that $a \in C^{m+2,\lambda}(\bar{\Omega} \times \mathbb{M}_n^+, \mathbb{M}_n)$, and that (3.5), (3.6), (3.7), and (3.12) hold. Then, for each $R \in \mathbb{O}_n^+$, mapping (7.8) is a (linear) homeomorphism of $\mathscr{V}_R^{m+2,\lambda}$ into $\mathscr{E}_R^{m,\lambda}$.*

CHAPTER IV

Boundary Problems of Place in Finite Elastostatics

This chapter is devoted to the boundary value problem of place in finite elastostatics. This problem corresponds to the assignment of the trace on $\partial\Omega$ of the unknown equilibrium deformation ϕ. Here we confine our attention to the case in which the trace of ϕ on $\partial\Omega$ is required to be the identity $\iota_{\partial\Omega}$. The case when the trace of ϕ on $\partial\Omega$ is assigned to be the trace on $\partial\Omega$ of a given deformation can obviously be reduced to the one studied here. We consider the case when the body forces are independent of the deformation (dead load), and the case when the body forces depend on the unknown deformation (live load).

Theorems on existence, uniqueness, and analytic dependence on the load are obtained in Sobolev spaces and in Schauder spaces. They are of a local type. In the case of dead loads some stronger results are also given (see Theorems 5.3 and 5.4). Note that the solutions whose existence we prove are actually C^1 diffeomorphisms of $\bar\Omega$ onto itself; the possibility that the solutions satisfy such a requirement is based on an important topological property of the set of admissible deformations (see Theorem 3.2).

The results presented are useful in clarifying the position of a linearized theory within the nonlinear theory. In §2 we prove, on the basis of Theorem 6.1 of Chapter II, that no linearization is "admissible" with respect to the pair $(W^{1,p}(\Omega, \mathbb{R}^n), W^{-1,p}(\Omega, \mathbb{R}^n))$ of spaces for solutions and data; note that this is a natural choice of (Sobolev) spaces for solutions and data in studying the formally linearized operator of the finite elastostatics operator, as we have seen in Chapter III.

The results exposed in this chapter are essentially improvements of the analogous results either stated or proved in VALENT [1978a, 1979, 1981, 1985b]. A local theorem on existence and uniqueness for a dead boundary value problem of place was obtained by VAN BUREN [1968] and summarized in WANG & TRUESDELL [1973].

§1. Formulation of the Problem

Let $(x, Z) \mapsto a(x, Z)$ be a given function of $\Omega \times \mathsf{M}_n^+$ into M_n, let $(x, y) \mapsto b(x, y)$ be a given function of $\Omega \times \mathbb{R}^n$ into $\mathbb{R}^n$, and let μ be a given function of Ω into $\mathbb{R}$.

We will study the problem of finding a function $\phi: \bar{\Omega} \to \mathbb{R}^n$ such that

ϕ is an orientation-preserving C^1 diffeomorphism of $\bar{\Omega}$
onto $\bar{\Omega}$ (see Chapter II, §1), $\qquad\qquad$ (1.1)

$$\begin{cases} \operatorname{div} A(D\phi) + \mu B(\phi) = 0 & \text{in } \Omega, \\ \phi|_{\partial\Omega} = \imath_{\partial\Omega}, \end{cases} \qquad (1.2)$$

where $A(D\phi): \Omega \to \mathsf{M}_n$ and $B(\phi): \Omega \to \mathbb{R}^n$ are the functions defined by

$$\begin{cases} A(D\phi)(x) = a(x, D\phi(x)), \\ B(\phi)(x) = b(x, \phi(x)), & x \in \Omega, \end{cases}$$

and where $\imath_{\partial\Omega}$ denotes, as always, the identity function of $\partial\Omega$ onto itself.

The physical meaning of the symbols is explained in Chapter I. We only recall that (if $n = 3$) Ω represents a fixed reference configuration of an elastic body, that ϕ is the unknown deformation of the body, that the function a defines the material response in the sense that $a(x, D\phi(x))$ is the first Piola–Kirchhoff stress at the point $x \in \Omega$ when the body is deformed by ϕ, and that the function b defines a body force in the sense that $b(x, \phi(x))$ is the body force per unit mass at the point $\phi(x)$ in the deformed configuration $\phi(\Omega)$. Note that the requirement (1.1) for the solutions of (1.2) is very strong. Actually, the set of functions within which we seek the solutions of problem (1.2) is neither a linear space nor a convex set.

We begin by dealing with the case when b is constant in the second variable. In this case b can be identified with a function defined in Ω, so that $B(\phi) = b$ for any ϕ and thus (putting $f = \mu b$) problem (1.2) reduces to

$$\begin{cases} \operatorname{div} A(D\phi) + f = 0 & \text{in } \Omega, \\ \phi|_{\partial\Omega} = \imath_{\partial\Omega}, \end{cases} \qquad (1.3)$$

with f a given $\mathbb{R}^n$-valued function defined on Ω. Let us consider the (finite) elastostatics operator

$$\phi \mapsto \operatorname{div} A(D\phi). \qquad (1.4)$$

Local existence and uniqueness for problem (1.3) may be found by applying the inverse function theorem to operator (1.4), provided there exist two Banach spaces $\mathcal{B}_u$ and $\mathcal{B}_f$ such that:

(a) each $u \in \mathcal{B}_u$ is a function of Ω into $\mathbb{R}^n$ somehow satisfying the condition $u|_{\partial\Omega} = 0$;
(b) there is a $u_0 \in \mathcal{B}_u$ such that

$$u \mapsto \operatorname{div} A(D(\iota_\Omega + u)) \qquad (1.5)$$

is a continuously differentiable mapping from an open neighborhood U of 0 in $\mathcal{B}_u$ into $\mathcal{B}_f$;
(c) the differential at u_0 of operator (1.5) is a bijection of $\mathcal{B}_u$ onto $\mathcal{B}_f$.

By observing that the formal differential of (1.4) at any point is a second-order linear partial differential operator, a natural choice of spaces $\mathcal{B}_u$ and $\mathcal{B}_f$ might seem to be the following: $\mathcal{B}_u = W_0^{1,2}(\Omega, \mathbb{R}^n)$, $\mathcal{B}_f = W^{-1,2}(\Omega, \mathbb{R}^n)$, or, more generally, the following: $\mathcal{B}_u = W_0^{1,p}(\Omega, \mathbb{R}^n)$, $\mathcal{B}_f = W^{-1,p}(\Omega, \mathbb{R}^n)$. (Cf. Theorem 4.2 of Chapter III.) Unfortunately, these choices are not possible, as we will discover in the next section. For conditions ((a), (b), (c)) to be satisfied it is necessary that $\mathcal{B}_u$ be a space of functions which are smoother than those of $W_0^{1,p}(\Omega, \mathbb{R}^n)$.

§2. Remarks on Admissibility of a Linearization

In this section we suppose the function $a: \Omega \times \mathbb{M}_n^+ \to \mathbb{M}_n$ to be such that, for almost all $x \in \Omega$, the function $Z \mapsto a(x, Z)$ is continuously differentiable, that, for all $Z \in \mathbb{M}_n^+$, the functions $x \mapsto a(x, Z)$ and $x \mapsto D_{Z_{ij}} a(x, Z)$ $(i, j = 1, \ldots, n)$ are measurable, and that

$$|a(x, Z)| \leq \varphi(x) + c|Z|, \quad |D_{Z_{ij}} a(x, Z)| \leq \varphi_{ij}(x) + c_{ij}|Z|, \quad \forall (x, Z) \in \Omega \times \mathbb{M}_n^+$$

with φ, $\varphi_{ij} \in L^p(\Omega)$ and c, c_{ij} real numbers > 0 independent of (x, Z). In this hypotheses, we have seen in Chapter II, §6 that $\phi \mapsto \operatorname{div} A(D\phi)$, together with its formal linearization (cf. Chapter III, §3) at any deformation ϕ_0 are continuous mappings of $W^{1,p}(\Omega, \mathbb{R}^n)$ into $W^{-1,p}(\Omega, \mathbb{R}^n)$.

Now we emphasize the following consequence of Theorem 6.1 of Chapter II.

Remark 2.1. *Let the hypotheses made above on the function a be satisfied and let $\phi_0 \in W_0^{1,p}(\Omega, \mathbb{R}^n)$. If there are two neighborhoods U of ϕ_0 in $W_0^{1,p}(\Omega, \mathbb{R}^n)$ and V of $\operatorname{div} A(D\phi_0)$ in $W^{-1,p}(\Omega, \mathbb{R}^n)$ such that (1.4) is a* *homeomorphism of U onto V differentiable at ϕ_0, then for almost all $x \in \Omega$ the function $Z \mapsto a(x, Z)$ is affine.*

Proof. Assume that there is a pair (U, V) with the property expressed in Remark 2.1. Since the mapping $\phi \mapsto A(D\phi)$ is continuous from $W_0^{1,p}(\Omega, \mathbb{R}^n)$ into $L^p(\Omega, \mathbb{M}^n)$ and the mapping $\phi \mapsto \operatorname{div} A(D\phi)$ induces a homeomorphism between U and V, the mapping $A(D\phi) \mapsto \operatorname{div} A(D\phi)$ is a homeomorphism of the subset $\{A(D\phi): \phi \in U\}$ of $L^p(\Omega, \mathbb{M}^n)$ onto V, and hence its inverse is differentiable at $\operatorname{div} A(D\phi_0)$. Then, since (1.4), regarded as a mapping from $W_0^{1,p}(\Omega, \mathbb{R}^n)$ into $W^{-1,p}(\Omega, \mathbb{R}^n)$, is differentiable at ϕ_0, the operator $\phi \mapsto A(D\phi)$ is also differentiable at ϕ_0, when it is regarded as a mapping from $W_0^{1,p}(\Omega, \mathbb{R}^n)$ into $L^p(\Omega, \mathbb{M}^n)$. Therefore, in view of Theorem 6.1 of Chapter II, for almost all $x \in \Omega$ the function $Z \mapsto a(x, D\phi_0(x) + Z) - a(x, D\phi_0(x))$ is linear. $\square$

Remark 2.1 suggests a definition of admissibility of a linearization of problem (1.3) with respect to a pair $(\mathscr{B}_u, \mathscr{B}_f)$ of Banach spaces (for solutions and data). Let $\mathscr{B}_u$ and $\mathscr{B}_f$ be Banach spaces such that each $u \in \mathscr{B}_u$ is a $\mathbb{R}^n$-valued function defined on Ω and somehow verifying the condition $u|_{\partial\Omega} = 0$, and that (1.5) is a homeomorphism of an open subset U of $\mathscr{B}_u$ onto an open subset V of $\mathscr{B}_f$. If $\phi_0 \in \iota_{\bar{\Omega}} + U$, we will say that the linear problem

$$\begin{cases} \operatorname{div} L_{D\phi_0}(Du) = f & \text{in } \Omega, \\ u|_{\partial\Omega} = 0, \end{cases} \tag{2.1}$$

where $L_{D\phi_0}(Du)$ is defined by (3.3) of Chapter III, viz. by

$$L_{D\phi_0}(Du)(x) = \sum_{h,k=1}^{n} D_{Z_{hk}} a(x, D\phi_0(x)) D_k u_h(x),$$

is obtained by (formal) *linearization at ϕ_0* of problem (1.3).

Moreover (following VALENT [1978b]), we will say that the *linearization at ϕ_0* of problem (1.3) is *admissible with respect to the pair $(\mathscr{B}_u, \mathscr{B}_f)$* provided the following two conditions are satisfied:

(i) The linearized problem (2.1) is well-posed with respect to $(\mathscr{B}_u, \mathscr{B}_f)$ in the sense that $u \mapsto \operatorname{div} L_{D\phi_0}(Du)$ is a continuous bijection (and hence a homeomorphism) of $\mathscr{B}_u$ onto $\mathscr{B}_f$.

(ii) Operator (1.4) is differentiable at ϕ_0 and its differential at ϕ_0 coincides with the mapping $u \mapsto \operatorname{div} L_{D\phi_0}(Du)$.

Since (1.5) is a homeomorphism of U onto V it is not difficult to prove (see, e.g., CARTAN [1967], Lemma on p. 55) that *the pair $((i), (ii))$ is equivalent to the pair $((i), (ii)')$, where $(ii)'$ is the following condition*

(ii)′
$$\lim_{f_0 + f \in V, \|f\|_{\mathscr{B}_f} \to 0} \frac{\|\hat{u}(f_0, f) - (u_0 + \hat{u}_L(f))\|_{\mathscr{B}_u}}{\|f\|_{\mathscr{B}_f}} = 0,$$

with $u_0 = \phi_0 - \iota_{\bar{\Omega}}$, $f_0 = \mathrm{div}\, A(D\phi_0)$, and $\hat{u}(f_0 + f)$ and $\hat{u}_L(f)$ defined by the following conditions:

$$\begin{cases} \hat{u}(f_0 + f) \in U, & \mathrm{div}\, A(D(\iota_\Omega + \hat{u}(f))) = f_0 + f, \\ \hat{u}_L(f) \in \mathscr{B}_u, & \mathrm{div}\, L_{D\phi_0}(D\hat{u}_L(f)) = f. \end{cases}$$

In Chapter III, Corollary 4.3, we have proved that if

$$a \in C^1(\bar{\Omega} \times \mathsf{M}_n^+, \mathsf{M}_n)$$

with Ω of class C^1 and if (3.5), (3.6), (3.7), (3.12) of Chapter III are satisfied, then for every $R \in \mathbb{O}_n^+$ and every real $p > 1$ the operator $u \mapsto L_R(Du)$ is a (linear) homeomorphism of $W_0^{1,p}(\Omega, \mathbb{R}^n)$ onto $W^{-1,p}(\Omega, \mathbb{R}^n)$. This fact might lead us to believe that, under the previous hypotheses on a and Ω, the linearization at any rigid deformation of problem (1.3) is admissible. On the contrary, Remark 2.1 expresses the fact that (excluding the trivial case when operator (1.4) is linear) *no linearization of operator (1.4) is admissible with respect to the pair* $(W_0^{1,p}(\Omega, \mathbb{R}^n)$, $W^{-1,p}(\Omega, \mathbb{R}^n))$.

On this subject we note that, in view of Theorems 4.1, 4.2 of Chapter II and 7.1, 7.2 of Chapter III, we immediately obtain—by use of the inverse function theorem—the following remarks.

Remark 2.2. Let $a \in C^{m+2}(\bar{\Omega} \times \mathsf{M}_n^+, \mathsf{M}_n)$ with Ω of class C^{m+2}, $m \geq 0$, and let $\phi_0 \in W^{m+2,p}(\Omega, \mathbb{R}^n)$ with p real number > 1. If

$$\int_\Omega \left(\sum_{h,k=1}^n D_{z_{hk}} a(x, D\phi_0(x)) D_k u_h \right) \cdot Du\, dx \geq c \|u\|_{1,2}^2, \qquad \forall u \in \mathscr{D}(\Omega, \mathbb{R}^n), \quad (2.2)$$

then the linearization of the operator (1.4) at ϕ_0 is admissible with respect to the pair $(W_0^{1,p}(\Omega, \mathbb{R}^n) \cap W^{m+2,p}(\Omega, \mathbb{R}^n), W^{m,p}(\Omega, \mathbb{R}^n))$.

Remark 2.3. Let $a \in C^{m+2,\lambda}(\bar{\Omega} \times \mathsf{M}_n^+, \mathsf{M}_n)$ with Ω of class $C^{m+2,\lambda}$, $m \geq 0$, and let $\phi_0 \in C^{m+2,\lambda}(\bar{\Omega}, \mathbb{R}^n)$ with $\lambda \in]0, 1]$. If (2.2) is satisfied, then the linearization of the operator (1.4) at ϕ_0 is admissible with respect to the pair $(C_0^0(\bar{\Omega}, \mathbb{R}^n) \cap C^{m+2,\lambda}(\bar{\Omega}, \mathbb{R}^n), C^{m,\lambda}(\bar{\Omega}, \mathbb{R}^n))$.

To conclude this section we recall that *condition (2.2) is satisfied with ϕ_0 any rigid deformation if* (3.5), (3.6), (3.7), *and* (3.12) *of Chapter* III *are satisfied.*

§3. A Topological Property of Sets of Admissible Deformations

A function $\phi: \bar{\Omega} \to \mathbb{R}^n$ will be called an *admissible deformation* (for the problems $((1.1), (1.2))$ and $((1.1), (1.3)))$ if ϕ satisfies the conditions (1.1) and $\phi|_{\partial\Omega} = \iota_{\partial\Omega}$. We denote by $\mathscr{A}_{m+2,p}$ and by $\mathscr{A}_{m+2,\lambda}$ the set of those admissible deformations that lie in $W^{m+2,p}(\Omega, \mathbb{R}^n)$ and in $C^{m+2,\lambda}(\bar{\Omega}, \mathbb{R}^n)$, respectively.

Our goal, in this section, is to prove Theorem 3.2. We begin by pointing out the following basic

Lemma 3.1. *Suppose that* Ω *is connected and coincides with the interior of* $\bar{\Omega}$. *Then a* C^1 *function* $\phi: \bar{\Omega} \to \mathbb{R}^n$ *such that* $\phi(x) = x, \, \forall x \in \partial\Omega$ *is an orientation-preserving* C^1 *diffeomorphism of* $\bar{\Omega}$ *onto* $\bar{\Omega}$ *if (and only if)* $\det D\phi(x) > 0, \, \forall x \in \bar{\Omega}$.

Proof. Let $\phi: \bar{\Omega} \to \mathbb{R}^n$ be a C^1 function such that $\phi(x) = x, \, \forall x \in \partial\Omega$ and that $\det D\phi(x) > 0, \, \forall x \in \bar{\Omega}$. If ψ is a C^1 $\mathbb{R}^n$-valued function defined on an open neighborhood of $\bar{\Omega}$ in $\mathbb{R}^n$ such that $\psi(x) = \phi(x), \, \forall x \in \bar{\Omega}$ (see Chapter II, §1), then $\det D\psi(x) > 0, \, \forall x \in \bar{\Omega}$ and $\psi(x) = x, \, \forall x \in \partial\Omega$. By the inverse function theorem, for each $x \in \bar{\Omega}$, there is an open neighborhood U_x of x in $\mathbb{R}^n$ and an open neighborhood V_x of $\phi(x)$ in $\mathbb{R}^n$ such that ψ induces a C^1 diffeomorphism of U_x onto V_x. Since, moreover, ϕ induces the identity on $\partial\Omega$ and Ω is of class C^1, from a theorem† proved by CIARLET [1986] (Theorem 5.5-2) it follows that $\phi: \bar{\Omega} \to \mathbb{R}^n$ is (globally) one-to-one and that $\phi(\bar{\Omega}) = \bar{\Omega}$. Now it is easy to conclude that ϕ is an orientation-preserving C^1 diffeomorphism of $\bar{\Omega}$ onto $\bar{\Omega}$. $\square$

Theorem 3.2. *Let* Ω *be of class* C^1 *and let* m *be any integer* ≥ 0. *Then, for* $p(m+1) > n$, $\mathscr{A}_{m+2,p}$ *is an open subset of the affine subspace* $(\iota_{\bar{\Omega}} + W_0^{1,p}(\Omega, \mathbb{R}^n)) \cap W^{m+2,p}(\Omega, \mathbb{R}^n)$ *of* $W^{m+2,p}(\Omega, \mathbb{R}^n)$, *while for any* $\lambda \in \,]0,1]$, $\mathscr{A}_{m+2,\lambda}$ *is an open subset of the affine subspace* $(\iota_{\bar{\Omega}} + C_0^0(\bar{\Omega}, \mathbb{R}^n)) \cap C^{m+2,\lambda}(\bar{\Omega}, \mathbb{R}^n)$ *of* $C^{m+2,\lambda}(\bar{\Omega}, \mathbb{R}^n)$.

Proof. First of all, we recall that if $p(m+1) > n$, $W^{m+2,p}(\Omega)$ can be continuously embedded in $C^1(\bar{\Omega})$ and that (see Chapter II, §1) for $u \in C^1(\bar{\Omega})$ the following equivalence holds:

$$v \in W_0^{1,p}(\Omega, \mathbb{R}^n) \iff v|_{\partial\Omega} = 0.$$

---

† The statement of this theorem is the following. "Let Ω be a bounded, open, connected subset of $\mathbb{R}^n$ such that the interior of $\bar{\Omega}$ coincides with Ω, let $\phi_0 \in C^0(\bar{\Omega}, \mathbb{R}^n)$ be an injective mapping, and let $\phi \in C^0(\bar{\Omega}, \mathbb{R}^n) \cap C^1(\Omega, \mathbb{R}^n)$ be a mapping that satisfies

$$\begin{cases} \det D\phi(x) > 0, & \forall x \in \Omega, \\ \phi(x) = \phi_0(x), & \forall x \in \partial\Omega. \end{cases}$$

Then the mapping $\phi: \bar{\Omega} \to \phi(\bar{\Omega})$ is a homeomorphism, the mapping $\phi: \Omega \to \phi(\Omega)$ is a C^1 diffeomorphism, and $\phi(\Omega) = \phi_0(\Omega)$, $\phi(\bar{\Omega}) = \phi_0(\bar{\Omega})$." The proof of this theorem is based on the properties of the topological degree.

Other very interesting theorems concerning global invertibility of a locally invertible mapping can be found in MEISTER & OLECH [1963] and in BALL [1981].

Thus, if $p(m + 1) > n$,

$$u \in W^{m+2,p}(\Omega, \mathbb{R}^n), \qquad u|_{\partial\Omega} = 0 \iff u \in W_0^{1,p}(\Omega, \mathbb{R}^n) \cap W^{m+2,p}(\Omega, \mathbb{R}^n),$$

so that

$$\mathscr{A}_{m+2,p} \subseteq (\iota_{\bar{\Omega}} + W_0^{1,p}(\Omega, \mathbb{R}^n)) \cap W^{m+2,p}(\Omega, \mathbb{R}^n).$$

We assume that $p(m + 1) > n$ and we will prove that $\mathscr{A}_{m+2,p}$ is open in $(\iota_{\bar{\Omega}} + W_0^{1,p}(\Omega, \mathbb{R}^n)) \cap W^{m+2,p}(\Omega, \mathbb{R}^n)$. Accordingly, we fix $\phi_0 \in \mathscr{A}_{m+2,p}$ and we show that there is a neighborhood U_0 of ϕ_0 in $W^{m+2,p}(\Omega, \mathbb{R}^n)$ such that $U_0 \cap (\iota_{\bar{\Omega}} + W_0^{1,p}(\Omega, \mathbb{R}^n)) \subseteq \mathscr{A}_{m+2,p}$. We put

$$\mu_0 = \inf_{x \in \Omega} \det D\phi_0(x).$$

From the hypotheses we have $\mu_0 > 0$. Clearly, the map $\phi \mapsto \det D\phi$ is continuous from $C^1(\bar{\Omega}, \mathbb{R}^n)$ into $C^0(\bar{\Omega})$, and therefore is continuous from $W^{m+2,p}(\Omega, \mathbb{R}^n)$ into $C^0(\bar{\Omega})$ because the embedding $W^{m+2,p}(\Omega) \subseteq C^1(\bar{\Omega})$ is continuous. Hence there is an open, convex neighborhood U_0 of ϕ_0 in $W^{m+2,p}(\Omega, \mathbb{R}^n)$ such that $\phi \in U_0$ implies

$$\sup_{x \in \Omega} |\det D\phi(x) - \det D\phi_0(x)| < \mu_0.$$

We immediately deduce that $\phi \in U_0$ implies

$$\inf_{x \in \Omega} \det D\phi(x) > 0.$$

Note that, under our hypotheses, Ω has a finite number of connected components which are open subsets of $\mathbb{R}^n$ (because, as is easily seen, Ω is locally connected) and the distance between any two connected components of Ω is strictly positive.

Moreover, each connected component Γ of Ω coincides with the interior of its closure (in $\mathbb{R}^n$). Therefore, from Lemma 3.1 it follows that, for each connected component Γ of Ω, every $\phi \in (\iota_{\bar{\Omega}} + W_0^{1,p}(\Omega, \mathbb{R}^n)) \cap U_0$ is an orientation-preserving C^1 diffeomorphism of $\bar{\Gamma}$ onto itself; then every $\phi \in (\iota_{\bar{\Omega}} + W_0^{1,p}(\Omega, \mathbb{R}^n)) \cap U_0$ is an orientation-preserving C^1 diffeomorphism of $\bar{\Omega}$ onto $\bar{\Omega}$, and hence $(\iota_{\bar{\Omega}} + W_0^{1,p}(\Omega, \mathbb{R}^n)) \cap U_0 \subseteq \mathscr{A}_{m+2,p}$. This completes the proof of the first part of the theorem.

Quite analogous arguments show that $\mathscr{A}_{m+2,\lambda}$ is an open subset of the affine subspace $(\iota_{\bar{\Omega}} + C_0^0(\bar{\Omega}, \mathbb{R}^n)) \cap C^{m+2,\lambda}(\bar{\Omega}, \mathbb{R}^n)$ of $C^{m+2,\lambda}(\bar{\Omega}, \mathbb{R}^n)$. □

§4. Local Theorems on Existence, Uniqueness, and Analytic Dependence on f for Problem $((1.1), (1.3))$

Let m be an integer ≥ 1. If σ is an $\mathbb{M}_n$-valued function defined on Ω we will denote by $A_{Z_{hk}}(\sigma)$ the $\mathbb{M}_n$-valued function defined on Ω by setting $\forall x \in \Omega$

$$A_{Z_{hk}}(\sigma)(x) = D_{Z_{hk}}a(x, \sigma(x)). \tag{4.1}$$

According to the convention assumed at the end of §1 in Chapter II, $(W^{m+2,p}(\Omega, \mathbb{R}^n))^+$ and $(C^{m+2,\lambda}(\bar{\Omega}, \mathbb{R}^n))^+$ stand for $\{\phi \in W^{m+2,p}(\Omega, \mathbb{R}^n): \det D\phi > 0\}$ and $\{\phi \in C^{m+2,\lambda}(\bar{\Omega}, \mathbb{R}^n): \det D\phi > 0\}$, respectively.

The following two lemmas are obvious consequences of Theorems 4.1 and 4.2 of Chapter II, combined with Remark 1.1 of Chapter II.

Lemma 4.1. *Assume that Ω has the cone property (in particular, Ω is of class C^1), that $a \in C^{m+2}(\bar{\Omega} \times \mathbb{M}_n^+, \mathbb{M}_n)$, and that $p(m+1) > n$. Then (1.4) is a continuously differentiable mapping from $(W^{m+2,p}(\Omega, \mathbb{R}^n))^+$ into $W^{m,p}(\Omega, \mathbb{R}^n)$ and its differential at any point $\phi \in (W^{m+2,p}(\Omega, \mathbb{R}^n))^+$ is the (continuous, linear) mapping*

$$v \mapsto \operatorname{div}\left(\sum_{h,k=1}^{n} A_{Z_{hk}}(D\phi)D_k v_h \right). \tag{4.2}$$

Lemma 4.2. *Assume that Ω is of class C^1 and that $a \in C^{m+3}(\bar{\Omega} \times \mathbb{M}_n^+, \mathbb{M}_n)$. Then (1.4) is a continuously differentiable mapping from $(C^{m+2,\lambda}(\bar{\Omega}, \mathbb{R}^n))^+$ into $C^{m,\lambda}(\bar{\Omega}, \mathbb{R}^n)$ and its differential at any point $\phi \in (C^{m+2,\lambda}(\bar{\Omega}, \mathbb{R}^n))^+$ is defined by (4.2).*

As far as the analyticity of (4.1) is concerned, Theorems 5.1 and 5.2 of Chapter II combined with Remark 1.1 of Chapter II yield the following two lemmas.

Lemma 4.3. *Assume that Ω has the cone property, that*

$$a \in C^\infty(\bar{\Omega} \times \mathbb{M}_n^+, \mathbb{M}_n),$$

and that the functions $(x, Z) \mapsto D_x^\alpha a(x, Z)$, $|\alpha| \leq m + 1$, are analytic in Z uniformly with respect to x. Then, for $p(m+1) > n$, (1.4) is an analytic mapping of $(W^{m+2,p}(\Omega, \mathbb{R}^n))^+$ into $W^{m,p}(\Omega, \mathbb{R}^n)$.

Lemma 4.4. *Assume that Ω is of class C^1, that $a \in C^\infty(\bar{\Omega} \times \mathbb{M}_n^+, \mathbb{M}_n)$, and that the functions $(x, Z) \mapsto D_x^\alpha a(x, Z)$, $|\alpha| \leq m + 1$, are analytic in Z uniformly with respect to x. Then (1.4) is an analytic mapping of $(C^{m+2,\lambda}(\bar{\Omega}, \mathbb{R}^n))^+$ into $C^{m,\lambda}(\bar{\Omega}, \mathbb{R}^n)$.*

We will denote by

$$\mathscr{S}_{m+2,p} \qquad [\text{respectively } \mathscr{S}_{m+2,\lambda}]$$

the set of $\phi \in \mathscr{A}_{m+2,p}$ [respectively $\phi \in \mathscr{A}_{m+2,\lambda}$] such that

$$\int_\Omega \sum_{h,k=1}^{n} (A_{Z_{hk}}(D\phi)D_k v_h) \cdot Dv \, dx \geq c(\phi)\|v\|_{1,2}^2, \qquad \forall v \in \mathscr{D}(\Omega, \mathbb{R}^n), \tag{4.3}$$

with $c(\phi)$ a suitable number > 0 independent of v.

In other words, $\mathcal{S}_{m+2,p}$ [respectively $\mathcal{S}_{m+2,\lambda}$] is the set of those admissible deformations that belong to $W^{m+2,p}(\Omega, \mathbb{R}^n)$ and are *uniformly Hadamard stable* (in the terminology of GURTIN & SPECTOR [1979]), or *infinitesimally strongly stable* (according to the definition given by FICHERA [1972b]). Note that, in view of Corollary 3.4 of Chapter III, *every rigid deformation belongs to $\mathcal{S}_{m+2,p}$* (and to $\mathcal{S}_{m+2,\lambda}$) *provided Ω has the cone property, $a \in C^1(\bar{\Omega} \times \mathbb{M}_n^+, \mathbb{M}_n)$, and (3.5), (3.6), (3.7), (3.12) of Chapter III are satisfied*.

Now let $a \in C^{m+2}(\bar{\Omega} \times \mathbb{M}_n^+, \mathbb{M}_n)$ with Ω of class C^{m+2} and consider the C^1 mapping (1.5) acting from the open subset $\mathcal{A}_{m+2,p} - \iota_{\bar{\Omega}}$ of $W_0^{1,p}(\Omega, \mathbb{R}^n) \cap W^{m+2,p}(\Omega, \mathbb{R}^n)$, with $p > n$, into $W^{m,p}(\Omega, \mathbb{R}^n)$ (see Theorem 3.2 and Lemma 4.1). In view of Theorem 7.1 of Chapter III and Lemma 4.1 the differential of such a mapping at any point $\phi_0 \in \mathcal{S}_{m+2,p}$ is a (linear) bijection of $W_0^{1,p}(\Omega, \mathbb{R}^n) \cap W^{m+2,p}(\Omega, \mathbb{R}^n)$ onto $W^{m,p}(\Omega, \mathbb{R}^n)$. Moreover, by Lemma 4.3, mapping (1.5) is analytic provided $a \in C^\infty(\bar{\Omega} \times \mathbb{M}_n^+, \mathbb{M}_n)$ and the functions $(x, Z) \mapsto D_x^\alpha a(x, Z)$, $|\alpha| \leq m + 1$, are analytic in Z uniformly with respect to x. Therefore, in view of the inverse function theorem (see Appendix I), for each $\phi_0 \in \mathcal{S}_{m+2,p}$ there is an open neighborhood U_0 of ϕ_0 in $W^{m+2,p}(\Omega, \mathbb{R}^n)$ contained in $\mathcal{S}_{m+2,p}$, and an open neighborhood V_0 of div $A(D\phi_0)$ in $W^{m,p}(\Omega, \mathbb{R}^n)$ such that (1.4) is a C^1 diffeomorphism of U_0 onto V_0; furthermore, (1.4) is analytic at ϕ_0 if $a \in C^\infty(\bar{\Omega} \times \mathbb{M}_n^+, \mathbb{M}_n)$ and the functions $(x, Z) \mapsto D_x^\alpha a(x, Z)$, $|\alpha| \leq m + 1$, are analytic in Z uniformly with respect to X. Thus we can state the following theorem.

Theorem 4.5. *Assume that $a \in C^{m+2}(\bar{\Omega} \times \mathbb{M}_n^+, \mathbb{M}_n)$ with Ω of class C^{m+2} and that $p > n$. Then for each $\phi_0 \in \mathcal{S}_{m+2,p}$ [in particular, for every rigid deformation ϕ_0 if (3.5), (3.6), (3.7), and (3.12) of Chapter III hold] there are two numbers $\xi > 0$ and $\eta > 0$ such that, for all $f \in W^{m,p}(\Omega, \mathbb{R}^n)$ with $\| f - \text{div } A(D\phi_0) \|_{m,p} < \xi$, problem ((1.1), (1.3)) has one and only one solution $\hat{\phi}(f)$ in $W^{m+2,p}(\Omega, \mathbb{R}^n)$ satisfying the condition $\| \hat{\phi}(f) - \phi_0 \|_{m+2,p} < \eta$. Moreover, the mapping $f \mapsto \hat{\phi}(f)$ is continuously differentiable; it is analytic at div $A(D\phi_0)$ if $a \in C^\infty(\bar{\Omega} \times \mathbb{M}_n^+, \mathbb{M}_n)$ and the functions $(x, Z) \mapsto D_x^\alpha a(x, Z)$, $|\alpha| \leq m$, are analytic in Z uniformly with respect to x.*

Analogously, combining Theorem 7.2 of Chapter III, Theorem 3.2 and Lemmas 4.2 and 4.3, we get

Theorem 4.6. *Assume that $a \in C^{m+3}(\bar{\Omega} \times \mathbb{M}_n^+, \mathbb{M}_n)$ with Ω of class $C^{m+2,\lambda}$. Then for each $\phi_0 \in \mathcal{S}_{m+2,\lambda}$, $\lambda \in]0, 1[$ [in particular, for every rigid deformation ϕ_0 if (3.5), (3.6), (3.7), and (3.12) of Chapter III hold] there are two numbers $\xi > 0$ and $\eta > 0$ such that, for all $f \in C^{m,\lambda}(\Omega, \mathbb{R}^n)$ with $\| f - \text{div } A(D\phi_0) \|_{m,\lambda} < \xi$, problem ((1.1), (1.3)) has one and only one solution $\hat{\phi}(f)$ in $C^{m+2,\lambda}(\bar{\Omega}, \mathbb{R}^n)$ satisfying the condition $\| \hat{\phi}(f) - \phi_0 \|_{m+2,p} < \eta$. Moreover, the mapping $f \mapsto \hat{\phi}(f)$ is continuously differentiable; it is analytic* at div $A(D\phi_0)$ if $a \in C^\infty(\bar{\Omega} \times \mathsf{M}_n^+, \mathsf{M}_n)$ and the functions $(x, Z) \mapsto D_x^\alpha a(x, Z)$, $|\alpha| \leq m + 1$, are analytic in Z uniformly with respect to x.

§5. Stronger Results on Existence and Uniqueness for Problem ((1.1), (1.3))

As in §4, m shall denote an integer ≥ 1.

Remark 5.1. Let Ω be of class C^1 and suppose that the derivatives $D_{Z_{hk}}a$ $(h, k = 1, \ldots, n)$, exist and are continuous functions on $\bar{\Omega} \times \mathsf{M}_n^+$. Then $\mathscr{S}_{m+2,\lambda}$ is an open subset of the affine subspace $(\iota_{\bar{\Omega}} + C_0^0(\bar{\Omega}, \mathbb{R}^n)) \cap C^{m+2,\lambda}(\bar{\Omega}, \mathbb{R}^n)$ of $C^{m+2,\lambda}(\bar{\Omega}, \mathbb{R}^n)$, and, for $p(m + 1) > n$, $\mathscr{S}_{m+2,p}$ is an open subset of the affine subspace $(\iota_{\bar{\Omega}} + W_0^{1,p}(\Omega, \mathbb{R}^n)) \cap W^{m+2,p}(\Omega, \mathbb{R}^n)$ of $W^{m+2,p}(\Omega, \mathbb{R}^n)$.

Proof. Let $\phi_0 \in \mathscr{S}_{m+2,\lambda}$ [respectively $\phi_0 \in \mathscr{S}_{m+2,p}$ with $p(m + 1) > n$]: then there is a number $c(\phi_0) > 0$ such that

$$\int_\Omega \sum_{h,k=1}^n (A_{Z_{hk}}(D\phi_0)D_k v_h) \cdot Dv \, dx \geq c(\phi_0)\|v\|_{1,2}^2, \qquad \forall v \in \mathscr{D}(\Omega, \mathbb{R}^n). \quad (5.1)$$

Since $D_{Z_{hk}} a_{ij}$ are continuous functions on $\bar{\Omega} \times \mathsf{M}_n^+$, it is easily seen that $\phi \mapsto A_{Z_{hk}}(D\phi)$ is a continuous mapping from $(C^1(\bar{\Omega}, \mathbb{R}^n))^+$ into $C^0(\bar{\Omega}, \mathsf{M}_n)$; therefore, it is continuous from $(C^{m+2,\lambda}(\bar{\Omega}, \mathbb{R}^n))^+$ into $C^0(\bar{\Omega}, \mathsf{M}_n)$ and, if $p(m + 1) > n$, from $(W^{m+2,p}(\Omega, \mathbb{R}^n))^+$ into $C^0(\bar{\Omega}, \mathsf{M}_n)$ in view of the Sobolev embedding theorem.

Then it is not difficult to see that for every number ε such that $0 < \varepsilon < c(\phi_0)$ there is a neighborhood U_0 of ϕ_0 in $(C^{m+2,\lambda}(\bar{\Omega}, \mathbb{R}^n))$ [respectively $(W^{m+2,p}(\Omega, \mathbb{R}^n))$] such that $\phi_0 \in U_0$ implies

$$\left| \int_\Omega \sum_{h,k=1}^n [(A_{Z_{hk}}(D\phi_0) - A_{Z_{hk}}(D\phi))D_h v_k] \cdot Dv \, dx \right| < \varepsilon\|v\|_{1,2}^2, \quad \forall v \in \mathscr{D}(\Omega, \mathbb{R}^n).$$

$$(5.2)$$

Combining (5.1) and (5.2) we obtain that if $\phi \in U_0$, then

$$\int_\Omega \sum_{h,k=1}^n (A_{Z_{hk}}(D\phi)D_k v_h) \cdot Dv \, dx \geq (c(\phi_0) - \varepsilon)\|v\|_{1,2}^2, \qquad \forall v \in \mathscr{D}(\Omega, \mathbb{R}^n).$$

Since, by Theorem 3.2, U_0 can be chosen such that $U_0 \subseteq \mathscr{A}_{m+2,\lambda}$ [respectively $U_0 \subseteq \mathscr{A}_{m+2,p}$], the proof is complete. $\square$

Results like those stated in Remark 5.1 and in the following lemma have been proved by GURTIN & SPECTOR [1979].

Lemma 5.2. *Assume that Ω and the function a are as in Lemma* 4.1 *[respectively in Lemma* 4.2]. *Then the mapping* (1.4) *of* $(W^{m+2,p}(\Omega, \mathbb{R}^n))^+$ *into* $W^{m,p}(\Omega, \mathbb{R}^n)$ *with* $p(m+1) > n$ *[respectively of* $(C^{m+2,\lambda}(\bar{\Omega}, \mathbb{R}^n))^+$ *into* $C^{m,\lambda}(\bar{\Omega}, \mathbb{R}^n)]$ *is one-to-one on each convex subset of* $\mathscr{S}_{m+2,p}$ *[respectively of* $\mathscr{S}_{m+2,\lambda}]$.

Proof. It suffices to prove that the result holds for the mapping (1.4) acting from $(W^{m+2,p}(\Omega, \mathbb{R}^n))^+$ into $W^{m,p}(\Omega, \mathbb{R}^n)$, because the proof in the case when (1.4) is considered as a mapping from $(C^{m+2,\lambda}(\bar{\Omega}, \mathbb{R}^n))^+$ into $C^{m,\lambda}(\bar{\Omega}, \mathbb{R}^n)$ is quite analogous. If N is any integer $\geqq 1$ we denote by $(\cdot, \cdot)_0$ the scalar product on $(L^2(\Omega))^N$ defined by

$$(f, g)_0 = \int_\Omega f(x) \cdot g(x) \, dx.$$

From Theorem 4.1 of Chapter II it follows that $\sigma \mapsto A(\sigma)$ is a continuously differentiable mapping from $(W^{m+1,p}(\Omega, \mathbb{R}^n))^+$ into $W^{m+1,p}(\Omega, \mathbb{M}^n)$ and that

$$A'(\sigma)(\tau) = \sum_{h,k=1}^n A_{Z_{hk}}(\sigma)\tau_{hk}, \qquad \forall \sigma, \tau \in (W^{m+1,p}(\Omega, \mathbb{R}^n))^+.$$

Therefore, (4.3) can be written in the form

$$(A'(D\phi)(v), Dv)_0 \geqq c(\phi)\|v\|_{1,2}^2, \qquad \forall v \in \mathscr{D}(\Omega, \mathbb{R}^n). \tag{5.3}$$

Now let $\mathscr{C}$ be a convex subset of $\mathscr{S}_{m+2,p}$ and let $\phi, \psi \in \mathscr{C}$ be such that $\phi \neq \psi$. Note that

$$A(D\phi) - A(D\psi) = \int_0^1 A'(D\psi + tD(\phi - \psi)) \, dt$$

and that, for all $t \in [0, 1]$, (5.3) holds with $\psi + t(\phi - \psi)$ instead of ϕ, because $\psi + t(\phi - \psi) \in \mathscr{C}$, $\forall t \in [0, 1]$ and $\mathscr{C} \subseteq \mathscr{S}_{m+2,p}$. Then, because plainly

$$(-\operatorname{div} A(D\phi), v)_0 = (A(D\phi), Dv)_0,$$

$$\forall v \in W_0^{1,p}(\Omega, \mathbb{R}^n) \cap W^{m+2,p}(\Omega, \mathbb{R}^n) \qquad \text{with} \quad p(m+1) > n,$$

we get

$$(\operatorname{div} A(D\psi) - \operatorname{div} A(D\phi), \phi - \psi)_0$$

$$= (A(D\phi) - A(D\psi), D\phi - D\psi)_0$$

$$= \int_0^1 (A'(D\psi + tD(\phi - \psi))(\phi - \psi), D\phi - D\psi)_0 \, dt$$

$$\geqq \int_0^1 c(\psi + t(\phi - \psi))(D\phi - D\psi, D\phi - D\psi)_0 \, dt > 0,$$

which implies div $A(D\phi) \neq$ div $A(D\psi)$. Thus, mapping (1.4) is one-to-one on $\mathscr{C}$ and the proof of the lemma is complete. $\square$

In view of Lemma 5.2, Theorem 4.5 [respectively Theorem 4.6] immediately gives the following Theorem 5.3 [respectively Theorem 5.4].

Theorem 5.3. *Assume that* $a \in C^{m+2}(\bar{\Omega} \times M_n^+, M_n)$ *with* Ω *of class* C^{m+2} *and that* $p > n$. *If* $\mathscr{C}$ *is a convex, open subset of the affine subspace* $(\iota_{\bar{\Omega}} + W_0^{1,p}(\Omega, \mathbb{R}^n)) \cap W^{m+2,p}(\Omega, \mathbb{R}^n)$ *of* $W^{m+2,p}(\Omega, \mathbb{R}^n)$ *such that* $\mathscr{C} \subseteq \mathscr{S}_{m+2,p}$,[†] *then for each* $\phi_0 \in \mathscr{S}$ *there is a number* $\xi > 0$ *such that, for every* $f \in W^{m,p}(\Omega, \mathbb{R}^n)$ *with* $\| f - \operatorname{div} A(D\phi_0)\|_{m,p} < \xi$, *problem* $((1.1),(1.3))$ *has one and only one solution* $\hat{\phi}(f)$ *in* $\mathscr{C}$. *Moreover, the mapping* $f \mapsto \hat{\phi}(f)$ *is continuously differentiable; it is analytic at* div $A(D\phi_0)$ *provided* $a \in C^{\infty}(\bar{\Omega} \times M_n^+, M_n)$ *and the functions* $(x, Z) \mapsto D_x^{\alpha} a(x, Z)$, $|\alpha| \leq m + 1$, *are analytic in* Z *uniformly with respect to* x.

Theorem 5.4. *Assume that* $a \in C^{m+3}(\bar{\Omega} \times M_n^+, M_n)$ *with* Ω *of class* $C^{m+2,\lambda}$ $(\lambda \in {]}0, 1[)$. *If* $\mathscr{C}$ *is a convex, open subset of the affine subspace* $(\iota_{\bar{\Omega}} + C_0^0(\bar{\Omega}, \mathbb{R}^n)) \cap C^{m+2,\lambda}(\bar{\Omega}, \mathbb{R}^n)$ *of* $C^{m+2,\lambda}(\bar{\Omega}, \mathbb{R}^n)$ *such that* $\mathscr{C} \subseteq \mathscr{S}_{m+2,\lambda}$,[‡] *then for each* $\phi_0 \in \mathscr{C}$ *there is a number* $\xi > 0$ *such that, for every* $f \in C^{m,\lambda}(\bar{\Omega}, \mathbb{R}^n)$ *with* $\| f - \operatorname{div} A(D\phi_0)\|_{m,\lambda} < \xi$, *problem* $((1.1),(1.3))$ *has one and only one solution* $\hat{\phi}(f)$ *in* $\mathscr{C}$. *Moreover, the mapping* $f \mapsto \hat{\phi}(f)$ *is continuously differentiable; it is analytic at* div $A(D\phi_0)$ *provided* $a \in C^{\infty}(\bar{\Omega} \times M_n^+, M_n)$ *and the functions* $(x, Z) \mapsto D_x^{\alpha} a(x, Z)$, $|\alpha| \leq m + 1$, *are analytic in* Z *uniformly with respect to* x.

Note that, by Remark 5.1, under the hypotheses of Theorem 5.3, $\mathscr{S}_{m+2,p}$ is open in $(\iota_{\bar{\Omega}} + W_0^{1,p}(\Omega, \mathbb{R}^n)) \cap W^{m+2,p}(\Omega, \mathbb{R}^n)$ and hence for each $\phi_0 \in \mathscr{S}_{m+2,p}$ there is a number $\eta > 0$ such that the (convex, open) subset $\{\phi \in (\iota_{\bar{\Omega}} + W_0^{1,p}(\Omega, \mathbb{R}^n)) \cap W^{m+2,p}(\Omega, \mathbb{R}^n): \|\phi - \phi_0\|_{m+2,p} < \eta\}$ of $(\iota_{\bar{\Omega}} + W_0^{1,p}(\Omega, \mathbb{R}^n)) \cap W^{m+2,p}(\Omega, \mathbb{R}^n)$ is contained in $\mathscr{S}_{m+2,p}$. Analogously, under the hypotheses of Theorem 5.4, for each $\phi_0 \in \mathscr{S}_{m+2,\lambda}$ there is an open ball of $(\iota_{\bar{\Omega}} + C_0^0(\bar{\Omega}, \mathbb{R}^n)) \cap C^{m+2,\lambda}(\bar{\Omega}, \mathbb{R}^n)$ contained in $\mathscr{S}_{m+2,\lambda}$. Thus the following remark holds.

Remark 5.5. *Theorem 5.3 implies Theorem 4.5; Theorem 5.4 implies Theorem 4.6.*

---

† Such a $\mathscr{C}$ exists because (in view of Remark 5.1) $\mathscr{S}_{m+2,p}$ is open in $(\iota_{\bar{\Omega}} + W_0^{1,p}(\Omega, \mathbb{R}^n)) \cap W^{m+2,p}(\Omega, \mathbb{R}^n)$.

‡ Such a $\mathscr{C}$ exists because of Remark 5.1.

§6. Local Theorems on Existence and Uniqueness for Problem ((1.1), (1.2))

In this section we deal with the (live) boundary problem ((1.1), (1.2)). For any smooth enough $a: \Omega \times M_n^+ \to M_n$, $\phi: \bar{\Omega} \to \mathbb{R}^n$ and any $f: \Omega \times \mathbb{R}^n \to \mathbb{R}^n$, $\mu: \Omega \to \mathbb{R}$ we set

$$\Lambda(f, \phi) = \operatorname{div} A(D\phi) + \mu B(\phi) \qquad (6.1)$$

with $A(D\phi)$ and $B(\phi)$ as defined in §1. We will obtain local results of existence and uniqueness for problem ((1.1), (1.2)) by applying the implicit function theorem to the equation $\Lambda(f, \phi) = 0$. Also in this section m shall denote an integer ≥ 1.

Recalling Theorem 4.5 and Corollary 2.3 of Chapter II and Lemma 4.1 we immediately obtain.

Lemma 6.1. *Assume that Ω has the cone property, that $a \in C^{m+2}(\bar{\Omega} \times M_n^+, M_n)$, that $\mu \in W^{m+1,p}(\Omega)$, and that $p(m + 1) > n$. Let G be a convex, bounded, open subset of $\mathbb{R}^n$ containing Ω. Then $(f, u) \mapsto \Lambda(f, \iota_{\bar{\Omega}} + u)$ is a continuously differentiable mapping from the open (see Theorem 3.2) subset $C^{m+1}(\bar{G}, \mathbb{R}^n) \times (\mathscr{A}_{m+2,p} - \iota_{\bar{\Omega}})$ of the Banach space $C^{m+1}(\bar{G}, \mathbb{R}^n) \times (W_0^{1,p}(\Omega, \mathbb{R}^n) \cap W^{m+2,p}(\Omega, \mathbb{R}^n))$ into $W^{m,p}(\Omega, \mathbb{R}^n)$, and its differential at any $(\bar{f}, \bar{u}) \in C^{m+1}(\bar{G}, \mathbb{R}^n) \times (\mathscr{A}_{m+2,p} - \iota_{\bar{\Omega}})$ is the (continuous, linear) mapping of $C^{m+1}(\bar{G}, \mathbb{R}^n) \times (W_0^{1,p}(\Omega, \mathbb{R}^n) \cap W^{m+2,p}(\Omega, \mathbb{R}^n))$ into $W^{m,p}(\Omega, \mathbb{R}^n)$ defined by*

$$(f, u) \mapsto \operatorname{div}\left(\sum_{h,k=1}^{n} A_{Z_{hk}}(D\bar{\phi}) D_k u_h \right) + \mu \sum_{j=1}^{n} F_{y_j}(\bar{\phi}) u_j + F(\phi_0), \qquad (6.2)$$

where $\bar{\phi} = \iota_{\bar{\Omega}} + \bar{u}$, $A_{Z_{hk}}(D\bar{\phi})$ are defined by (4.1) and $\bar{F}_{y_j}(\bar{\phi})$ is the $\mathbb{R}^n$-valued function defined on Ω by

$$\bar{F}_{y_j}(\bar{\phi})(x) = D_{y_j}\bar{f}(x, \bar{\phi}(x)).$$

Combining Theorem 4.6 and Remark 1.1 with Lemma 4.2, we readily deduce.

Lemma 6.2. *Assume that Ω is of class C^1, that $a \in C^{m+3}(\bar{\Omega} \times M_n^+, M_n)$, and that $\mu \in C^{m+1,\lambda}(\bar{\Omega})$. Let G be a convex, bounded, open subset of $\mathbb{R}^n$ containing Ω. Then $(f, u) \mapsto \Lambda(f, \iota_{\bar{\Omega}} + u)$ is a continuously differentiable mapping from the open (see Theorem 3.2) subset $C^{m+2}(\bar{G}, \mathbb{R}^n) \times \mathscr{A}_{m+2,\lambda}$ of $C^{m+2}(\bar{G}, \mathbb{R}^n) \times (C_0^0(\bar{\Omega}, \mathbb{R}^n) \cap C^{m+2,\lambda}(\bar{\Omega}, \mathbb{R}^n))$ into $C^{m,\lambda}(\bar{\Omega}, \mathbb{R}^n)$, and its differential at any $(\bar{f}, \bar{u}) \in C^{m+2}(\bar{G}, \mathbb{R}^n) \times \mathscr{A}_{m+2,\lambda}$ is the (continuous, linear) mapping of $C^{m+2}(\bar{G}, \mathbb{R}^n) \times (C_0^0(\bar{\Omega}, \mathbb{R}^n) \cap C^{m+2,\lambda}(\bar{\Omega}, \mathbb{R}^n))$ into $C^{m,\lambda}(\bar{\Omega}, \mathbb{R}^n)$ defined by (6.2).*

Note that, under the hypotheses of Lemma 6.1 [respectively Lemma 6.2], for any $f \in C^{m+1}(\bar{G}, \mathbb{R}^n)$ [respectively $f \in C^{m+2}(\bar{G}, \mathbb{R}^n)$] the mapping $u \mapsto \Lambda(f, \iota_{\bar{\Omega}} + u)$ is a C^1 operator from $\mathscr{A}_{m+2,p} - \iota_{\bar{\Omega}}$ into $W^{m,p}(\Omega, \mathbb{R}^n)$ [respectively from $\mathscr{A}_{m+2,\lambda} - \iota_{\bar{\Omega}}$ into $C^{m,\lambda}(\bar{\Omega}, \mathbb{R}^n)$] and its differential at $\bar{u}$ is

$$u \mapsto \text{div}\left(\sum_{h,k=1}^{n} A_{Z_{hk}}(D\bar{\phi})D_k u_h \right) + \sum_{j=1}^{n} F_{y_j}(\bar{\phi})u_j,$$

with $\bar{\phi} = \bar{u} + \iota_{\bar{\Omega}}$, $A_{Z_{hk}}(D\bar{\phi})$ defined by (4.1) and $F_{y_j}(\bar{\phi})$ the $\mathbb{R}^n$-valued function defined on Ω by $F_{y_j}(\bar{\phi})(x) = D_{y_j}f(x, \bar{\phi}(x))$.

Then, in view of Lemma 6.1 [respectively Lemma 6.2] and Theorem 3.2, a straightforward application of the implicit function theorem (see Appendix I), gives the following Theorem 6.3 [respectively Theorem 6.4].

Theorem 6.3. *Assume that Ω is of class C^1, that $a \in C^{m+2}(\bar{\Omega} \times \mathsf{M}_n^+, \mathsf{M}_n)$, that $\mu \in W^{m+1,p}(\Omega)$, and that $p(m+1) > n$. Let G be a convex, bounded, open subset of $\mathbb{R}^n$ containing Ω and let $(\bar{f}, \bar{\phi}) \in C^{m+1}(\bar{G}, \mathbb{R}^n) \times \mathscr{A}_{m+2,p}$. If $\Lambda(\bar{f}, \bar{\phi}) = 0$ and the linear equation*

$$\text{div}\left(\sum_{h,k=1}^{n} A_{Z_{hk}}(D\bar{\phi})D_k u_h \right) + \mu \sum_{j=1}^{n} F_{y_j}(\bar{\phi})u_j = g \qquad (6.3)$$

has a unique solution u in $W_0^{1,p}(\Omega, \mathbb{R}^n) \cap W^{m+2,p}(\Omega, \mathbb{R}^n)$ for every $g \in W^{m,p}(\Omega, \mathbb{R}^n)$, then there are numbers $\xi > 0$ and $\eta > 0$ such that, for each $f \in C^{m+1}(\bar{G}, \mathbb{R}^n)$ with $|f - \bar{f}|_{m+1} < \xi$, problem $((1.1), (1.2))$ has one and only one solution $\hat{\phi}(f)$ in $W^{m+2,p}(\Omega, \mathbb{R}^n)$ satisfying the condition $\|\hat{\phi}(f) - \bar{\phi}\|_{m+2,p} < \eta$. Moreover (using the Landau symbol $\mathcal{O}$)

$$\hat{\phi}(f) = \bar{\phi} + \hat{u}_L(\bar{f}, \bar{\phi})(f - \bar{f}) + \mathcal{O}(|f - \bar{f}|_{m+1}), \qquad (6.4)$$

where $\hat{u}_L(\bar{f}, \bar{\phi})(f - \bar{f})$ denotes the solution in $W_0^{1,p}(\Omega, \mathbb{R}^n) \cap W^{m+2,p}(\Omega, \mathbb{R}^n)$ of (6.3) when $g = \mu(f - \bar{f}) \circ \bar{\phi}$.

Theorem 6.4. *Assume that Ω is of class C^1, that $a \in C^{m+3}(\bar{\Omega} \times \mathsf{M}_n^+, \mathsf{M}_n)$, and that $u \in C^{m,\lambda}(\bar{\Omega})$. Let G be a convex, bounded, open subset of $\mathbb{R}^n$ containing Ω and let $(\bar{f}, \bar{\phi}) \in C^{m+2}(\bar{G}, \mathbb{R}^n) \times \mathscr{A}_{m+2,\lambda}$. If $\Lambda(\bar{f}, \bar{\phi}) = 0$ and the linear equation (6.3) has a unique solution u in $C_0^0(\bar{\Omega}, \mathbb{R}^n) \cap C^{m+2,\lambda}(\bar{\Omega}, \mathbb{R}^n)$ for every $g \in C^{m,\lambda}(\bar{\Omega}, \mathbb{R}^n)$, then there are numbers $\xi > 0$ and $\eta > 0$ such that, for each $f \in C^{m+2}(\bar{G}, \mathbb{R}^n)$ with $|f - \bar{f}|_{m+2} < \xi$, problem $((1.1), (1.2))$ has one and only one solution $\hat{\phi}(f)$ in $C^{m+2,\lambda}(\bar{\Omega}, \mathbb{R}^n)$ satisfying the condition $\|\hat{\phi}(f) - \bar{\phi}\|_{m+2,\lambda} < \eta$. Moreover, (6.4) holds where $\hat{u}(\bar{f}, \bar{\phi})(f - \bar{f})$ denotes the solution in $C_0^0(\bar{\Omega}, \mathbb{R}^n) \cap C^{m+2,\lambda}(\bar{\Omega}, \mathbb{R}^n)$ of (6.3) when $g = \mu(f - \bar{f}) \circ \bar{\phi}$.*

By arguments like those used in proving Theorems 6.1, 6.2, 6.4, 7.1, and 7.2 of Chapter III it is possible to prove the spontaneous generalization of Theorem 7.1, 7.3 of Chapter III contained in the following remarks.

Remark 6.5. *Let Ω be of class C^{m+2} and let the assumptions of Theorem 6.3 be satisfied. Then for every $g \in W^{m,p}(\Omega, \mathbb{R}^n)$ with $p > n$, equation (6.3) has one and only one solution u in $W_0^{1,p}(\Omega, \mathbb{R}^n) \cap W^{m+2,p}(\Omega, \mathbb{R}^n)$ provided*

$$\int_\Omega \sum_{h,k=1}^n (D_{z_{hk}} a(x, D\bar{\phi}(x)) D_k u_h(x)) \cdot Du(x)\, dx + \int_\Omega \sum_{j=1}^n \mu(x) (D_{y_j} \bar{f}(x, \bar{\phi}(x)) u_j(x))$$

$$\cdot u(x)\, dx \geqq c \|u\|_{1,2}^2, \qquad \forall u \in \mathscr{D}(\Omega, \mathbb{R}^n), \quad (6.5)$$

with c a number > 0 independent of u.

Remark 6.6. *Let Ω be of class $C^{m+2,\lambda}$ and let the assumptions of Theorem 6.4 be satisfied. Then for every $g \in C^{m,\lambda}(\bar{\Omega}, \mathbb{R}^n)$ with $\lambda \in\,]0, 1[$, equation (6.3) has one and only one solution u in $C_0^0(\bar{\Omega}, \mathbb{R}^n) \cap C^{m+2,\lambda}(\bar{\Omega}, \mathbb{R}^n)$ provided (6.5) holds with $c > 0$ independent of u.*

We conclude by recalling that (see Corollary 3.4 of Chapter III) inequality (6.5) holds, with $\bar{\phi}$ any rigid deformation and $\bar{f} = 0$, provided Ω has the cone property, $a \in C^1(\bar{\Omega} \times \mathbb{M}_n^+, \mathbb{M}_n)$, and (3.4), (3.6), (3.7), (3.12) of Chapter III are satisfied.

CHAPTER V

Boundary Problems of Traction in Finite Elastostatics. An Abstract Method. The Special Case of Dead Loads

The first part of this chapter is devoted to the general traction problem in finite elastostatics. Unlike that which occurs in dealing with boundary problems of place, a direct application of the implicit function theorem cannot be made in studying local existence and uniqueness for a traction problem. Here an abstract method is presented to obtain a modification of the boundary problem which leads to another problem having the following requisites:

 (i) it is equivalent, in a certain sense, to the starting problem;
(ii) the implicit function theorem applies to it.

Our method is founded upon Lemmas 3.2 and 4.3, and it applies to various kinds of traction problems.

In the first part of Chapter VI we will obtain from this method local theorems of existence, uniqueness, and analytic dependence of a parameter for the boundary problems of pressure type when the load is invariant under translations or rotations; while in the second part of the present chapter the same method is used to study the case of dead loads. In this case we establish two general theorems of existence, uniqueness, and analytic dependence on a parameter (see Theorems 6.7 and 6.8), and we point out some significant consequences (see Corollaries 6.9 and 6.10). Note that the n-dimensional framework (generalizing three-dimensional elasticity) enables us to analyze a wider variety of cases and to penetrate more deeply into the matter.

We remark that there are some interesting types of loads for which the abstract method does not work: this is the case, e.g., of the load acting upon a heavy elastic body submerged in a heavy liquid—this will be treated in the second part of Chapter VI where we shall devise a sharper procedure.

§1. Generality on the Traction Problem in Finite Elastostatics

The boundary ondition of traction for the equilibrium equation $\operatorname{div} S_\phi + \mu_\phi b_\phi = 0$ of a body (see (4.7) of Chapter I) consists of prescribing $S_\phi v_\phi$ on the boundary $\partial\phi(\Omega)$ of $\phi(\Omega)$. As we saw in §4 of Chapter I, this corresponds (in the case $n = 3$) to the assignment of the contact forces acting upon the body by its environment in the deformation ϕ.

In the case of an elastic body this leads to the problem of finding an orientation-preserving, one-to-one, smooth function $\phi: \bar\Omega \to \mathbb{R}^n$ such that

$$\begin{cases} \operatorname{div} S_\phi + \mu_\phi b_\phi = 0 & \text{in } \phi(\Omega), \\ S_\phi v_\phi = \sigma_\phi & \text{on } \partial\phi(\Omega), \end{cases} \tag{1.1}$$

where v_ϕ is the unit outward normal to $\partial\phi(\Omega)$, $b_\phi: \phi(\Omega) \to \mathbb{R}^n$ and $\sigma_\phi: \partial\phi(\Omega) \to \mathbb{R}^n$ are prescribed functions, and $S_\phi: \phi(\Omega) \to \operatorname{Sym}_n$, $\mu_\phi: \phi(\Omega) \to \mathbb{R}$ are the functions defined by putting $\forall x \in \Omega$

$$S_\phi(\phi(x)) = s(x, D\phi(x)), \qquad \mu_\phi(\phi(x)) = \frac{\mu(x)}{\det D\phi(x)},$$

with $s: \Omega \times \mathbb{M}_n^+ \to \operatorname{Sym}_n$ and $\mu: \Omega \to \mathbb{R}$ given function.† Recall that (when $n = 3$) s is the response function of the elastic material and μ is the density of the body in the reference configuration Ω.

In Chapter I, §2 and §4, we have remarked that, putting $\forall x \in \Omega$

$$S(x) = s(x, D\phi(x)) \operatorname{cof} D\phi(x),$$

(1.1) reduces to

$$\begin{cases} \operatorname{div} S + \mu b_\phi \circ \phi = 0 & \text{in } \Omega, \\ Sv = |(\operatorname{cof} D\phi)v| \sigma_\phi \circ \phi & \text{on } \partial\Omega, \end{cases}$$

where v is the unit outward normal to $\partial\Omega$. Let us set

$$\begin{cases} B(\phi) = b_\phi \circ \phi, \\ G(\phi) = |(\operatorname{cof} D\phi)|_{\partial\Omega} v| \sigma_\phi \circ \phi, \end{cases}$$

and let us denote by a the function of $\Omega \times \mathbb{M}_n^+$ into $\mathbb{M}_n$ related with s by

$$a(x, Z) = s(x, Z) \operatorname{cof} Z,$$

so that $S(x) = a(x, D\phi(x))$.

---

† Recall that we always use the subscript ϕ to designate a function defined on $\phi(\Omega)$ or $\partial\phi(\Omega)$. (See Chapter I, §2.)

Thus our problem is to find a one-to-one, smooth function $\phi: \bar{\Omega} \to \mathbb{R}^n$ such that $\det D\phi > 0$ and

$$\begin{cases} \operatorname{div} A(D\phi) + \mu B(\phi) = 0 & \text{in } \Omega, \\ -A(D\phi)\nu + G(\phi) = 0 & \text{on } \partial\Omega, \end{cases} \tag{1.2}$$

where $\mu: \Omega \to \mathbb{R}$ is a given function, $A(D\phi): \Omega \to M_n$ is the function defined by

$$A(D\phi)(x) = a(x, D\phi(x)),$$

with $a: \Omega \times M_n^+ \to M_n$ a given function, $\phi \mapsto B(\phi)$ is a given mapping taking its values in a space of functions of Ω into $\mathbb{R}^n$, and $\phi \mapsto G(\phi)$ is a given mapping taking its values in a space of functions of $\partial\Omega$ into $\mathbb{R}^n$. The pair $(\mu B, G)$ shall be called the *load*.

In the case when there are two functions $(x, y, Z) \mapsto b(x, y, Z)$ from $\Omega \times \mathbb{R}^n \times M_n$ into $\mathbb{R}^n$ and $(x, y, z) \mapsto g(x, y, z)$ from $\partial\Omega \times \mathbb{R}^n \times \mathbb{R}^n$ into $\mathbb{R}^n$ such that

$$\begin{cases} B(\phi)(x) = b(x, \phi(x), D\phi(x)), & \forall x \in \Omega, \\ G(\phi)(x) = g(x, \phi(x), (\operatorname{cof} D\phi(x))\nu(x)), & \forall x \in \partial\Omega, \end{cases}$$

the load $(\mu B, G)$ will be said to be *simple*. This definition of a simple load is more restrictive than that given by SPECTOR [1980], but it includes many important cases of physical interest.

Chapter VI will be devoted to the study of problem (1.2) for the class of those simple loads such that $g(x, y, z) = h(x, y)z$ with h a real-valued function defined on $\Omega \times \mathbb{R}^n$. In this case the surface traction is prescribed to be parallel to the normal to the boundary of the unknown deformed equilibrium configuration.

In §6 of this chapter we will study the case when the load is simple with b independent of (y, Z) and g independent of (y, z): in this case, it is customary to say that the load is *dead* (see SEWELL [1967]).

§2. Preliminary Discussion

Consider problem (1.2) in the case of a *general load*, i.e., when $\phi \mapsto B(\phi)$ and $\phi \mapsto G(\phi)$ are general operators. Let us put $F(\phi) = \mu B(\phi)$ and introduce a real parameter ε, so that (1.2) takes the form

$$\begin{cases} \operatorname{div} A(D\phi) + \varepsilon F(\phi) = 0 & \text{in } \Omega, \\ -A(D\phi)\nu + \varepsilon G(\phi) = 0 & \text{on } \partial\Omega. \end{cases} \tag{2.1}$$

We will consider the following hypotheses on the function a (see Chapter

III, §3):

$$a(x, RZ) = Ra(x, Z), \qquad \forall(x, Z, R) \in \Omega \times \mathbb{M}_n^+ \times \mathbb{O}_n^+, \quad (2.2)$$

$$a(x, Z)Z^\mathsf{T} \in \text{Sym}_n, \qquad \forall(x, Z) \in \Omega \times \mathbb{M}_n^+, \quad (2.3)$$

$$a(x, I) = 0, \qquad \forall x \in \Omega, \quad (2.4)$$

$$Z \cdot \left(\sum_{h,k=1}^n D_{Z_{hk}} a(x, I) Z_{hk} \right) > 0, \qquad \forall(x, Z) \in \bar{\Omega} \times (\text{Sym}_n \backslash \{0\}), \quad (2.5)$$

where, as always, $D_{Z_{hk}}$ is the partial derivative operator with respect to the coordinate Z_{hk} of Z.

We recall that (2.2) follows from the independence of the observer of the elastic response (see Chapter I, §6), that (2.3) is a consequence of the balance of angular momentum (see Chapter I, §5), that (2.4) means that the reference configuration Ω is unstressed, and that the hypothesis (2.5) on the elastic response a, at the reference configuration, is usually assumed when such a configuration is unstressed. Let us set

$$\Lambda(\phi, \varepsilon) = (\text{div } A(D\phi) + \varepsilon F(\phi), -A(D\phi)v + \varepsilon G(\phi)).$$

Combining Theorems 4.1, 4.2, 5.1, 5.2 of Chapter II with Remark 6.5 of Chapter III we easily deduce

Lemma 2.1. *Let m be an integer ≥ 0, ϕ_0 a rigid deformation of $\bar{\Omega}$, and R_0 the value of the gradient of ϕ_0 (at any point $x \in \bar{\Omega}$). Assume that Ω is of class C^{m+2} and that $a \in C^{m+2}(\bar{\Omega} \times \mathbb{M}_n^+, \mathbb{M}_n)$ [respectively Ω is of class $C^{m+2,\lambda}$ and $a \in C^{m+3}(\bar{\Omega} \times \mathbb{M}_n^+, \mathbb{M}_n)$. Then*

$$\phi \mapsto (\text{div } A(D\phi), -A(D\phi)|_{\partial\Omega} v) \quad (2.6)$$

is a C^1 mapping from $(W^{m+2,p}(\Omega, \mathbb{R}^n))^+$ into

$$W^{m,p}(\Omega, \mathbb{R}^n) \times W^{m+1-1/p,p}(\partial\Omega, \mathbb{R}^n)$$

with $p(m+1) > n$ [respectively from $(C^{m+2,\lambda}(\bar{\Omega}, \mathbb{R}^n))^+$ into $C^{m,\lambda}(\bar{\Omega}, \mathbb{R}^n) \times C^{m+1,\lambda}(\partial\Omega, \mathbb{R}^n)$], and its differential at ϕ_0 is the mapping E_{R_0} defined by

$$E_{R_0}(v) = (\text{div } L_{R_0}(Dv), -L_{R_0}(Dv)|_{\partial\Omega} v), \quad (2.7)$$

where $L_{R_0}(Dv)$ is the function of $\bar{\Omega}$ into $\mathbb{M}_n$ defined by

$$L_{R_0}(Dv)(x) = \sum_{h,k=1}^n D_{Z_{hk}} a(x, R_0) D_k v_h(x), \qquad x \in \bar{\Omega}.$$

If, furthermore, we suppose that $a \in C^\infty(\bar{\Omega} \times \mathbb{M}_n^+, \mathbb{M}_n)$ and that the functions $(x, Z) \mapsto D_x^\alpha a(x, Z)$, $|\alpha| \leq m+1$, are analytic in Z, at R_0, uniformly with respect to x, then mapping (2.6) is analytic at ϕ_0.

Recall that (by Theorems 7.6 and 7.7 of Chapter III), *if the assumptions of the first part of Lemma 2.1 are satisfied and (2.2), (2.3), (2.4), (2.5)* *hold, then (2.7) is a (linear) homeomorphism of $\mathscr{V}_{R_0}^{m+2,p}$ onto $\mathscr{E}_{R_0}^{m,p}$ with $p(m+1) > n$ [respectively of $\mathscr{V}_{R_0}^{m+2,\lambda}$ onto $\mathscr{E}_{R_0}^{m,\lambda}$ with $0 < \lambda < 1$], where*

$$\begin{cases} \mathscr{V}_{R_0}^{m+2,p} = \left\{ v \in W^{m+2,p}(\Omega, \mathbb{R}^n) \colon \int_{\Omega} v \, dx = 0, \; R_0^{\mathrm{T}} \int_{\Omega} Dv \, dx \in \mathrm{Sym}_n \right\}, \\ \mathscr{E}_{R_0}^{m,p} = \{ (f, g) \in W^{m,p}(\Omega, \mathbb{R}^n) \times W^{m+1-1/p,p}(\partial\Omega, \mathbb{R}^n) \colon \\ \qquad\qquad (R_0^{\mathrm{T}} f, R_0^{\mathrm{T}} g) \text{ is equilibrated} \}, \end{cases} \tag{2.8}$$

$$\begin{cases} \mathscr{V}_{R_0}^{m+2,\lambda} = \left\{ v \in C^{m+2,\lambda}(\bar{\Omega}, \mathbb{R}^n) \colon \int_{\Omega} v \, dx = 0, \; R_0^{\mathrm{T}} \int_{\Omega} Dv \in \mathrm{Sym}_n \right\}, \\ \mathscr{E}_{R_0}^{m,\lambda} = \{ (f, g) \in C^{m,\lambda}(\bar{\Omega}, \mathbb{R}^n) \times C^{m+1,\lambda}(\partial\Omega, \mathbb{R}^n) \colon \\ \qquad\qquad (R_0^{\mathrm{T}} f, R_0^{\mathrm{T}} g) \text{ is equilibrated} \}. \end{cases} \tag{2.9}$$

We also recall (see Chapter III, §5) that $(R_0^{\mathrm{T}} f, R_0^{\mathrm{T}} g)$ is equilibrated (with respect to $\iota_{\bar{\Omega}}$) if and only if (f, g) is equilibrated with respect to any rigid deformation of $\bar{\Omega}$ having R_0 as the gradient at any $x \in \bar{\Omega}$.

These facts suggest two possible choices for solutions and data in order to obtain—by use of the implicit function theorem—local theorems of existence, uniqueness, and analytic dependence on ε of solutions to problem (2.1): $W^{m+2,p}(\Omega, \mathbb{R}^n)$, $W^{m,p}(\Omega, \mathbb{R}^n) \times W^{m+1-1/p,p}(\partial\Omega, \mathbb{R}^n)$ with $p(m+1) > n$, and $C^{m+2,\lambda}(\bar{\Omega}, \mathbb{R}^n)$, $C^{m,\lambda}(\bar{\Omega}, \mathbb{R}^n) \times C^{m+1,\lambda}(\partial\Omega, \mathbb{R}^n)$ with $0 < \lambda < 1$. Accordingly, we will suppose that $\phi \mapsto (F(\phi), G(\phi))$ is a C^1 mapping from an open subset U of $(W^{m+2,p}(\Omega, \mathbb{R}^n))^+$ [respectively $(C^{m+2,\lambda}(\bar{\Omega}, \mathbb{R}^n))^+$] into $W^{m,p}(\Omega, \mathbb{R}^n) \times W^{m+1-1/p,p}(\partial\Omega, \mathbb{R}^n)$ [respectively $C^{m,\lambda}(\bar{\Omega}, \mathbb{R}^n) \times C^{m+1,\lambda}(\partial\Omega, \mathbb{R}^n)$], so that, under the assumptions of the first part of Lemma 2.1, $(\phi, \varepsilon) \mapsto \Lambda(\phi, \varepsilon)$ shall be a C^1 operator from $U \times \mathbb{R}$ into $W^{m,p}(\Omega, \mathbb{R}^n) \times W^{m+1-1/p,p}(\partial\Omega, \mathbb{R}^n)$ [respectively $C^{m,\lambda}(\bar{\Omega}, \mathbb{R}^n) \times C^{m+1,\lambda}(\partial\Omega, \mathbb{R}^n)$].

We must remark that *a direct application of the implicit function theorem to the equation*

$$\Lambda(\phi, \varepsilon) = 0,$$

in order to express ϕ as a function of ε near $(\phi_0, 0)$ with ϕ_0 a rigid deformation of $\bar{\Omega}$, *is not possible* because the values of (the partial differential) $\Lambda'_{\phi}(\phi_0, 0)$, namely of operator E_{R_0}, are equilibrated with respect to ϕ_0, while the values taken by Λ are not equilibrated with respect to ϕ_0. Note that, under the assumptions on Ω, a, and ϕ of Lemma 2.1, the divergence theorem gives

$$\int_{\Omega} \operatorname{div} A(D\phi) \, dx - \int_{\partial\Omega} A(D\phi)v \, d\sigma = 0 \tag{2.10}$$

and, for any $\chi \in C^1(\bar{\Omega})$,

$$\int_{\Omega} \chi \otimes \operatorname{div} A(D\phi) \, dx - \int_{\partial\Omega} \chi \otimes A(D\phi)v \, d\sigma = -\int_{\Omega} (D\chi)A(D\phi)^{\mathrm{T}} \, dx. \tag{2.11}$$

Thus $\Lambda(\phi, \varepsilon)$ is equilibrated with respect to ϕ_0 if and only if

$$
\begin{cases}
\varepsilon\left(\displaystyle\int_\Omega F(\phi)\, dx + \int_{\partial\Omega} G(\phi)\, d\sigma \right) = 0, \\[2mm]
\varepsilon\left(\displaystyle\int_\Omega \phi_0 \wedge F(\phi)\, dx + \int_{\partial\Omega} \phi_0 \wedge G(\phi)\, d\sigma \right) = \int_\Omega (D\phi_0 A(D\phi)^{\mathrm{T}} \\[4mm]
\hspace{6cm} - A(D\phi)(D\phi_0)^{\mathrm{T}})\, dx.
\end{cases}
$$

Note also that (2.3) implies

$$
\int_\Omega A(D\phi)(D\phi)^{\mathrm{T}}\, dx \in \mathrm{Sym}_n; \tag{2.12}
$$

hence from ((2.10), (2.11)) it follows that, *for any $\varepsilon \neq 0$, $\Lambda(\phi, \varepsilon)$ is equilibrated with respect to ϕ if and only if the pair $(F(\phi), G(\phi))$ is equilibrated with respect to ϕ.*

We emphasize the fact that *the condition expressing that, for $\varepsilon \neq 0$, $\Lambda(\phi, \varepsilon)$ is equilibrated with respect to ϕ does not involve ε, while the equilibratedness of $\Lambda(\phi, \varepsilon)$ with respect to ϕ_0 ($\neq \phi$) depends on ε.*

*   *   *   *   *

Let ϕ_0 be a rigid deformation of $\bar{\Omega}$ and let R_0 be the value of its gradient. If the assumptions of the first part of Lemma 2.1 are satisfied and (2.2), (2.3), (2.4), (2.5) hold, then we will show that, under suitable hypotheses on the (load) mapping $\phi \mapsto (F(\phi), G(\phi))$, a (modified) operator $(\psi, \varepsilon) \mapsto \bar{\Lambda}(\psi, \varepsilon)$ exists such that:

(i) $\bar{\Lambda}$ is defined on $V \times \mathbb{R}$, with V a suitable open neighborhood of ϕ_0 in $\mathscr{V}_{R_0}^{m+2,p}$ [respectively $\mathscr{V}_{R_0}^{m+2,\lambda}$];

(ii) the implicit function theorem applies to the equation $\bar{\Lambda}(\psi, \varepsilon) = 0$ at the point $(\phi_0, 0)$;

(iii) there is a homeomorphism, say $\hat{\phi}$, of V onto the set of ϕ belonging to a suitable neighborhood of ϕ_0 in $W^{m+2,p}(\Omega, \mathbb{R}^n)$ [respectively $C^{m+2,\lambda}(\bar{\Omega}, \mathbb{R}^n)$] and satisfying certain conditions (see (5.2), (5.3), and (5.4)), such that, for $\psi \in V$ and $\varepsilon \neq 0$, we have

$$
\bar{\Lambda}(\psi, \varepsilon) = 0 \iff \Lambda(\hat{\phi}(\psi), \varepsilon) = 0.
$$

The key to the proof of the existence of an operator Λ with the properties (i), (ii), (iii) consists of Lemmas 3.2 and 4.3. We begin by a simple, but crucial remark.

Remark 2.2. Let $(c, R) \in \mathbb{R}^n \times \mathbb{O}_n^+$. If (2.2) holds, then a C^1 function $\psi \colon \bar{\Omega} \to \mathbb{R}^n$ is a solution of problem†

$$
\begin{cases}
\mathrm{div}\, A(D\psi) + \varepsilon R^{\mathrm{T}} F(R\psi + c) = 0 & in\ \Omega, \\[2mm]
-A(D\psi)v + \varepsilon R^{\mathrm{T}} G(R\psi + c) = 0 & on\ \partial\Omega,
\end{cases} \tag{2.13}
$$

---

† To make our notation simple, whenever we deal with a constant function $x \mapsto c$ defined on $\bar{\Omega}$, such a function will be denoted by c (thus we will identify a constant function with its value).

if and only if, putting

$$\phi = c + R\psi, \tag{2.14}$$

ϕ is a solution of problem (2.1).

Proof. It suffices to observe that if ϕ and ψ are related by (2.14) and (2.2) holds, then $Ra(x, D\psi(x)) = a(x, RD\psi(x)) = a(x, D(R\psi(x) + c)) = a(x, D\phi(x))$, $\forall x \in \Omega$, whence

$$RA(D\psi) = A(D\phi). \quad \square$$

§3. A Basic Lemma

Let ρ be a rigid deformation of $\bar{\Omega}$ and let R be the value of the gradient of ρ (at any $x \in \bar{\Omega}$). As we have done in Chapter III, §1, we denote by $T_\rho(\mathcal{R}_{\bar{\Omega}})$ the tangent space at ρ to the manifold $\mathcal{R}_{\bar{\Omega}}$ of all rigid deformations of $\bar{\Omega}$. We recall that $T_\rho(\mathcal{R}_{\bar{\Omega}})$ is the space of functions $r_\rho: \bar{\Omega} \to \mathbb{R}^n$ of the type

$$r_\rho(x) = c + WRx, \qquad x \in \bar{\Omega}, \tag{3.1}$$

with $c \in \mathbb{R}^n$ and $W \in \text{Skew}_n$. In particular, the elements of $T_{i_{\bar{\Omega}}}(\mathcal{R}_{\bar{\Omega}})$ are the functions $r: \bar{\Omega} \to \mathbb{R}^n$ of the type

$$r(x) = c + Wx, \qquad x \in \bar{\Omega},$$

with $c \in \mathbb{R}^n$ and $W \in \text{Skew}_n$: they are the *infinitesimal rigid displacements* of $\bar{\Omega}$.

Let us denote by $\mathcal{E}_\rho$ the set of pairs $(f, g) \in L^2(\Omega, \mathbb{R}^n) \times L^2(\partial\Omega, \mathbb{R}^n)$ that are equilibrated with respect to ρ (see Chapter III, §5), i.e., such that

$$\int_\Omega f \, dx + \int_{\partial\Omega} g \, d\sigma = 0, \qquad \int_\Omega \rho \wedge f \, dx + \int_{\partial\Omega} \rho \wedge g \, d\sigma = 0.$$

Remark 3.1. *For each* $(f, g) \in L^2(\Omega, \mathbb{R}^n) \times L^2(\partial\Omega, \mathbb{R}^n)$ *there exist, and are uniquely determined by* R, $\hat{r}_\rho(f, g) \in T_\rho(\mathcal{R}_{\bar{\Omega}})$ *and* $\hat{e}_\rho(f, g) \in \mathcal{E}_\rho$ *such that*

$$(f, g) = \hat{e}_\rho(f, g) + (\hat{r}_\rho(f, g)|_\Omega, \hat{r}_\rho(f, g)|_{\partial\Omega}).$$

Moreover, the linear mappings

$$(f, g) \mapsto \hat{e}_\rho(f, g), \qquad (f, g) \to (\hat{r}_\rho(f, g)|_\Omega, \hat{r}_\rho(f, g)|_{\partial\Omega}) \tag{3.2}$$

are continuous from $L^2(\Omega, \mathbb{R}^n) \times L^2(\partial\Omega, \mathbb{R}^n)$ *into itself, with respect to any normable topology on* $L^2(\Omega, \mathbb{R}^n) \times L^2(\partial\Omega, \mathbb{R}^n)$ *finer than the usual topology of* $L^2(\Omega, \mathbb{R}^n) \times L^2(\partial\Omega, \mathbb{R}^n)$ *or equal to the latter topology.*

Proof. Let $\mathscr{L} = L^2(\Omega, \mathbb{R}^n) \times L^2(\partial\Omega, \mathbb{R}^n)$ with the inner product defined by

$$(f, g) \cdot (u, v) = \int_\Omega f \cdot u \, dx + \int_{\partial\Omega} g \cdot v \, d\sigma \tag{3.3}$$

and let $\mathscr{L}_\rho = \{(r_{\rho|\Omega}, r_{\rho|\partial\Omega}): r_\rho \in T_\rho(\mathscr{R}_{\bar{\Omega}})\}$. The subspace $\mathscr{L}_\rho$ of the Hilbert space $\mathscr{L}$ is complete because its dimension is finite. Moreover, the subspace of $\mathscr{L}$ orthogonal to $\mathscr{L}_\rho$ is just $\mathscr{E}_\rho$; this is easily seen after observing that if $(f, g) \in \mathscr{L}$, $r_\rho: \bar{\Omega} \to \mathbb{R}^n$ is defined by (3.1), and $c_1 = \rho(x) - Rx$, we have

$$\int_\Omega f \cdot r_\rho \, dx + \int_{\partial\Omega} g \cdot r_\rho \, d\sigma = (c - Wc_1) \cdot \left(\int_\Omega f \, dx + \int_{\partial\Omega} g \, d\sigma \right)$$
$$- W \cdot \left(\int_\Omega \rho \wedge f \, dx + \int_{\partial\Omega} \rho \wedge g \, d\sigma \right).$$

Then, the following direct orthogonal sum decomposition

$$\mathscr{L} = \mathscr{E}_\rho \oplus \mathscr{L}_\rho$$

holds and the projections (3.2) are continuous.

To conclude the proof it suffices to remark that any normable topology on $\mathscr{L}$ induces, on the linear subspace $\mathscr{L}_\rho$, the topology defined by the inner product (3.3), because any two norms on a finite dimensional linear space are equivalent; thus, the projection $(3.2)_2$—and consequently the projection $(3.2)_1$—remains continuous if $\mathscr{L}$ is endowed with any normable topology finer than those defined by the inner product (3.3). $\square$

We will say that $\hat{e}_\rho(f, g)$ is the *equilibrated part* of (f, g) *relative to the rigid deformation* ρ *of* $\bar{\Omega}$.

Lemma 3.2. *For any rigid deformation* ρ *of* $\bar{\Omega}$ *there is a neighborhood* U *of* ρ *in* $C^0(\bar{\Omega}, \mathbb{R}^n)$ *having the following property: if* $\phi \in U$ *and* (f, g) *is an element of* $L^2(\Omega, \mathbb{R}^n) \times L^2(\partial\Omega, \mathbb{R}^n)$ *equilibrated with respect to* ϕ, *then*

$$\hat{e}_\rho(f, g) = (0, 0) \Rightarrow (f, g) = (0, 0).$$

Proof. For any $\phi \in C^0(\bar{\Omega}, \mathbb{R}^n)$ let $M_\rho(\phi)$ be the (continuous, linear) mapping from $T_\rho(\mathscr{R}_{\bar{\Omega}})$ into $\mathbb{R}^n \times \text{Skew}_n$ defined by putting

$$M_\rho(\phi)(r_\rho) = \left(\int_\Omega r_\rho \, dx + \int_{\partial\Omega} r_\rho \, d\sigma, \int_\Omega \phi \wedge r_\rho \, dx + \int_{\partial\Omega} \phi \wedge r_\rho \, d\sigma \right),$$

$\forall r_\rho \in T_\rho(\mathscr{R}_{\bar{\Omega}})$. From Remark 3.1 it follows that $M_\rho(\rho)$ is one-to-one (and hence a homeomorphism). Moreover, straightforward calculations based on the Hölder inequality show that $\phi \mapsto M_\rho(\phi)$ is a continuous mapping from $C^0(\bar{\Omega}, \mathbb{R}^n)$ into the space of continuous, linear mappings of $T_\rho(\mathscr{R}_{\bar{\Omega}})$ into $\mathbb{R}^n \times \text{Skew}_n$ endowed with the bounded convergence topology.

Therefore (since the set of linear homeomorphisms of $T_\rho(\mathcal{R}_{\bar{\Omega}})$ onto $\mathbb{R}^n \times \text{Skew}_n$ is an open subset of the space of all continuous, linear mappings of $T_\rho(\mathcal{R}_{\bar{\Omega}})$into $\mathbb{R}^n \times \text{Skew}_n$ equipped with the bounded convergence topology) there is a neighborhood U of ρ such that, for each $\phi \in U$, $M_\rho(\phi)$ is one-to-one. Now let $\phi \in U$ and let (f, g) be an element of $L^2(\Omega, \mathbb{R}^n) \times L^2(\partial\Omega, \mathbb{R}^n)$ equilibrated with respect to ϕ. If $\hat{e}_\rho(f, g) = (0, 0)$, then by Remark 3.1 we have

$$f = \hat{r}_\rho(f, g)|_\Omega, \qquad g = \hat{r}_\rho(f, g)|_{\partial\Omega},$$

with $\hat{r}_\rho(f, g) \in T_\rho(\mathcal{R}_{\bar{\Omega}})$; hence, the pair $(\hat{r}_\rho(f, g)|_\Omega, \hat{r}_\rho(f, g)|_{\partial\Omega})$ is equilibrated with respect to ϕ and thus $\hat{r}_\rho(f, g) = 0$ because $M_\rho(\phi)$ is one-to-one; then $(f, g) = (0, 0)$. $\square$

§4. Critical Infinitesimal Rigid Displacements for a Load

We will regard the set $\mathcal{R}_{\bar{\Omega}}$ of all rigid deformations of $\bar{\Omega}$ (see §3) as a submanifold of the (real) topological linear space of the restrictions to $\bar{\Omega}$ of all affine functions of $\mathbb{R}^n$ into itself (equipped with the topology of bounded convergence).

Suppose that the (load) mapping $\phi \mapsto (F(\phi), G(\phi))$ is such that $(F(\phi), G(\phi))$ belongs to $L^1(\Omega, \mathbb{R}^n) \times L^1(\partial\Omega, \mathbb{R}^n)$ for every $\phi \in \mathcal{R}_{\bar{\Omega}}$. Let $M: \mathcal{R}_{\bar{\Omega}} \to \mathbb{R}^n \times \text{Skew}_n$ be the mapping defined by

$$M(\phi) = \left(\int_\Omega F(\phi)\, dx + \int_{\partial\Omega} G(\phi)\, d\sigma, \int_\Omega \phi \wedge F(\phi)\, dx + \int_{\partial\Omega} \phi \wedge G(\phi)\, d\sigma \right).$$
(4.1)

Moreover, let $\phi_0 \in \mathcal{R}_{\bar{\Omega}}$ be such that $(F(\phi_0), (G(\phi_0))$ is equilibrated with respect to ϕ_0, viz

$$M(\phi_0) = (0, 0),$$

let $(c_0, W_0) \in (\mathbb{R}^n \times \text{Skew}_n)\backslash\{(0, 0)\}$ and let r_0 be the (associated) infinitesimal rigid displacement $y \mapsto c_0 + W_0 y$ of $\mathbb{R}^n$.

We will say that the *infinitesimal rigid displacement* r_0 (or the pair (c_0, W_0)) is *critical at* ϕ_0 *for the load* (F, G) if, for every C^1 curve $t \mapsto \phi_t$ in $\mathcal{R}_{\bar{\Omega}}$ having the fixed rigid deformation ϕ_0 as the value at 0 and such that

$$\frac{d}{dt}(t \mapsto \phi_t)|_{t=0} = r_0 \circ \phi_0,$$
(4.2)

the derivative at 0 of the function $t \mapsto M(\phi_t)$ is zero, i.e.,

$$\begin{cases} \lim \dfrac{1}{t} \left(\displaystyle\int_\Omega F(\phi_t)\, dx + \int_{\partial\Omega} G(\phi_t)\, d\sigma \right) = 0, \\[2mm] \lim_{t \to 0} \dfrac{1}{t} \left(\displaystyle\int_\Omega \phi_t \wedge F(\phi_t)\, dx + \int_{\partial\Omega} \phi_t \wedge G(\phi_t)\, d\sigma \right) = 0. \end{cases}$$

Here the symbol $(d/dt)(t \mapsto \phi_t)|_{t=0}$ denotes the derivative at 0 of the curve $t \mapsto \phi_t$.

Remark 4.1. *Let the load* (F, G) *be such that the mapping* $M: \mathscr{R}_{\bar{\Omega}} \to \mathbb{R}^n \times$ Skew$_n$ *is differentiable at* ϕ_0. *Then an infinitesimal rigid displacement* $r_0 \neq 0$ *of* $\mathbb{R}^n$ *is critical for* (F, G) *at* ϕ_0 *if and only if the differential of* M *at* ϕ_0 *vanishes at the element* $r_0 \circ \phi_0$ *of* $T_{\phi_0}(\mathscr{R}_{\bar{\Omega}})$.

Proof. Let $\theta: T_{\phi_0}(\mathscr{R}_{\bar{\Omega}}) \to \mathbb{R}^n \times$ Skew$_n$ and $m: \mathbb{R}^n \times$ Skew$_n \to \mathbb{R}^n \times$ Skew$_n$ be the mappings defined by putting, for any $(c, W) \in \mathbb{R}^n \times$ Skew$_n$,

$$\theta(r \circ \phi_0) = (c, W), \qquad m(c, W) = M((\exp W)\phi_0 + c),$$

where r is the infinitesimal rigid displacement $y \mapsto c + Wy$ of $\mathbb{R}^n$. The differential of M at ϕ_0 is the linear mapping

$$m'(0, 0) \circ \theta: T_{\phi_0}(\mathscr{R}_{\bar{\Omega}}) \to \mathbb{R}^n \times \text{Skew}_n.$$

Hence the value at $r_0 \circ \phi_0 \in T_{\phi_0}(\mathscr{R}_{\bar{\Omega}})$ of the differential of M at ϕ_0 is the derivative at 0 of the function

$$\mathbb{R} \ni t \mapsto M(\phi_t),$$

where ϕ_t is the rigid deformation of $\bar{\Omega}$ defined by

$$\phi_t(x) = (\exp(tW_0))\phi_0(x) + tc_0, \qquad x \in \bar{\Omega}. \tag{4.3}$$

Then, as $t \mapsto \phi_t$ is a curve in $\mathscr{R}_{\bar{\Omega}}$ having the given rigid deformation ϕ_0 as the value at 0 and verifying (4.2), the proof is concluded. □

Henceforth, we suppose that the (load) mapping $\phi \mapsto (F(\phi), G(\phi))$ is defined on an open subset U of a Banach space $\mathscr{X}$, that $\mathscr{R}_{\bar{\Omega}} \subseteq U$, that $\rho \circ \phi \in U$, $\forall(\rho, \phi) \in \mathscr{R}_{\bar{\Omega}} \times U$, that $T_\rho(\mathscr{R}_{\bar{\Omega}}) \subseteq \mathscr{X}$, $\forall \rho \in \mathscr{R}_{\bar{\Omega}}$, and that the integrals appearing in (4.1) make sense $\forall \phi \in U$. In §5 we will consider the following two choices of $\mathscr{X}$ and U: $\mathscr{X} = W^{m+2,p}(\Omega, \mathbb{R}^n)$, $U = (W^{m+2,p}(\Omega, \mathbb{R}^n))^+$ with $p(m + 1) > n$ and $\mathscr{X} = C^{m+2,\lambda}(\bar{\Omega}, \mathbb{R}^n)$, $U = (C^{m+2,\lambda}(\bar{\Omega}, \mathbb{R}^n))^+$.

For $\phi_0 \in \mathscr{R}_{\bar{\Omega}}$ we will denote by

$$(\mathscr{K}_{\phi_0}(F, G), \mathscr{W}_{\phi_0}(F, G))$$

and pair, with $\mathscr{K}_{\phi_0}(F, G)$ a linear subspace of $\mathbb{R}^n$ and $\mathscr{W}_{\phi_0}(F, G)$ a linear subspace of Skew$_n$, such that $\mathscr{K}_{\phi_0}(F, G) \times \mathscr{W}_{\phi_0}(F, G)$ is *a maximal element of the set of all linear subspaces of* $\mathbb{R}^n \times$ Skew$_n$ *of the type* $\mathscr{K} \times \mathscr{W}$, *with* $\mathscr{K}$ *a linear subspace of* $\mathbb{R}^n$ *and* $\mathscr{W}$ *a linear subspace of* Skew$_n$, *that have the following property:* "if $(c, W) \in \mathscr{K} \times \mathscr{W}$ and

$$c \cdot \left(\int_\Omega F(\phi)\, dx + \int_{\partial\Omega} G(\phi)\, d\sigma \right)$$

$$+ W \cdot \left(\int_\Omega \phi \wedge F(\phi)\, dx + \int_{\partial\Omega} \phi \wedge G(\phi)\, d\sigma \right) = 0 \tag{4.4}$$

for every ϕ belonging to some neighborhood of ϕ_0 in $\mathscr{X}$ contained in U, then $(c, W) = (0, 0)$."

Remark that, having fixed a base $(a_1, \ldots, a_n)$ of $\mathbb{R}^n$ and a base $(W_1, \ldots, W_{\binom{n}{2}})$ of Skew_n, if $\bar{J} \times \bar{H}$ is a maximal element of the family of all subsets of $\{1, \ldots, n\} \times \{1, \ldots, \binom{n}{2}\}$ of the type $J \times H$, with J a subset of $\{1, \ldots, n\}$ and H a subset of $\{1, \ldots, \binom{n}{2}\}$, such that the set of functions

$$\begin{cases} \phi \mapsto a_j \cdot \left(\int_\Omega F(\phi) \, dx + \int_{\partial\Omega} G(\phi) \, d\sigma \right), & j \in J, \\ \phi \mapsto W_h \cdot \left(\int_\Omega \phi \wedge F(\phi) \, dx + \int_{\partial\Omega} \phi \wedge G(\phi) \, d\sigma \right), & h \in H, \end{cases}$$

is linearly independent on every neighborhood of ϕ_0 in $\mathscr{X}$ contained in U, then

$$\left(\sum_{j \in \bar{J}} \mathbb{R}a_j, \sum_{h \in \bar{H}} \mathbb{R}W_h \right)$$

is a pair $(\mathscr{K}_{\phi_0}(F, G), \mathscr{W}_{\phi_0}(F, G))$. Conversely, any pair $(\mathscr{K}_{\phi_0}(F, G), \mathscr{W}_{\phi_0}(F, G))$ is of this type for a suitable choice of the bases $(a_1, \ldots, a_n)$ of $\mathbb{R}^n$ and $(W_1, \ldots, W_{\binom{n}{2}})$ of Skew_n. These assertions are easily proved after observing that all linear subspaces $(\sum_{j \in \bar{J}} \mathbb{R}a_j, \sum_{h \in \bar{H}} \mathbb{R}W_h)$ of $\mathbb{R}^n \times \text{Skew}_n$ considered above have the same dimension and that such a dimension is also independent of the choice of the bases $(a_1, \ldots, a_n)$ and $(W_1, \ldots, W_{\binom{n}{2}})$.

Of course, the set of pairs $(\mathscr{K}_{\phi_0}(F, G), \mathscr{W}_{\phi_0}(F, G))$ depends on the topology of $\mathscr{X}$, though, to avoid overly heavy notation, this dependence does not appear explicitly. Observe that, if the mapping $\phi \mapsto (F(\phi), G(\phi))$ is constant (dead loading), then $(\mathscr{K}_{\phi_0}(F, G), \mathscr{W}_{\phi_0}(F, G))$ is independent of the topology of $\mathscr{X}$: indeed, if $\mathscr{X} \ni \phi \mapsto (F(\phi), G(\phi))$ is constant and (4.4) vanishes for every ϕ belonging to some neighborhood of ϕ_0 in $\mathscr{X}$, then (4.4) vanishes for all $\phi \in \mathscr{X}$. Various examples of pairs $(\mathscr{K}_{\phi_0}(F, G), \mathscr{W}_{\phi_0}(F, G))$, arising from the study of special types of (load) mappings $\phi \mapsto (F(\phi), G(\phi))$, will be discussed in §6 of this chapter and in Chapter VI.

The following lemma, together with Lemma 3.2, are the basic tools in proving Theorem 5.1, which is the main result of this chapter.

Lemma 4.2. *Let $\mathscr{K}$ be a linear subspace of $\mathbb{R}^n$ and let $\mathscr{W}$ be a linear subspace of Skew_n such that $\mathscr{K} \times \mathscr{W} \supseteq \mathscr{K}_{\phi_0}(F, G) \times \mathscr{W}_{\phi_0}(F, G)$ for some pair $(\mathscr{K}_{\phi_0}(F, G), \mathscr{W}_{\phi_0}(F, G))$. Suppose that no element $\neq (0, 0)$ of $\mathscr{K} \times \mathscr{W}$ is critical for (F, G) at ϕ_0. Then an open neighborhood $V (\subseteq U)$ of ϕ_0 in $\mathscr{X}$ and a number $\beta > 0$ exist such that for each $\psi \in V$ there is one and only one element $(\hat{c}(\psi), \hat{R}(\psi))$ of $\mathscr{K} \times \exp(\mathscr{W})$ such that†*

$$M(\hat{R}(\psi)\psi + \hat{c}(\psi)) = 0$$

---

† See the footnote on p. 107.

and

$$|\hat{c}(\psi)| + |\hat{R}(\psi) - I| \leq \beta,$$

where M is the mapping from U into $\mathbb{R}^n \times \text{Skew}_n$ defined by (4.1). V and β can be chosen such that the mappings $\psi \mapsto \hat{c}(\psi)$ and $\psi \mapsto \hat{R}(\psi)$ are of class C^1. Furthermore, the mappings $\psi \mapsto \hat{c}(\psi)$ and $\psi \mapsto \hat{R}(\psi)$ are analytic at ϕ_0 provided the mapping $M: U \mapsto \mathbb{R}^n \times \text{Skew}_n$ defined by (4.1) is analytic at ϕ_0.

Proof. Let π be the projection of $\mathbb{R}^n \times \text{Skew}_n$ into $\mathcal{X} \times \mathcal{W}$. For any $(c, W, \psi) \in \mathcal{X} \times \mathcal{W} \times \mathcal{X}$ such that $(\exp W)\psi + c \in U$ we put

$$\Gamma(c, W, \psi) = \pi(M(\exp W)\psi + c).$$

In accordance with the convention of the footnote on p. 107, here c denotes both an element of $\mathbb{R}^n$ and the constant function $x \mapsto c$ of $\bar{\Omega}$ into $\mathbb{R}^n$. In our hypotheses Γ is a C^1 mapping from an open neighborhood of $(0, 0, \phi_0)$ in $\mathcal{X} \times \mathcal{W} \times \mathcal{X}$ into $\mathcal{X} \times \mathcal{W}$; moreover, it is analytic at ϕ_0 provided $M: U \rightarrow \mathbb{R}^n \times \text{Skew}_n$ is analytic at ϕ_0. Since $M(\phi_0) = (0, 0)$ we have $\Gamma(0, 0, \phi_0) = (0, 0)$. It is not difficult to recognize that the differential at $(0, 0)$ of the mapping $(c, W) \mapsto \Gamma(c, W, \phi_0)$ is the linear mapping

$$(c, W) \mapsto \pi \circ (M'(\phi_0)(r \circ \phi_0)),$$

where r is the infinitesimal rigid displacement of $\mathbb{R}^n$ defined by $r(y) = c + Wy$. Note that from the definition of $\mathcal{X}_{\phi_0}(F, G)$ and $\mathcal{W}_{\phi_0}(F, G)$ it follows that there is a neighborhood $U_0 (\subseteq U)$ of ϕ_0 in $\mathcal{X}$ such that

$$M(\phi) = L((\pi \circ M)(\phi)), \qquad \forall \phi \in U_0, \tag{4.5}$$

where L is a suitable linear function of $\mathcal{X} \times \mathcal{W}$ into $\mathbb{R}^n \times \text{Skew}_n$, and so

$$M'(\phi_0)(r \circ \phi_0) = L((\pi \circ M'(\phi_0))(r \circ \phi_0), \qquad \forall r \in T_{i_{\bar{\Omega}}}(\mathcal{R}_{\bar{\Omega}}).$$

Consequently, if the differential at $(0, 0)$ of the mapping $(c, W) \mapsto \Gamma(c, W, \phi_0)$ vanishes at $(c_0, W_0) \in \mathcal{X} \times \mathcal{W}$, then $M'(\phi_0)(r \circ \phi_0) = 0$, where r_0 is the infinitesimal rigid displacement $y \mapsto c_0 + W_0 y$ of $\mathbb{R}^n$, and hence $(c_0, W_0) = (0, 0)$ because, otherwise, by Remark 4.1, (c_0, W_0) would be critical for (F, G) at ϕ_0. Therefore the differential at $(0, 0)$ of the mapping $(c, W) \mapsto \Gamma(c, W, \phi_0)$ is an one-to-one (linear) function of $\mathcal{X} \times \mathcal{W}$ onto itself. Then, by the implicit function theorem applied to the equation

$$\Gamma(c, W, \psi) = (0, 0),$$

an open neighborhood V_0 of ϕ_0 in $\mathcal{X}$ and a number $\eta > 0$ exist such that for each $\psi \in V_0$ there is one and only one element $(\hat{c}(\psi), \hat{W}(\psi))$ of $\mathcal{X} \times \mathcal{W}$ such that $|\hat{c}(\psi)| + |\hat{W}(\psi)| \leq \eta$ and that

$$\Gamma(\hat{c}(\psi), \hat{W}(\psi), \psi) = (0, 0). \tag{4.6}$$

Moreover, the mappings $\psi \mapsto \hat{c}(\psi)$ and $\psi \mapsto \hat{W}(\psi)$ are of class C^1 for a suitable choice of V and η: they are analytic at ϕ_0 provided $M: U \to \mathbb{R}^n \times \text{Skew}_n$ is analytic at ϕ_0.

To conclude the proof it suffices to remark that if V is a neighborhood of ϕ_0 in $\mathcal{X}$ contained in V_0 and such that

$$\psi \in V \Rightarrow (\exp \hat{W}(\psi))\psi + \hat{c}(\psi) \in U_0,$$

then, by (4.5), for $\psi \in V$ (4.6) implies

$$M((\exp \hat{W}(\psi))\psi + \hat{c}(\psi)) = (0, 0). \quad \square$$

An evident consequence of Lemma 4.2 is

Remark 4.3. *Let the assumptions of Lemma 4.2 be satisfied and let V, β, $\hat{c}$, $\hat{R}$ be as in the statement of Lemma 4.2. If (2.13) holds for $(\psi, \varepsilon, c, R) \in V \times (\mathbb{R}\backslash\{0\}) \times \mathcal{X} \times \exp \mathcal{W}$ with $|c| + |R - I| \leq \beta$, then $c = \hat{c}(\psi)$ and $R = \hat{R}(\psi)$. Hence, for $(\psi, \varepsilon, c, R) \in V \times (\mathbb{R}\backslash\{0\}) \times \mathcal{X} \times \exp \mathcal{W}$ with $|c| + |R - I| \leq \beta$, (2.13) holds if and only if*

$$\begin{cases} \text{div } A(D\psi) + \varepsilon \hat{R}(\psi)^{\mathrm{T}} F(\hat{R}(\psi)\psi + \hat{c}(\psi)) = 0 & \text{in } \Omega, \\ -A(D\psi)\nu + \varepsilon \hat{R}(\psi)^{\mathrm{T}} G(\hat{R}(\psi)\psi + \hat{c}(\psi)) = 0 & \text{on } \partial\Omega, \end{cases} \quad (4.7)$$

where $\hat{c}(\psi)$ is identified with the constant function $x \mapsto \hat{c}(\psi)$ of $\bar{\Omega}$ into R^n.

The following lemma will be useful in that part of the proof of Theorem 5.1 concerning uniqueness.

Lemma 4.4. *Let $\mathcal{X}$ be a real Banach space continuously embedded in $C^1(\bar{\Omega}, \mathbb{R}^n)$, Let $\phi_0 \in \mathcal{X}$, let $\mathcal{K}$ be a linear subspace of $\mathbb{R}^n$, and let $\mathcal{W}$ be a linear subspace of Skew_n. Suppose that $\phi \in \mathcal{X}$ satisfies the conditions*

$$\begin{cases} \det \displaystyle\int_\Omega D\phi \, dx > 0, \\[2mm] R_\phi \in \{(\exp W) R_{\phi_0}: W \in \mathcal{W}\}, \\[2mm] \displaystyle\int_\Omega (\phi - R_\phi R_{\phi_0}^{\mathrm{T}} \phi_0) \, dx \in \mathcal{K}, \end{cases} \quad (4.8)$$

where R_ϕ [respectively R_{ϕ_0}] is the element of $\mathbb{O}_n^+$ such that the matrix

$$R_\phi^{\mathrm{T}} \int_\Omega D\phi \, dx \quad \left[\text{respectively } R_{\phi_0}^{\mathrm{T}} \int_\Omega D\phi \, dx\right] \quad (4.9)$$

is symmetric and positive definite. Then there is a unique $(\psi, c, R) \in \mathcal{X} \times \mathcal{K} \times \exp \mathcal{W}$ such that

$$\int_\Omega \psi \, dx = \int_\Omega \phi_0 \, dx, \quad (4.10)$$

$$R_{\phi_0}^{\mathrm{T}} \int_\Omega D\psi \, dx \text{ is symmetric and positive definite}, \quad (4.11)$$

$$\phi(x) = c + R\psi(x), \qquad \forall x \in \bar{\Omega}; \quad (4.12)$$

c, R, and ψ are related to ϕ by

$$R = R_\phi R_{\phi_0}^T, \qquad c = \frac{1}{\text{vol}(\Omega)} \int_\Omega (\phi - R_\phi R_{\phi_0}^T \phi_0)\, dx, \qquad (4.13)$$

and

$$\psi(x) = R_{\phi_0} R_\phi^T \left(\phi(x) - \frac{1}{\text{vol}(\Omega)} \int_\Omega \phi\, dx \right) + \frac{1}{\text{vol}(\Omega)} \int_\Omega \phi_0\, dx. \qquad (4.14)$$

Moreover, the mapping $(\psi, c, R) \mapsto \phi$, with ϕ as defined in (4.12), is a homeomorphism of the subset $\{(\psi, c, R) \in \mathcal{X} \times \mathcal{K} \times \exp \mathcal{W}: (4.11) \text{ holds}\}$ of $\mathcal{X} \times \mathbb{R}^n \times \mathbb{M}_n$ onto the subset $\{\phi \in \mathcal{X}: (4.8) \text{ holds}\}$ of $\mathcal{X}$.

Proof. Some straightforward checks show that if R, c, ψ are defined by (4.13), (4.14), then from (4.8) it follows that $(\psi, c, R) \in \mathcal{X} \times \mathcal{K} \times \exp \mathcal{W}$ and that (4.10), (4.11), and (4.12) hold. As regards the uniqueness of (ψ, c, R) satisfying conditions (4.10), (4.11), and (4.12), we first remark that (4.12) yields

$$R_{\phi_0}^T \int_\Omega D\psi\, dx = (RR_{\phi_0})^T \int_\Omega D\phi\, dx$$

and that, in view of (4.11), this implies $R_\phi = RR_{\phi_0}$, namely, implies (4.13)$_1$. On the other hand, from (4.10) and (4.12) it follows that

$$\text{vol}(\Omega)c = \int_\Omega (\phi - R\phi_0)\, dx,$$

which, combined with the equality $R = R_\phi R_{\phi_0}^T$, gives (4.13)$_2$. Finally, (4.14) is a consequence of (4.12) and (4.13). Thus, if $(\psi, c, R) \in \mathcal{X} \times \mathcal{K} \times \exp \mathcal{W}$ and (4.10), (4.11), (4.12) are satisfied, then c, R, ψ are related to ϕ by ((4.13), (4.14)).

Note that if $(\psi, c, R) \in \mathcal{X} \times \mathcal{K} \times \exp \mathcal{W}$ and (4.10), (4.11) hold, then the function $x \mapsto c + R\psi(x)$ belongs to the set $\{\phi \in \mathcal{X}: (4.13) \text{ holds}\}$; this is easy to see by recalling that if $\phi(x) = c + R\psi(x)$ we have just verified that ((4.10), (4.11)) implies $R_\phi = RR_{\phi_0}$ and $\int_\Omega (\phi - R\phi_0)\, dx = \text{vol}(\Omega)c$. Therefore the mapping $(\psi, c, R) \mapsto c + R\psi$ (where c is identified with the constant function $x \mapsto c$ from $\bar{\Omega}$ into $\mathbb{R}^n$), is a (continuous) bijection of $\{(\psi, c, R) \in \mathcal{X} \times \mathcal{K} \times \exp \mathcal{W}: (4.10) \text{ and } (4.11) \text{ hold}\}$ onto $\{\phi \in \mathcal{X}: (4.8) \text{ holds}\}$. In order to recognize that the inverse of such a mapping is continuous it suffices to bear in mind (4.13) and (4.14) and observe that, as

$$\left(R_\phi \int_\Omega D\phi\, dx \right)^2 = \left(\int_\Omega D\phi\, dx \right)^T \left(\int_\Omega (D\phi)\, dx \right)$$

(see Chapter I, §6), the mapping $\phi \mapsto R_\phi$ is continuous from $C^1(\bar{\Omega}, \mathbb{R}^n)$ into $\mathbb{M}_n$. Thus the proof is concluded. $\square$

§5. A Local Theorem on Existence, Uniqueness, and Analytic Dependence on a Parameter

We recall that by a *deformation* of $\bar{\Omega}$ we mean an one-to-one C^1 function $\phi: \bar{\Omega} \to \mathbb{R}^n$ such that $\det D\phi > 0$, so that a deformation of $\bar{\Omega}$ is an orientation-preserving C^1 diffeomorphism of $\bar{\Omega}$ onto a subset of $\mathbb{R}^n$. (See Chapter I, §2.)

Theorem 5.1. *Let m be an integer > 0. Assume that Ω is of class C^{m+2} [respectively $C^{m+2,\lambda}$], that $a \in C^{m+2}(\bar{\Omega} \times \mathsf{M}_n^+, \mathsf{M}_n)$ [respectively $C^{m+2,\lambda}(\bar{\Omega} \times \mathsf{M}_n^+, \mathsf{M}_n)$], and that $\phi \mapsto (F(\phi), G(\phi))$ is a C^1 mapping of $(W^{m+2,p}(\Omega, \mathbb{R}^n))^+$ [respectively $(C^{m+2,\lambda}(\bar{\Omega}, \mathbb{R}^n))^+$] into $W^{m,p}(\Omega, \mathbb{R}^n) \times W^{m+1-1/p,p}(\partial\Omega, \mathbb{R}^n)$ [respectively $C^{m,\lambda}(\bar{\Omega}, \mathbb{R}^n) \times C^{m+1,\lambda}(\partial\Omega, \mathbb{R}^n)$] with $p(m+1) > n$ [respectively $0 < \lambda < 1$]. Let ϕ_0 be a rigid deformation of $\bar{\Omega}$ and let R_0 be the value of $D\phi_0$ (at any $x \in \bar{\Omega}$). Let $\mathscr{K}$ be a linear subspace of $\mathbb{R}^n$ and let $\mathscr{W}$ be a linear subspace of Skew_n such that $\mathscr{K}_{\phi_0}(F, G) \subseteq \mathscr{K}$ and $\mathscr{W}_{\phi_0}(F, G) \subseteq \mathscr{W}$ for some pair $(\mathscr{K}_{\phi_0}(F, G), \mathscr{W}_{\phi_0}(F, G))$.† Suppose that (2.2), (2.3), (2.4), (2.5) hold, that $(F(\phi_0), G(\phi_0))$ is equilibrated with respect to ϕ_0, and that no element of $(\mathscr{K} \times \mathscr{W}) \setminus \{(0,0)\}$ is critical for (F, G) at ϕ_0. Then, there are numbers $\bar{\varepsilon} > 0$ and $\delta > 0$ such that for each $\varepsilon \in [-\bar{\varepsilon}, \bar{\varepsilon}] \setminus \{0\}$ there is one and only one deformation ϕ_ε of $\bar{\Omega}$ belonging to $W^{m+2,p}(\Omega, \mathbb{R}^n)$ [respectively $C^{m+2,\lambda}(\bar{\Omega}, \mathbb{R}^n)$] and satisfying the following conditions:*

$$\begin{cases} \mathrm{div}\, A(D\phi_\varepsilon) + \varepsilon F(\phi_\varepsilon) = 0 & in\ \Omega, \\ -A(D\phi_\varepsilon)v + \varepsilon G(\phi_\varepsilon) = 0 & on\ \partial\Omega, \end{cases} \tag{5.1}$$

$$R_\varepsilon \in \{(\exp W)R_0 : W \in \mathscr{W}\}, \tag{5.2}$$

$$\int_\Omega (\phi_\varepsilon - R_\varepsilon R_0^T \phi_0)\, dx \in \mathscr{K}, \tag{5.3}$$

and

$$\|\phi_\varepsilon - \phi_0\|_{m+2,p} \leqq \delta \qquad [\text{respectively } \|\phi_\varepsilon - \phi_0\|_{m+2,\lambda} \leqq \delta], \tag{5.4}$$

where R_ε is the element of $\mathbb{O}_n^+$ such that the matrix

$$R_\varepsilon^T \int_\Omega D\phi_\varepsilon\, dx$$

is symmetric and positive definite. The mapping $\varepsilon \mapsto \phi_\varepsilon$ (extended to $[-\bar{\varepsilon}, \bar{\varepsilon}]$ by assuming that its value at $\varepsilon = 0$ is the fixed rigid deformation ϕ_0 of $\bar{\Omega}$)

---

† As, here, we are dealing with a load mapping $\phi \mapsto (F(\phi), G(\phi))$ defined on $(W^{m+2,p}(\Omega, \mathbb{R}^n))^+$ [respectively $(C^{m+2,\lambda}(\bar{\Omega}, \mathbb{R}^n))^+$], $\mathscr{K}_{\phi_0}(F, G) \times \mathscr{W}_{\phi_0}(F, G)$ is a maximal element of the set of all products $\mathscr{K} \times \mathscr{W}$, with $\mathscr{K}$ a linear subspace of $\mathbb{R}^n$ and $\mathscr{W}$ a linear subspace of Skew_n, having the following property: "if $(c, W) \in \mathscr{K} \times \mathscr{W}$ and (4.4) holds for each ϕ belonging to some neighborhood of ϕ_0 in $(W^{m+2,p}(\Omega, \mathbb{R}^n))^+$ [respectively $(C^{m+2,\lambda}(\bar{\Omega}, \mathbb{R}^n))^+$], then $(c, W) = (0, 0)$."

is of class C^1 in a suitable neighborhood of 0. *If, in addition, we suppose that* $a \in C^\infty(\overline{\Omega} \times M_n^+, M_n)$, *that the functions* $(x, Z) \mapsto D_x^\alpha a(x, Z)$, $|\alpha| \leq m + 1$, *are analytic in* Z, *at* R_0, *uniformly with respect to* x *(see Chapter II, §5), and that the (load) mapping* $\phi \mapsto (F(\phi), G(\phi))$ *is analytic at* ϕ_0, *then the mapping* $\varepsilon \mapsto \phi_\varepsilon$ *is analytic at* 0.

Proof. We consider only the case of Sobolev spaces, because in the case of Schauder spaces we should proceed likewise. Let V, $\hat{R}$, $\hat{c}$, and β be as in the statement of Lemma 4.2 with $\mathscr{X} = W^{m+2,p}(\Omega, \mathbb{R}^n)$ and $U = (W^{m+2,p}(\Omega, \mathbb{R}^n))^+$. For $(\psi, \varepsilon) \in V \times \mathbb{R}$ we put

$$\overline{\Lambda}(\psi, \varepsilon) = \hat{e}_{\phi_0}(\text{div } A(D\psi) + \varepsilon\hat{R}(\psi)^T F(\hat{R}(\psi)\psi + \hat{c}(\psi)), -A(D\psi)|_{\partial\Omega}\nu$$

$$+ \varepsilon\hat{R}(\psi)^T G(\hat{R}(\psi)\psi + \hat{c}(\psi))),$$

where $\hat{e}_{\phi_0}$ is the mapping defined in §3 and $\hat{c}(\psi)$ is identified with the constant function $x \mapsto \hat{c}(\psi)$ of $\overline{\Omega}$ into $\mathbb{R}^n$ (see the footnote on p. 107).

Combining Lemma 2.1 with Lemma 4.2 and Remark 3.1 we deduce that $(\psi, \varepsilon) \mapsto \overline{\Lambda}(\psi, \varepsilon)$ is a C^1 mapping from the open subset $V \times \mathbb{R}$ of $W^{m+2,p}(\Omega, \mathbb{R}^n) \times \mathbb{R}$ into the closed subspace $\mathscr{E}_{R_0}^{m,p}$ of $W^{m,p}(\Omega, \mathbb{R}^n) \times W^{m+1-1/p,p}(\partial\Omega, \mathbb{R}^n)$ defined by (2.8), and that the differential at ϕ_0 of the mapping $\psi \mapsto \overline{\Lambda}(\psi, 0)$ is the mapping E_{R_0} defined by (2.7); moreover, if we suppose that $a \in C^\infty(\overline{\Omega} \times M_n^+, M_n)$, that the functions $(x, Z) \mapsto D_x^\alpha a(x, Z)$, $|\alpha| \leq m + 1$, are analytic in Z, at R_0, uniformly with respect to x, and that the mapping $\phi \mapsto (F(\phi), G(\phi))$ is analytic at ϕ_0, then the mapping $(\psi, \varepsilon) \mapsto \overline{\Lambda}(\psi, \varepsilon)$ is analytic at $(\phi_0, 0)$.

By Theorem 7.6 of Chapter III, mapping E_{R_0} is a bijection of $\mathscr{V}_{R_0}^{m+2,p}$ onto $\mathscr{E}_{R_0}^{m,p}$, with $\mathscr{V}_{R_0}^{m+2,p}$ as defined by (2.8). Note that $\overline{\Lambda}(\phi_0, 0) = (0, 0)$, because, in view of Remark 3.1 of Chapter III, hypothesis (2.4) gives $A(D\phi_0) = 0$. Note also that, by (2.10), (2.11), (2.12), from Lemma 4.2 it follows that for any $(\psi, \varepsilon) \in V \times \mathbb{R}$, the pair

$$(\text{div } A(D\psi) + \varepsilon\hat{R}(\psi)^T F(\hat{R}(\psi)\psi + \hat{c}(\psi)),$$

$$-A(D\psi)|_{\partial\Omega}\nu + \varepsilon\hat{R}(\psi)^T G(\hat{R}(\psi)\psi + \hat{c}(\psi)))$$

is equilibrated with respect to ψ. Thus, in view of Lemma 3.2, the neighborhood V of ϕ_0 can be chosen such that, for $(\psi, \varepsilon) \in V \times \mathbb{R}$, (4.7) holds if and only if $\overline{\Lambda}(\psi, \varepsilon) = (0, 0)$. Then, using the implicit function theorem for the equation $\Gamma(v, \varepsilon) = (0, 0)$ with $\Gamma(v, \varepsilon) = \overline{\Lambda}(\phi_0 + v, \varepsilon)$, we arrive at the following result.

Two numbers $\varepsilon_0 > 0$ and $\eta > 0$ exist such that for each $\varepsilon \in [-\varepsilon_0, \varepsilon_0]$ there is one and only one $\psi_\varepsilon \in W^{m+2,p}(\Omega, \mathbb{R}^n)$ satisfying the conditions

$$\psi_\varepsilon \in (\phi_0 + \mathscr{V}_{R_0}^{m+2,p}) \cap V,$$

$$\begin{cases} \text{div } A(D\psi_\varepsilon) + \varepsilon\hat{R}(\psi_\varepsilon)^T F(\hat{R}(\psi_\varepsilon)\psi_\varepsilon + \hat{c}(\psi_\varepsilon)) = 0 & \text{in } \Omega, \\ -A(D\psi_\varepsilon)\nu + \varepsilon\hat{R}(\psi_\varepsilon)^T G(\hat{R}(\psi_\varepsilon)\psi_\varepsilon + \hat{c}(\psi_\varepsilon)) = 0 & \text{on } \partial\Omega, \end{cases} \quad (5.5)$$

and

$$\|\psi_\varepsilon - \phi_0\|_{m+2,p} \leq \eta. \tag{5.6}$$

The mapping $\varepsilon \mapsto \psi_\varepsilon$, from $[-\varepsilon_0, \varepsilon_0]$ into $W^{m+2,p}(\Omega, \mathbb{R}^n)$, is of class C^1 in a suitable neighborhood of 0; furthermore,

$$\psi_0 = \phi_0.$$

If, in addition, we suppose that $a \in C^\infty(\bar{\Omega} \times \mathbb{M}_b^+, \mathbb{M}_n)$, that the functions $(x, Z) \mapsto D_x^\alpha a(x, Z)$, $|\alpha| \leq m + 1$, are analytic in Z, at R_0, uniformly with respect to x, and that the mapping $\phi \mapsto (F(\phi), G(\phi))$ is analytic at ϕ_0, then the mapping $\varepsilon \mapsto \psi_\varepsilon$ is analytic at 0.

Now, for $|\varepsilon| \leq \varepsilon_0$ and $x \in \bar{\Omega}$, we set

$$\phi_\varepsilon(x) = \hat{c}(\psi_\varepsilon) + \hat{R}(\psi_\varepsilon)(\psi_\varepsilon(x)). \tag{5.7}$$

Observe that, from Lemma 4.2 and the properties of ψ_ε, the value of the mapping $\varepsilon \mapsto \phi_\varepsilon$ at $\varepsilon = 0$ is just the fixed rigid deformation ϕ_0. In view of Remark 2.2, from (5.5) it follows that (5.1) is satisfied when the function ϕ_ε is defined by (5.7). Remark that, evidently,

$$R_{\phi_0} = R_0; \tag{5.8}$$

thus, as $\psi_\varepsilon - \psi_0$ belongs to $\mathscr{V}_{R_0}^{m+2,p}$, we have

$$\int_\Omega \psi_\varepsilon \, dx = \int_\Omega \phi_0 \, dx, \qquad R_0^T \int_\Omega D\psi_\varepsilon \, dx \in \mathrm{Sym}_n.$$

Let ε_1 be a real number such that $0 < \varepsilon_1 \leq \varepsilon_0$, and that for every $\varepsilon \in [-\varepsilon_1, \varepsilon_1]$ the symmetric matrix

$$R_0^T \int_\Omega D\psi_\varepsilon \, dx$$

is positive definite, $\det D\phi_\varepsilon > 0$ and

$$\sum_{|\alpha|=1} \sup_{x \in \Omega} |D^\alpha(\phi_\varepsilon - \phi_0)(x)| < \frac{1}{c_\Omega},$$

where c_Ω is a number > 0 such that inequality (1.1) of Chapter II holds (see Remark 1.1 of Chapter II); ε_1 exists because the mappings $\varepsilon \mapsto \psi_\varepsilon$ and $\varepsilon \mapsto \phi_\varepsilon$ are continuous from a suitable neighborhood of 0 in $\mathbb{R}$ into $C^1(\bar{\Omega}, \mathbb{R}^n)$. Note that, for $|\varepsilon| \leq \varepsilon_1$, the function ϕ_ε defined by (5.7) is one-to-one in view of Remark 1.2 of Chapter II and thus ϕ_ε is a deformation of $\bar{\Omega}$. Moreover, for $|\varepsilon| \leq \varepsilon_1$, conditions (5.2) and (5.3) are satisfied when ϕ_ε is defined by (5.7); this follows from Lemma 4.4, after remarking that, by Lemma 4.2, we have

$$\hat{c}(\psi_\varepsilon) \in \mathscr{K}, \qquad \hat{R}(\psi_\varepsilon) \in \exp \mathscr{W}, \qquad \forall \varepsilon \in [-\varepsilon_1, \varepsilon_1].$$

Furthermore, bearing in mind Lemma 4.2 and the properties of ψ_ε we easily realize that the mapping $\varepsilon \mapsto \phi_\varepsilon$, with ϕ_ε defined by (5.7), is of class C^1 in a suitable neighborhood of 0; and that it is analytic at 0 provided we suppose that $\phi \mapsto (F(\phi), G(\phi))$ is analytic at ϕ_0, that the function a belongs to $C^\infty(\bar{\Omega} \times M_n^+, M_n)$, and the functions $(x, Z) \mapsto D_x^\alpha a(x, Z)$, $|\alpha| \leq m + 1$, are analytic in Z, at R_0, uniformly with respect to x.

Let us now prove the uniqueness of ϕ_ε for $|\varepsilon|$ small enough, under the conditions listed in the statement of the theorem. We begin by observing that since (under our hypotheses) $W^{m+2,p}(\Omega)$ is continuously embedded in $C^1(\bar{\Omega})$, in view of (5.8), we can deduce from Lemma 4.4 that a number $\delta > 0$ exists such that if

$$
\begin{cases}
\|\phi - \phi_0\|_{m+2,p} \leq \delta, \\[2mm]
\det \int_\Omega D\phi \, dx > 0, \\[2mm]
R_\phi \in \{(\exp W)R_0 : W \in \mathscr{W}\}, \quad \int_\Omega (\phi - R_\phi R_0^T \phi_0) \, dx \in \mathscr{K}
\end{cases}
\tag{5.9}
$$

hold, with R_ϕ the element of $\mathbb{O}_n^+$ such that matrix (4.9) is symmetric and positive definite, then, putting

$$
R = R_\phi R_0^T, \qquad c = \frac{1}{\mathrm{vol}(\Omega)} \int_\Omega (\phi - R_\phi R_0^T \phi_0) \, dx, \tag{5.10}
$$

and

$$
\psi(x) = R^T \phi(x) - R^T c, \qquad \forall x \in \bar{\Omega}, \tag{5.11}
$$

we have

$$
\psi \in (\phi_0 + \mathscr{V}_{R_0}^{m+2,p}) \cap V, \tag{5.12}
$$

and

$$
\|\psi - \phi_0\|_{m+2,p} \leq \eta, \qquad |c| + |R - I| \leq \beta. \tag{5.13}
$$

Now let $\bar{\varepsilon} \in \mathbb{R}$ such that $0 < \bar{\varepsilon} \leq \varepsilon_1$ and $\|\phi_\varepsilon - \phi_0\|_{m+2,p} \leq \delta$ (with ϕ_ε the function defined by (5.7)) whenever $|\varepsilon| \leq \bar{\varepsilon}$; such a number exists because the mapping $\varepsilon \mapsto \phi_\varepsilon$ is continuous from a neighborhood of 0 in $\mathbb{R}$ into $W^{m+2,p}(\Omega, \mathbb{R}^n)$. We prove that, if $0 < |\varepsilon| \leq \varepsilon_0$, the function ϕ_ε defined by (5.7) is the only element of $W^{m+2,p}(\Omega, \mathbb{R}^n)$ which satisfies (5.1), (5.2), (5.3), and (5.5). We fix $\varepsilon \in [-\bar{\varepsilon}, \bar{\varepsilon}] \setminus \{0\}$. Let $\phi \in W^{m+2,p}(\Omega, \mathbb{R}^n)$ be a solution of problem (2.1) satisfying (5.9) and let $(\psi, c, R) \in W^{m+2,p}(\Omega, \mathbb{R}^n) \times \mathscr{K} \times \exp \mathscr{W}$ be related to ϕ by (5.10) and (5.11). Then (5.12) and (5.13) hold; thus, by Remarks 2.2 and 4.3, (4.7) holds and we have $R = \hat{R}(\psi)$ and $c = \hat{c}(\psi)$. Therefore $\psi = \psi_\varepsilon$, because ψ_ε is the only element of $(\phi_0 + \mathscr{V}_{R_0}^{m+2,p}) \cap V$ satisfying (5.5) and (5.6). Hence we have $\phi = \phi_\varepsilon$ with ϕ_ε defined by (5.7). $\square$

§6. The Case of Dead Loads

In this section we deal with the case in which the (load) mapping $\mathcal{X} \ni \phi \mapsto (F(\phi), G(\phi))$ is constant. We denote by (f, g) the value taken by $\phi \mapsto (F(\phi), G(\phi))$ at any $\phi \in \mathcal{X}$; thus, f and g are $\mathbb{R}^n$-valued functions defined on Ω and $\partial\Omega$, respectively. We suppose that $(f, g) \in (L^1(\Omega, \mathbb{R}^n) \times L^1(\partial\Omega, \mathbb{R}^n))\backslash\{(0, 0)\}$ and $C^\infty(\bar{\Omega}, \mathbb{R}^n) \subseteq \mathcal{X} \subseteq C^0(\bar{\Omega}, \mathbb{R}^n)$, and we assume that

$$\int_\Omega f\, dx + \int_{\partial\Omega} g\, d\sigma = 0; \tag{6.1}$$

this equality is a necessary condition for problem (2.1) to have solutions.

Let ϕ_0 be an element of $\mathcal{R}_{\bar{\Omega}}$ such that

$$\int_\Omega \phi_0 \wedge f\, dx + \int_{\partial\Omega} \phi_0 \wedge g\, d\sigma = 0. \tag{6.2}$$

We write $(\mathcal{K}_{\phi_0}(f, g), \mathcal{W}_{\phi_0}(f, g))$ instead of $(\mathcal{K}_{\phi_0}(F, G), \mathcal{W}_{\phi_0}(F, G))$ and we say that an *infinitesimal rigid displacement* r is *critical for* (f, g) when r is critical for (F, G). As we remarked in §4, the set of pairs $(\mathcal{K}_{\phi_0}(f, g), \mathcal{W}_{\phi_0}(f, g))$ is independent of the topology of $\mathcal{X}$. Actually, *the set of pairs* $(\mathcal{K}_{\phi_0}(f, g), \mathcal{W}_{\phi_0}(f, g))$ *is the same for each real Banach space* $\mathcal{X}$ *such that* $C^\infty(\bar{\Omega}, \mathbb{R}^n) \subseteq \mathcal{X} \subseteq C^0(\bar{\Omega}, \mathbb{R}^n)$; in order to realize that it suffices to recall that $C^\infty(\bar{\Omega}, \mathbb{R}^n)$ is dense in $C^0(\bar{\Omega}, \mathbb{R}^n)$ and observe that $W \cdot (\int_\Omega \phi \wedge f\, dx + \int_{\partial\Omega} \phi \wedge g\, d\sigma)$ vanishes for all $\phi \in \mathcal{X}$ whenever it vanishes for every ϕ belonging to some neighborhood of ϕ_0 in $\mathcal{X}$. Moreover, it is evident that, for any pair $(\mathcal{K}_{\phi_0}(f, g), \mathcal{W}_{\phi_0}(f, g))$, we have

$$\mathcal{K}_{\phi_0}(f, g) = \{0\}.$$

We also note that, clearly, *each constant function from* $\mathbb{R}^n$ *into* $\mathbb{R}^n$ *is critical for* (f, g) *at* ϕ_0.

The following remark gives a significant characterization of those infinitesimal rigid displacements of $\mathbb{R}^n$ that are critical for (f, g).

Remark 6.1. *Let* $W \in \text{Skew}_n\backslash\{0\}$ *and let* r *be the infinitesimal rigid displacement* $y \mapsto Wy$ *of* $\mathbb{R}^n$. *A necessary and sufficient condition for* r *to be critical for* (f, g) *at* ϕ_0 *is that* (f, g) *is equilibrated with respect to all rigid deformations* ϕ_t *of* $\bar{\Omega}$, *with* $t \in \mathbb{R}$ *and* ϕ_t *defined by putting*

$$\phi_t(x) = (\exp(tW))\phi_0(x), \tag{6.3}$$

$\forall x \in \bar{\Omega}$. *Furthermore,* r *is critical for* (f, g) *at* ϕ_0 *if and only if*

$$\int_\Omega (W\phi_0) \wedge f\, dx + \int_{\partial\Omega} (W\phi_0) \wedge g\, d\sigma = 0. \tag{6.4}$$

Proof. Consider the mapping $M: \mathcal{R}_{\bar{\Omega}} \to \mathbb{R}^n \times \text{Skew}_n$ defined by (4.1). Note that, in view of (6.1) and (6.2), (6.4) holds if and only if $M(\phi_t) = 0$, $\forall t \in \mathbb{R}$, namely, if and only if the function

$$\mathbb{R} \ni t \mapsto \int_{\Omega} \phi_t \otimes f \, dx + \int_{\partial\Omega} \phi_t \otimes g \, d\sigma \tag{6.5}$$

takes symmetric values. To realize that observe that (6.4) holds if the skew part of the derivative of (6.5) at $t = 0$ vanishes and that ((6.2), (6.4)) implies that each term of the power series expansion of (6.5) is a symmetric matrix. To conclude the proof it suffices to recall Remark 4.1, and observe that the differential of M at ϕ_0 vanishes at $r \circ \phi_0$ if and only if the derivative of function $t \mapsto M(\phi_t)$ vanishes at 0. $\square$

It is well known that, for any $W \in \mathrm{Skew}_3\backslash\{0\}$, the set $\{y \in \mathbb{R}^3 \colon Wy = 0\}$ is an one-dimensional linear subspace of $\mathbb{R}^3$: this subspace is called the *axis* of W. From the definition of exp W, it easily follows that the axis of W coincides with the axis of the rotation $y \mapsto (\exp W)y$ of $\mathbb{R}^3$; i.e., with the set $\{y \in \mathbb{R}^3 \colon (\exp W)y = y\}$. (See Appendix II.)

Let us emphasize the fact that, in the case $n = 3$, Remark 6.1 states that *the infinitesimal rigid displacement $y \mapsto Wy$ of $\mathbb{R}^3$ is critical for (f, g) at ϕ_0 if and only if (f, g) is equilibrated with respect to any rigid deformation of $\bar{\Omega}$ obtained from ϕ_0 by an arbitrary rotation about the axis of W.* Whenever, for $n = 3$, it occurs that (f, g) is equilibrated with respect to any rigid deformation of $\bar{\Omega}$ obtained from ϕ_0 by an arbitrary rotation about a one-dimensional linear subspace of $\mathbb{R}^3$, this subspace is called an *axis of equilibrium* for (f, g) at ϕ_0 (see SIGNORINI [1949], Chapter I, §2). After this definition, Remark 6.1, in the case $n = 3$, can be expressed by saying that *the infinitesimal rigid displacement $y \mapsto Wy$ $(W \in \mathrm{Skew}_n\backslash\{0\})$, of $\mathbb{R}^3$ is critical for (f, g) at ϕ_0 if and only if the axis of W is an axis of equilibrium for (f, g) at ϕ_0.*

Consider the matrix K_{ϕ_0} defined by

$$K_{\phi_0} = \int_{\Omega} \phi_0 \otimes f \, dx + \int_{\partial\Omega} \phi_0 \otimes g \, d\sigma.$$

K_{ϕ_0} is symmetric, because of (6.2).

Lemma 6.2. *Let $\omega_1, \ldots, \omega_n$ be the eigenvalues of K_{ϕ_0}. For no infinitesimal rigid displacement of $\mathbb{R}^n$ of the type $y \mapsto Wy$, with $0 \neq W \in \mathrm{Skew}_n$, to be critical for (f, g) at ϕ_0, it is necessary and sufficient that*

$$\prod_{\substack{i,j=1,\ldots,n \\ i \neq j}} (\omega_i + \omega_j) \neq 0. \tag{6.6}$$

Proof. By Remark 6.1 the infinitesimal rigid displacement $y \mapsto Wy$ of $\mathbb{R}^n$, with $0 \neq W \in \mathrm{Skew}_n$, is critical for (f, g) at ϕ_0 if and only if (6.4) holds. Then, since (6.4) takes the form $WK_{\phi_0} \in \mathrm{Sym}_n$, the statement of Remark

6.2 can be reformulated by saying that (6.6) holds if and only if

$$W \in \text{Skew}_n, \qquad WK_{\phi_0} \in \text{Sym}_n \Rightarrow W = 0. \tag{6.7}$$

In order to prove the equivalence of (6.6) and (6.7) we observe that, as $K_{\phi_0} \in \text{Sym}_n$, there is an $\bar{R} \in \mathbb{O}_n^+$ such that $\bar{R}^T K_{\phi_0} \bar{R}$ is the diagonal matrix whose diagonal is $(\omega_1, \ldots, \omega_n)$. We note that the condition $WK_{\phi_0} \in \text{Sym}_n$ is equivalent to the condition $\bar{R}^T WK_{\phi_0} \bar{R} \in \text{Sym}_n$, namely, equivalent to $W'\bar{R}^T K_{\phi_0} \bar{R} \in \text{Sym}_n$, where W' is the skew $n \times n$ matrix related to W by $W' = \bar{R}^T W \bar{R}$. We also note that the condition $W'\bar{R}^T K_{\phi_0} \bar{R} \in \text{Sym}_n$ can be written in the form $W'_{ij}(\omega_i + \omega_j) = 0$, $\forall i, j = 1, \ldots, n$. Then (6.6) is equivalent to (6.7), because (6.6) is a necessary and sufficient condition for $W'\bar{R}^T K_{\phi_0} \bar{R} \in \text{Sym}_n$ implying $W' = 0$, namely, implying $W = 0$. $\square$

Remark 6.2 concerns the case when $\mathscr{W}_{\phi_0}(F, G) = \text{Skew}_n$, while the following result concerns the case when $\mathscr{W}_{\phi_0}(F, G)$ is a proper subspace of Skew_n.

Lemma 6.3. *Let* Π_0 *be the smallest linear subspace of* $\mathbb{R}^n$ *such that*

$$\begin{cases} f(x) \in \Pi_0 & \text{for (almost) all } x \in \Omega, \\ g(x) \in \Pi_0 & \text{for (almost) all } x \in \partial\Omega, \end{cases} \tag{6.8}$$

and let $(a_1, \ldots, a_{k_0})$ *be an orthonormal base of* Π_0. *Then* $\mathscr{W}_{\phi_0}(F, G)$ *is the linear subspace of* Skew_n *spanned by*

$$\{a_i \wedge a_j : i = 1, \ldots, k_0; i < j \leq n\}. \tag{6.9}$$

Therefore

$$\dim \mathscr{W}_{\phi_0}(F, G) = \binom{n}{2} - \binom{n-k_0}{2} \qquad (= (n-1) + \cdots + (n - k_0)).$$

Proof. Let $(a_1, \ldots, a_{k_0}, \ldots, a_n)$ be an orthonormal base of $\mathbb{R}^n$. For every $W \in \text{Skew}_n$ we have

$$2W = \sum_{\substack{i=1,\ldots,n-1 \\ i<j\leq n}} ((a_i \wedge a_j) \cdot W)(a_i \wedge a_j);$$

hence (if $k_0 < n + 1$)

$$2W = \sum_{\substack{i=1,\ldots,k_0 \\ i<j\leq n}} ((a_i \wedge a_j) \cdot W)(a_i \wedge a_j) + \sum_{\substack{i=k_0+1,\ldots,n \\ i<j\leq n}} ((a_i \wedge a_j) \cdot W)(a_i \wedge a_j). \tag{6.10}$$

From (6.8) it follows that

$$\begin{cases} f(x) = \sum_{j=1}^{k_0} (f(x) \cdot a_j)a_j & \text{for (almost) all } x \in \Omega, \\ g(x) = \sum_{j=1}^{k_0} (g(x) \cdot a_j)a_j & \text{for (almost) all } x \in \partial\Omega. \end{cases} \tag{6.11}$$

Observing that for any $y, z \in \mathbb{R}^n$

$$\tfrac{1}{2}(a_i \wedge a_j) \cdot (y \wedge z) = (y \cdot a_i)(z \cdot a_j) - (y \cdot a_j)(z \cdot a_i), \tag{6.12}$$

from (6.11) we immediately deduce

$$(a_i \wedge a_j) \cdot \left(\int_\Omega \phi \wedge f \, dx + \int_{\partial\Omega} \phi \wedge g \, d\sigma \right) = 0 \tag{6.13}$$

$$\text{for} \quad i = k_0 + 1, \ldots, n; \quad j = k_0 + 1, \ldots, n.$$

In view of (6.10) and (6.13) the linear subspace of Skew$_n$ spanned by (6.9) contains $\mathscr{W}_{\phi_0}(f, g)$. Then, to prove that the subspace of Skew$_n$ spanned by (6.9) coincides with $\mathscr{W}_{\phi_0}(f, g)$, it suffices to show that if $W \in$ Skew$_n$ and

$$W \cdot \left(\int_\Omega \phi \wedge f \, dx + \int_{\partial\Omega} \phi \wedge g \, d\sigma \right) = 0, \tag{6.14}$$

then

$$W \cdot (a_i \wedge a_j) = 0 \qquad \text{for} \quad i = 1, \ldots, k_0; \quad j = 2, \ldots, n. \tag{6.15}$$

Accordingly, suppose that (6.14) holds $\forall \phi \in \mathscr{X}$, and hence $\forall \phi \in C^\infty(\bar{\Omega}, \mathbb{R}^n)$. Using (6.10), (6.11), (6.12), and (6.13) we obtain

$$2W \cdot \left(\int_\Omega \phi(x) \wedge f(x) \, dx + \int_{\partial\Omega} \phi(x) \wedge g(x) \, d\sigma \right)$$

$$= \sum_{\substack{i=1,\ldots,n \\ j=1,\ldots,k_0}} W \cdot (a_i \wedge a_j) \left(\int_\Omega (f(x) \cdot a_j)(\phi(x) \cdot a_i) \, dx \right.$$

$$\left. + \int_{\partial\Omega} (g(x) \cdot a_j)(\phi(x) \cdot a_i) \, d\sigma \right).$$

Then, from the fact that (6.14) holds for every $\phi \in C^\infty(\bar{\Omega}, \mathbb{R}^n)$ it follows that, for $i = 1, \ldots, n$,

$$\int_\Omega \sum_{j=1}^{k_0} (W \cdot (a_i \wedge a_j))(f(x) \cdot a_j)(\phi(x) \cdot a_i) \, dx$$

$$+ \int_{\partial\Omega} \sum_{j=1}^{k_0} (W \cdot (a_i \wedge a_j))(g(x) \cdot a_j)(\phi(x) \cdot a_i) \, dx = 0, \qquad \forall \phi \in C^\infty(\bar{\Omega}, \mathbb{R}^n). \tag{6.16}$$

Recalling that $\mathscr{D}(\Omega)$ is dense in $L^2(\Omega)$ and that (from the Stone–Weierstrass theorem) the restrictions to $\partial\Omega$ of all polynomial functions of $\mathbb{R}^n$ into $\mathbb{R}$ constitute a dense set in $C(\partial\Omega)$, we easily see that (6.16) implies, for $i = 1, \ldots, n$,

$$\sum_{j=1}^{k_0} (W \cdot (a_i \wedge a_j))(f(x) \cdot a_j) = 0 \qquad \text{for (almost) all} \quad x \in \Omega,$$

$$\sum_{j=1}^{k_0} (W \cdot (a_i \wedge a_j))(g(x) \cdot a_j) = 0 \qquad \text{for (almost) all} \quad x \in \partial\Omega.$$

Thus (6.15) is true, because, otherwise, the subspace of $\mathbb{R}^n$ spanned by $(a_1, \ldots, a_{k_0})$ would not be minimal within the set of linear subspaces of $\mathbb{R}^n$ containing $f(x)$ for (almost) all $x \in \Omega$ and $g(x)$ for (almost) all $x \in \partial\Omega$. $\square$

The following lemma generalizes Lemma 4.2.

Lemma 6.4. *Let Π be a linear subspace of $\mathbb{R}^n$ containing $f(x)$ for (almost) all $x \in \Omega$ and $g(x)$ for (almost) all $x \in \partial\Omega$, let $(a_1, \ldots, a_k)$ be an orthonormal base of Π, let $(a_1, \ldots, a_k, \ldots, a_n)$ be an orthonormal base of $\mathbb{R}^n$, let $\mathscr{W}_\Pi$ be the linear subspace of Skew_n spanned by*

$$\{a_i \wedge a_j : i = 1, \ldots, k, i < j \leq n\},$$

and let $\omega_1, \ldots, \omega_k$ be the eigenvalues of the matrix

$$\left(\int_\Omega (\phi_0(x) \cdot a_i)(f(x) \cdot a_r)\, dx + \int_{\partial\Omega} (\phi_0(x) \cdot a_i)(g(x) \cdot a_r)\, d\sigma \right)_{i,r=1,\ldots,k}. \quad (6.17)$$

For no infinitesimal rigid displacement of $\mathbb{R}^n$ of the type $y \mapsto Wy$, with $0 \neq W \in \mathscr{W}_\Pi$, to be critical for (f, g) at ϕ_0, it is necessary and sufficient that

$$\prod_{\substack{i,r=1,\ldots,k \\ i \neq r}} (\omega_i + \omega_r) \neq 0 \quad (6.18)$$

and, if $k \neq n$, that matrix (6.17) is nonsingular.

Proof. Evidently, if $k = n$, we have $\mathscr{W}_\Pi = \mathrm{Skew}_n$ and condition (6.18) coincides with condition (6.6); thus, if $k = n$, Lemma 6.4 reduces to Lemma 6.2. Therefore, we can suppose that $k \neq n$. Let $W \in \mathscr{W}_\Pi$. Then

$$2W = \sum_{\substack{i=1,\ldots,k \\ i < j \leq n}} (W \cdot (a_i \wedge a_j)) a_i \wedge a_j. \quad (6.19)$$

Because of Remark 6.1, we must prove that the implication

$$\int_\Omega (W\phi_0) \wedge f\, dx + \int_{\partial\Omega} (W\phi_0) \wedge g\, d\sigma = 0 \Rightarrow W = 0 \quad (6.20)$$

holds if and only if matrix (6.17) is nonsingular and (6.18) holds. Since for any three elements x, y, z of $\mathbb{R}^n$ we have

$$(x \wedge y)z = (y \cdot z)x - (x \cdot z)y,$$

from (6.19) it follows that

$$2W\phi_0(x) = \sum_{\substack{i=1,\ldots,k \\ i < j \leq n}} (W \cdot (a_i \wedge a_j))(\phi_0(x) \cdot a_j) a_i$$

$$- \sum_{\substack{i=1,\ldots,k \\ i < j \leq n}} (W \cdot (a_i \wedge a_j))(\phi_0(x) \cdot a_i) a_j.$$

Note that, for $s = k + 1, \ldots, n$,

$$
\begin{cases}
f(x) \cdot a_s = 0 & \text{for (almost) all} \quad x \in \Omega, \\
g(x) \cdot a_s = 0 & \text{for (almost) all} \quad x \in \partial\Omega.
\end{cases}
$$

Then, for (almost) every $x \in \Omega$, we have

$$
2W\phi_0(x) \wedge f(x) = \sum_{\substack{i,r=1,\ldots,k \\ i<j\leq n}} (W \cdot (a_i \wedge a_j))(\phi_0(x) \cdot a_j)(f(x) \cdot a_r)a_i \wedge a_r
$$

$$
- \sum_{\substack{i,r=1,\ldots,k \\ i<j\leq n}} (W \cdot (a_i \wedge a_j))(\phi_0(x) \cdot a_i)(f(x) \cdot a_r)a_j \wedge a_r,
$$

which gives without difficulty

$$
2W\phi_0(x) \wedge f(x) = \sum_{\substack{r=1,\ldots,k \\ j=k+1,\ldots,n}} \left(\sum_{i=1}^{k} (W \cdot (a_i \wedge a_j))(\phi_0(x) \cdot a_i) \right)(f(x) \cdot a_r)a_r \wedge a_j
$$

$$
+ \sum_{i,r=1,\ldots,k} \left(\sum_{j=1}^{n} (W \cdot (a_i \wedge a_j))(\phi_0(x) \cdot a_j) \right)(f(x) \cdot a_r)a_i \wedge a_r.
$$

Likewise, we get, for (almost) every $x \in \partial\Omega$

$$
2W\phi_0(x) \wedge g(x) = \sum_{\substack{r=1,\ldots,k \\ j=k+1,\ldots,n}} \left(\sum_{i=1}^{k} (W \cdot (a_i \wedge a_j))(\phi_0(x) \cdot a_i) \right)(g(x) \cdot a_r)a_r \wedge a_j
$$

$$
+ \sum_{i,r=1,\ldots,k} \left(\sum_{j=1}^{n} (W \cdot (a_i \wedge a_j))(\phi_0(x) \cdot a_j) \right)(g(x) \cdot a_r)a_i \wedge a_r.
$$

Thus

$$
2\left(\int_\Omega (W\phi_0) \wedge f \, dx + \int_{\partial\Omega} (W\phi_0) \wedge g \, d\sigma \right)
$$

$$
= \sum_{\substack{r=1,\ldots,k \\ j=k+1,\ldots,n}} \left[\sum_{i=1}^{k} (W \cdot (a_i \wedge a_j)) \left(\int_\Omega (\phi_0(x) \cdot a_i)(f(x) \cdot a_r) \, dx \right. \right.
$$

$$
\left. + \int_{\partial\Omega} (\phi_0(x) \cdot a_i)(g(x) \cdot a_r) \, d\sigma \right) \Big] a_r \wedge a_j
$$

$$
+ \sum_{1 \leq i < r \leq k} \left[\sum_{j=1}^{n} (W \cdot (a_i \wedge a_j)) \left(\int_\Omega (\phi_0(x) \cdot a_i)(f(x) \cdot a_r) \, dx \right. \right.
$$

$$
\left. + \int_{\partial\Omega} (\phi_0(x) \cdot a_i)(g(x) \cdot a_r) \, d\sigma \right)
$$

$$
- \sum_{j=1}^{n} (W \cdot (a_r \wedge a_j)) \left(\int_\Omega (\phi_0(x) \cdot a_r)(f(x) \cdot a_i) \, dx \right.
$$

$$
\left. + \int_{\partial\Omega} (\phi_0(x) \cdot a_r)(g(x) \cdot a_i) \, d\sigma \right) \Big] a_i \wedge a_r,
$$

which shows that the left-hand side of implication (6.20) is true if and only if

$$\sum_{i=1}^{k} (W \cdot (a_i \wedge a_s)) \left(\int_{\Omega} (\phi_0(x) \cdot a_i)(f(x) \cdot a_r) \, dx \right.$$

$$\left. + \int_{\partial\Omega} (\phi_0(x) \cdot a_i)(g(x) \cdot a_r) \, d\sigma \right) = 0,$$

$$\forall r = 1, \ldots, k \text{ and } \forall s = k + 1, \ldots, n, \quad (6.21)$$

and

$$\sum_{i=1}^{k} (W \cdot (a_i \wedge a_j)) \left(\int_{\Omega} (\phi_0(x) \cdot a_i)(f(x) \cdot a_r) \, dx + \int_{\partial\Omega} (\phi_0(x) \cdot a_i)(g(x) \cdot a_r) \, d\sigma \right)$$

$$= \sum_{j=1}^{n} (W \cdot (a_r \wedge a_j)) \left(\int_{\Omega} (\phi_0(x) \cdot a_r)(f(x) \cdot a_i) \, dx \right.$$

$$\left. + \int_{\partial\Omega} (\phi_0(x) \cdot a_r)(g(x) \cdot a_i) \, d\sigma \right), \quad \forall i, r = 1, \ldots, k. \quad (6.22)$$

Now to conclude our proof it suffices to remark that, in view of (6.19), the set of conditions (6.21) and (6.22) implies $W = 0$ if and only if matrix (6.17) is not singular and (6.18) holds. As a justification of the last assertion we observe that:

(i) the set of conditions (6.21) implies $W \cdot (a_i \wedge a_j) = 0$, $\forall i = 1, \ldots, k$ and $\forall j = k + 1, \ldots, n$ if and only if matrix (6.17) is not singular;
(ii) the set of conditions (6.22) means that the product of the matrix

$$\sum_{i,j=1}^{k} (W \cdot (a_i \wedge a_j)) a_i \wedge a_j, \quad (6.23)$$

by matrix (6.17), is a symmetric matrix, and, as we saw in proving Remark 6.2, this fact implies that (6.23) is the zero-matrix if and only if (6.18) is satisfied.   □

The following result, which particularizes Lemmas 6.3 and 6.4, has interest, especially in the case $n = 3$.

Corollary 6.5. *Suppose that there is an $e \in \mathbb{R}^n \backslash \{0\}$ such that*

$$\begin{cases} f(x) \wedge e = 0 & \text{for (almost) all } x \in \Omega, \\ g(x) \wedge e = 0 & \text{for (almost) all } x \in \partial\Omega, \end{cases}$$

and let $(a_1, \ldots, a_n)$ be an orthonormal base of $\mathbb{R}^n$ with $a_1 = e$. Then $\mathscr{W}_{\phi_0}(f, g)$ is the linear subspace of Skew_n spanned by $(e \wedge a_j)_{j=2,\ldots,n}$. Moreover, for no infinitesimal rigid displacement of $\mathbb{R}^n$ of the type $y \mapsto Wy$, with $0 \neq W \in \mathscr{W}_{\phi_0}(f, g)$, to be critical for (f, g) at ϕ_0, it is necessary *and sufficient that*

$$\int_{\Omega} \phi_0 \cdot f \, dx + \int_{\partial\Omega} \phi_0 \cdot g \, d\sigma \neq 0.$$

The next remark, whose proof is obvious, characterizes the pair of conditions appearing in the statement of Lemma 6.4.

Remark 6.6. *Let the matrix* $(B_{rs})_{r,s=1,\ldots,k+1}$ *be obtained from the real matrix* $(A_{ij})_{i,j=1,\ldots,k}$ *by putting* $B_{ij} = A_{ij}$ *for* $i, j = 1, \ldots, k$ *and* $B_{r,k+1} = B_{k+1,s} = 0$ *for* $r, s = 1, \ldots, k+1$. *If* $\alpha_1, \ldots, \alpha_k$ *are the eigenvalues of* $(A_{ij})_{i,j=1,\ldots,k}$, *then* $\alpha_1, \ldots, \alpha_k$, 0 *are the eigenvalues of* $(B_{rs})_{r,s=1,\ldots,k+1}$. *Hence, the following two conditions* (i) *and* (ii) *are equivalent:*

(i) $\det(A_{ij})_{i,j=1,\ldots,k} \neq 0$ *and* $\displaystyle\prod_{\substack{i,j=1,\ldots,k \\ i \neq j}} (\alpha_i + \alpha_j) \neq 0$;

(ii) $\displaystyle\sum_{\substack{r,s=1,\ldots,k \\ r \neq s}} (\beta_r + \beta_s) \neq 0$, *with* $\beta_1, \ldots, \beta_{k+1}$ *the eigenvalues of* $(B_{rs})_{r,s=1,\ldots,k+1}$.

From Remark 6.6 it follows that *if* (*in Lemma* 6.4) $k = n - 1$, *then for matrix* (6.17) *to be not singular and* (6.18) *holds, it is necessary and sufficient that* (6.6) *holds*. Note that this is in accordance with the fact that, by Lemma 6.3, if $k = n - 1$ (as well as if $k = n$) we have $\mathscr{W}_{\phi_0}(f, g) = \text{Skew}_n$.

$$* \quad * \quad * \quad * \quad *$$

After our discussion of the feature of $\mathscr{W}_{\phi_0}(f, g)$ and the existence of critical infinitesimal rigid displacements of $\mathbb{R}^n$ of the type $y \mapsto Wy$ with $W \in \mathscr{W}_{\phi_0}(f, g) \setminus \{0\}$, we are now in a position to state some consequences of Theorem 5.1 in the special case when the (load) mapping $\phi \mapsto (F(\phi), G(\phi))$ is constant.

Taking into account that $\mathscr{K}_{\phi_0}(f, g) = 0$ for any pair $(\mathscr{K}_{\phi_0}(f, g), \mathscr{W}_{\phi_0}(f, g))$, from Theorem 5.1 we deduce

Theorem 6.7. *Let* m *be an integer* ≥ 0. *Assume that* Ω *is of class* C^{m+2}, *that* $(f, g) \in W^{m,p}(\Omega, \mathbb{R}^n) \times W^{m+1-1/p,p}(\partial\Omega, \mathbb{R}^n)$ *with* $p(m + 1) > n$, *and that* $a \in C^{m+2}(\bar{\Omega} \times \mathbb{M}_n^+, \mathbb{M}_n)$ [*respectively that* Ω *is of class* $C^{m+2,\lambda}$ *with* $0 < \lambda < 1$, *that* $(f, g) \in C^{m,\lambda}(\bar{\Omega}, \mathbb{R}^n) \times C^{m+1,\lambda}(\partial\Omega, \mathbb{R}^n)$, *and that* $a \in C^{m+3}(\bar{\Omega} \times \mathbb{M}_n^+, \mathbb{M}_n)$]. *Let* ϕ_0 *be a rigid deformation of* $\bar{\Omega}$ *and let* R_0 *be the value of the gradient of* ϕ_0 (*at any* $x \in \bar{\Omega}$). *Let* $\mathscr{W}$ *be a linear subspace of* Skew_n *containing some* $\mathscr{W}_{\phi_0}(f, g)$.† *Suppose that* (2.2), (2.3), (2.4), (2.5), (6.1), (6.2) *hold and that no infinitesimal rigid displacement* $y \mapsto Wy$ *of* $\mathbb{R}^n$, *with* $W \in \mathscr{W} \setminus \{0\}$, *is critical for* (f, g) *at* ϕ_0. *Then, there*

---

† As $\mathscr{K}_{\phi_0}(f, g) = 0$, here $\mathscr{W}_{\phi_0}(f, g)$ is a maximal element of the set of all linear subspaces $\mathscr{W}$ of Skew_n having the following property: "if $W \in \mathscr{W}$ and $W \cdot (\int_\Omega \phi \wedge f \, dx + \int_{\partial\Omega} \phi \wedge g \, d\sigma) = 0$ for each $\phi \in C^\infty(\bar{\Omega}, \mathbb{R}^n)$, then $W = 0$."

are numbers $\bar{\varepsilon} > 0$ and $\delta > 0$ such that for each $\varepsilon \in [-\bar{\varepsilon}, \bar{\varepsilon}] \setminus \{0\}$ there is one and only one deformation† ϕ_ε of $\bar{\Omega}$ belonging to $W^{m+2,p}(\Omega, \mathbb{R}^n)$ [respectively $C^{m+2,\lambda}(\bar{\Omega}, \mathbb{R}^n)$] and satisfying the following conditions:

$$\begin{cases} \text{div } A(D\phi_\varepsilon) + \varepsilon f = 0 & \text{in } \Omega, \\ -A(D\phi_\varepsilon)\nu + \varepsilon g = 0 & \text{on } \partial\Omega, \end{cases} \tag{6.24}$$

$$\|\phi_\varepsilon - \phi_0\|_{m+2,p} \leqq \delta \qquad [\text{respectively } \|\phi_\varepsilon - \phi_0\|_{m+2,\lambda} \leqq \delta], \tag{6.25}$$

$$\int_\Omega (\phi_\varepsilon - R_\varepsilon R_0^{\mathrm{T}} \phi_0) \, dx = 0, \tag{6.26}$$

$$R_\varepsilon \in \{(\exp W) R_0 : W \in \mathscr{W}\}, \tag{6.27}$$

where R_ε is the element of $\mathbb{O}_n^+$ such that the matrix

$$R_\varepsilon^{\mathrm{T}} \int_\Omega D\phi_\varepsilon \, dx \tag{6.28}$$

is symmetric and positive definite. The mapping $\varepsilon \mapsto \phi_\varepsilon$ (extended to $[-\bar{\varepsilon}, \bar{\varepsilon}]$ by assuming that its value at $\varepsilon = 0$ is the given rigid deformation ϕ_0 of $\bar{\Omega}$) is of class C^1 in a suitable neighborhood of 0. If, in addition, $a \in C^\infty(\bar{\Omega} \times \mathbb{M}_n^+, \mathbb{M}_n)$ and the functions $(x, Z) \mapsto D_x^\alpha a(x, Z)$, $|\alpha| \leqq m + 1$, are analytic in Z, at R_0, uniformly with respect to x (see Chapter II, §5), then the mapping $\varepsilon \mapsto \phi_\varepsilon$ is analytic at 0.

Theorem 6.7, combined with Lemma 6.4, yields the next result.

Theorem 6.8. *Let m be an integer $\geqq 0$. Assume that Ω is of class C^{m+2}, that $(f, g) \in W^{m,p}(\Omega, \mathbb{R}^n) \times W^{m+1-1/p,p}(\partial\Omega, \mathbb{R}^n)$ with $p(m + 1) > n$, and that $a \in C^{m+2}(\bar{\Omega} \times \mathbb{M}_n^+, \mathbb{M}_n)$ [respectively that Ω is of class $C^{m+2,\lambda}$ with $0 < \lambda < 1$, that $(f, g) \in C^{m,\lambda}(\bar{\Omega}, \mathbb{R}^n) \times C^{m+1,\lambda}(\partial\Omega, \mathbb{R}^n)$, and that $a \in C^{m+3}(\bar{\Omega} \times \mathbb{M}_n^+, \mathbb{M}_n)]$. Let ϕ_0 be a rigid deformation of $\bar{\Omega}$ and R_0 the value of $D\phi_0$ (at any $x \in \bar{\Omega}$). Moreover, let Π be a linear subspace of $\mathbb{R}^n$ containing $f(x)$ for (almost) all $x \in \Omega$ and $g(x)$ for (almost) $x \in \partial\Omega$, let $(a_1, \ldots, a_k)$ be an orthonormal base of Π, let $(a_1, \ldots, a_k, \ldots, a_n)$ be an orthonormal base of $\mathbb{R}^n$, let $\mathscr{W}_\Pi$ be the linear subspace of Skew_n spanned by*

$$\{a_i \wedge a_j : i = 1, \ldots, k; i < j \leqq n\},$$

and let $\omega_1, \ldots, \omega_k$ be the eigenvalues of matrix (6.17). Suppose that (2.2), (2.3), (2.4), (2.5), (6.1), (6.2), (6.18) are satisfied and, if $k \neq n$, matrix (6.17) is nonsingular. Then, there are numbers $\bar{\varepsilon} > 0$ and $\delta > 0$ such that for each $\varepsilon \in [-\bar{\varepsilon}, \bar{\varepsilon}] \setminus \{0\}$ there is one and only one deformation ϕ_ε of $\bar{\Omega}$ belonging to $W^{m+2,p}(\Omega, \mathbb{R}^n)$ [respectively $C^{m+2,\lambda}(\bar{\Omega}, \mathbb{R}^n)$] and satisfying the conditions

---

† By a deformation of $\bar{\Omega}$ we mean an one-to-one C^1 function $\phi: \bar{\Omega} \to \mathbb{R}^n$ such that $\det D\phi > 0$. (See Chapter I, §2.)

(6.24), (6.25), (6.26) *and*

$$R_\varepsilon \in \{(\exp W)R_0 : W \in \mathcal{W}_\Pi\},$$

where R_ε is the element of $\mathbb{O}_n^+$ such that matrix (6.28) is symmetric and positive definite. The mapping $\varepsilon \mapsto \phi_\varepsilon$ (extended to $[-\bar{\varepsilon}, \bar{\varepsilon}]$ by assuming that its value at $\varepsilon = 0$ is the given rigid deformation ϕ_0 of $\bar{\Omega}$) is of class C^1 in a suitable neighborhood of 0. If, in addition, $a \in C^\infty(\bar{\Omega} \times \mathbb{M}_n^+, \mathbb{M}_n)$ and the functions $(x, Z) \mapsto D_x^\alpha a(x, Z)$, $|\alpha| \le m + 1$, are analytic in Z, at R_0, uniformly with respect to x, then the mapping $\varepsilon \mapsto \phi_\varepsilon$ is analytic at 0.

We emphasize the following two immediate consequences of Theorem 6.8 which correspond to the following two choices of the dimension k of Π: $k = n$ and $k = 1$. (More easily, Corollaries 6.9 and 6.10 can be obtained combined Theorem 6.8 with Lemma 6.2 and Corollary 6.5, respectively.)

Corollary 6.9. *Let Ω, ϕ_0, R_0, and m be as in Theorem 6.8. Assume that the function a belongs to $C^{m+2}(\bar{\Omega} \times \mathbb{M}_n^+, \mathbb{M}_n)$ [respectively $C^{m+2,\lambda}(\bar{\Omega} \times \mathbb{M}_n^+, \mathbb{M}_n)$] and satisfies (2.2), (2.3), (2.4), (2.5); moreover, assume that (f, g) belongs to $W^{m,p}(\Omega, \mathbb{R}^n) \times W^{m+1-1/p,p}(\partial\Omega, \mathbb{R}^n)$ with $p(m + 1) > n$ [respectively $C^{m,\lambda}(\bar{\Omega}, \mathbb{R}^n) \times C^{m+1,\lambda}(\bar{\Omega}, \mathbb{R}^n)$ with $0 < \lambda < 1$] and satisfies (6.1), (6.2), (6.6). Then, there exist numbers $\bar{\varepsilon} > 0$ and $\delta > 0$ such that for each $\varepsilon \in [-\bar{\varepsilon}, \bar{\varepsilon}] \backslash \{0\}$ there is one and only one deformation ϕ_ε belonging to $W^{m+2,p}(\Omega, \mathbb{R}^n)$ [respectively $C^{m+2,\lambda}(\Omega, \mathbb{R}^n)$] and satisfying conditions (6.24), (6.25), and (6.26). The mapping $\varepsilon \mapsto \phi_\varepsilon$ (extended to $[-\bar{\varepsilon}, \bar{\varepsilon}]$ by assuming that its value at $\varepsilon = 0$ is the given rigid deformation ϕ_0 of $\bar{\Omega}$) is of class C^1 in a suitable neighborhood of 0. If, moreover, $a \in C^\infty(\bar{\Omega} \times \mathbb{M}_n^+, \mathbb{M}_n)$ and the functions $(x, Z) \mapsto D_x^\alpha a(x, Z)$, $|\alpha| \le m + 1$, are analytic in Z, at R_0, uniformly with respect to x, then the mapping $\varepsilon \mapsto \phi_\varepsilon$ is analytic at 0.*

Corollary 6.10. *Let Ω, ϕ_0, R_0, and m be as in Theorem 6.8. Assume that the function a belongs to $C^{m+2}(\bar{\Omega} \times \mathbb{M}_n^+, \mathbb{M}_n)$ [respectively $C^{m+2,\lambda}(\bar{\Omega} \times \mathbb{M}_n^+, \mathbb{M}_n)$] and satisfies (2.2), (2.3), (2.4), (2.5). As to (f, g), assume that (f, g) belongs to $W^{m,p}(\Omega, \mathbb{R}^n) \times W^{m+1-1/p,p}(\partial\Omega, \mathbb{R}^n)$ with $p(m + 1) > n$ [respectively $C^{m,\lambda}(\bar{\Omega}, \mathbb{R}^n) \times C^{m+1,\lambda}(\partial\Omega, \mathbb{R}^n)$ with $0 < \lambda < 1$], that there is an $e \in \mathbb{R}^n \backslash \{0\}$ such that*

$$\begin{cases} f(x) \wedge e = 0 & \text{for (almost) all} \quad x \in \Omega, \\ g(x) \wedge e = 0 & \text{for (almost) all} \quad x \in \partial\Omega, \end{cases}$$

that (6.1), (6.2) hold, and that

$$\int_\Omega \phi_0 \cdot f \, dx + \int_{\partial\Omega} \phi_0 \cdot g \, d\sigma \ne 0.$$

Let $\mathcal{W}_e$ be the linear subspace of Skew_n spanned by $(e \wedge a_j)_{j=1,\dots,n}$, where $(a_1, \dots, a_n)$ is any orthonormal base of $\mathbb{R}^n$ with $a_1 = e$. Then, numbers $\bar{\varepsilon} > 0$ and $\delta > 0$ exist such that for each $\varepsilon \in [-\bar{\varepsilon}, \bar{\varepsilon}] \backslash \{0\}$ there is one and *only one deformation ϕ_ε of $\bar{\Omega}$ belonging to $W^{m+2,p}(\Omega, \mathbb{R}^n)$ [respectively $C^{m+2,\lambda}(\bar{\Omega}, \mathbb{R}^n)$] and satisfying the conditions (6.24), (6.25), (6.26) and*

$$R_\varepsilon \in \{(\exp W)R_0 : W \in \mathscr{W}_e\},$$

where R_ε is the element of $\mathbb{O}_n^+$ such that matrix (6.29) is symmetric and positive definite. The mapping $\varepsilon \mapsto \phi_\varepsilon$ (extended to $[-\bar{\varepsilon}, \bar{\varepsilon}]$ by assuming that its value at $\varepsilon = 0$ is the given rigid deformation ϕ_0 of $\bar{\Omega}$) is of class C^1 in a suitable neighborhood of 0. If, in addition, $a \in C^\infty(\bar{\Omega} \times \mathbb{M}_n^+, \mathbb{M}_n)$ and the functions $(x, Z) \mapsto D_x^\alpha a(x, Z)$, $|\alpha| \leqq m + 1$, are analytic in Z, at R_0, uniformly with respect to x, then the mapping $\varepsilon \mapsto \phi_\varepsilon$ is analytic at 0.

§7. Some Historical Notes

STOPPELLI, in a series of fundamental papers (see STOPPELLI [1954, 1955, 1957a, 1957b, 1958]), was able to prove a local theorem of existence, uniqueness, and analytic dependence on a parameter for the traction problem in the special case of a dead load having no axis of equilibrium, and he also studied the existence of solutions and their analytic dependence on a parameter when a dead load has an axis of equilibrium. The case of dead loads, although rather unrealistic, is mathematically very interesting. For a summary and a discussion of STOPPELLI's works we refer to GRIOLI [1962], TRUESDELL & NOLL [1965], WANG & TRUESDELL [1973], and MARSDEN & HUGHES [1978, 1983].

Recently, CHILLINGWORTH, MARSDEN, and WAN in three basic articles (see CHILLINGWORTH, MARSDEN, & WAN [1982, 1983] and WAN & MARSDEN [1984]) completed and extended STOPPELLI's existence results by means of geometrical techniques and singularity theory. I further mention the papers of SPECTOR [1982], LANZA DE CRISTOFORIS & VALENT [1982], BALL & SCHAEFFER [1983], MARSDEN & WAN [1983], VALENT [1986], and LE DRET [1986].

STOPPELLI's works arose within SIGNORINI's school. From the 1930s onward, SIGNORINI [1949, 1950] devised a clever perturbation method for investigating uniqueness and the position of the classical linear theory within the nonlinear theory of the traction problem with dead loads for hyperelastic bodies near an unstressed configuration. In this context he discovered some surprising facts which stimulated interest on the subject and led to various works. In particular, besides the already-mentioned memoirs of STOPPELLI, I mention paper by TOLOTTI [1942] and the crucial contributions of CAPRIZ & PODIO-GUIDUGLI [1974, 1979, 1981, 1982a, 1982b], who gave a conclusive answer to the question posed by SIGNORINI about compatibility of the linear and nonlinear theories in elasticity, and showed how SIGNORINI's perturbation scheme could be extended so as to cover a very large class of traction problems. On this subject see also BHARATHA & LEVINSON [1978].

A very important role, in view of a bifurcation analysis of solutions of the traction problem in finite elasticity under deformation-dependent loads, is played by the study, initiated by CAPRIZ & PODIO-GUIDUGLI [1982a] and pursued by PODIO-GUIDUGLI & VERGARA CAFFARELLI [1984] and PODIO-GUIDUGLI, VERGARA CAFFARELLI, & VIRGA [1984, 1986], of boundary value problems of linearized elasticity with deformation-dependent loads: those authors established a Green formula which allowed them to define formal adjoints; moreover, they provided a very interesting analysis of the role of ellipticity and normality assumptions and of the "complementing condition" of AGMON, DOUGLIS, & NIRENBERG [1964].

Finally, I refer to the interesting works of GRIOLI [1983a, 1983b], who proposed and studied a new perturbation procedure for traction problems of finite elasticity, using a suitable constitutive perturbation parameter.

Boundary Problems of Pressure Type in Finite Elastostatics

In this chapter we deal with the most spontaneous and physically reasonable class of traction problems in finite elastostatics. This class is characterized by the fact that the prescribed surface traction is parallel to the normal to the boundary of the unknown deformed equilibrium configuration. Within this class we consider the following three types of loads:

(a) loads invariant under translations;
(b) loads invariant under rotations;
(c) loads invariant under the group of isometries $y \mapsto c + Ry$ ($c \in \mathbb{R}^n$, $R \in \mathbb{O}_n^+$), of $\mathbb{R}^n$ such that $c \neq 0$, $c \cdot e = 0$, and $Re = e$, e being a fixed unit vector of $\mathbb{R}^n$.

In case (a) the pressure is constant. Case (b) includes, for example, the load consisting of a central body force proportional to the mass density, and a surface pressure proportional to the distance from a fixed sphere whose center coincides with the center of the body force. Finally, case (c) includes the load acting upon a *heavy body submerged in a heavy liquid*. The interest in the equilibrium problem of a heavy elastic body submerged in a liquid has been stimulated recently, in Italy, by GRIOLI [1982]. He, followed by CAPRIZ & PODIO-GUIDUGLI [1982a, 1982b], applied a perturbation process of the SIGNORINI type to such a problem, thus emphasizing some substantial differences between this traction problem and one with dead loads.

Here we will see that in cases (a) and (b) (as well as in the case of

dead loads) local theorems of existence, uniqueness, and analytic dependence on a parameter can be proved by using the abstract method
described in Chapter V, while such a method does not always apply to
case (c): for example, the load acting upon a heavy body submerged in
a homogeneous heavy liquid satisfies condition (c), but the method of
Chapter V does not apply to it. The bulk of the present chapter is
devoted to establishing a local theorem of existence, uniqueness, and
analytic dependence on a parameter for the (n-dimensional version of
the) traction problem with such a load (see Theorem 6.17). To prove
Theorem 6.17 we seek, as in Chapter V, a (modified) boundary problem
equivalent, in a certain sense, to the starting problem and such that the
implicit function theorem applies to it. But, here, the attainment of such
a modified problem is far more difficult and requires a subtler strategy.
Certainly, Theorem 6.17 is the main result of this chapter and of the
book. Note that the deformation ϕ_0 near which we find existence, uniqueness, and analytic dependence on the parameter ε in Theorem 4.17 differs
from the deformation $R_0 \iota_{\bar{\Omega}}$ by a translation which depends on the densities μ and ρ, besides the differential at $R_0 \iota_{\bar{\Omega}}$ of the finite elastostatics
operator. An anticipation, without proofs, of results described in this
chapter has been given in VALENT [1986].

§1. Preliminaries

We recall (see Chapter V, §1) that the traction problem for the equilib-
rium equation div $S_\phi + \mu_\phi b_\phi = 0$ consists in prescribing $S_\phi v_\phi$ on $\partial\Omega$.† In
this chapter we will study the class of traction problems of pressure
type for an elastic body: it corresponds to prescribe $S_\phi v_\phi$ parallel to v_ϕ
on $\partial\phi(\Omega)$. Precisely, we consider the problem of finding a deformation
$\phi: \bar{\Omega} \to \mathbb{R}^n$ of $\bar{\Omega}$ such that

$$\begin{cases} \operatorname{div} S_\phi + \mu_\phi b_\phi = 0 & \text{in } \phi(\Omega), \\ S_\phi v_\phi = p_\phi v_\phi & \text{on } \partial\phi(\Omega), \end{cases} \tag{1.1}$$

where v_ϕ is the unit outward normal to $\partial\phi(\Omega)$ and the functions $S_\phi: \phi(\Omega) \to$
$\mathbb{M}_n$, $\mu_\phi: \phi(\Omega) \to \mathbb{R}$, $b_\phi: \phi(\Omega) \to \mathbb{R}^n$, and $p_\phi: \partial\phi(\Omega) \to \mathbb{R}$ are defined by putting $S_\phi(\phi(x)) = s(x, D\phi(x))$, $\mu_\phi(\phi(x)) = \mu(x)/\det D\phi(x)$, $b_\phi(\phi(x)) = b(x, \phi(x))$,
$\forall x \in \Omega$, and $p_\phi(\phi(x)) = \pi(\phi(x))$, $\forall x \in \partial\Omega$, with $s: \Omega \times \mathbb{M}_n \to \mathbb{M}_n$, $\mu: \Omega \to \mathbb{R}$,
$b: \Omega \times \mathbb{R}^n \to \mathbb{R}^n$, and $\pi: \mathbb{R}^n \to \mathbb{R}$ given functions. Evidently, the meaning
of p_ϕ is that of a pressure upon the boundary of $\phi(\Omega)$.

Setting $S(x) = S_\phi(\phi(x)) \operatorname{cof} D\phi(x)$ (see (2.5) of Chapter I), and in view of
(2.3) and (2.6) of Chapter I, (when Ω, ϕ, s are sufficiently smooth) (1.1)

---

† Recall once more that we use the subscript ϕ to designate a function defined on $\phi(\Omega)$
or $\partial\phi(\Omega)$. (See Chapter I, §2.)

becomes

$$\begin{cases} \text{div } S + \mu b_\phi \circ \phi = 0 & \text{in } \Omega, \\ Sv = (\pi \circ \phi)(\text{cof } D\phi)v & \text{on } \partial\Omega, \end{cases}$$

where v is the unit outward normal to $\partial\Omega$. Since $S(x) = a(x, D\phi(x))$, $\forall x \in \Omega$, with the function $a: \Omega \times \mathbb{M}_n^+ \to \mathbb{M}_n$ related to the function s by $a(x, Z) = s(x, Z) \text{ cof } Z$, our problem is of type

$$\begin{cases} \text{div } A(D\phi) + \mu B(\phi) = 0 & \text{in } \Omega, \\ (-A(D\phi) + (\pi \circ \phi) \text{ cof } D\phi)v = 0 & \text{on } \partial\Omega, \end{cases} \tag{1.2}$$

where $\mu: \Omega \to \mathbb{R}$ and $\pi: \mathbb{R} \to \mathbb{R}$ are given functions and $A(D\phi): \bar{\Omega} \to \mathbb{M}_n$ and $B(\phi): \Omega \to \mathbb{R}^n$ are the functions defined by putting

$$\begin{cases} A(D\phi)(x) = a(x, D\phi(x)), & \forall x \in \Omega, \\ B(\phi)(x) = b(x, \phi(x)), & \forall x \in \partial\Omega, \end{cases} \tag{1.3}$$

with $a: \Omega \times \mathbb{M}_n^+ \to \mathbb{M}_n$ and $b: \Omega \times \mathbb{R}^n \to \mathbb{R}^n$ given functions.

We note that, in the context of the three-dimensional elasticity, to give the functions b and π means to prescribe the external forces (load) acting on the body, while to give the function a means to prescribe a relation between stress and deformation.

As we have done in Chapter V, we will make the assumptions:

$$a(x, RZ) = Ra(x, Z), \qquad \forall(x, Z, R) \in \Omega \times \mathbb{M}_n^+ \times \mathbb{O}_n^+, \tag{1.4}$$

$$a(x, Z)Z^T \in \text{Sym}_n, \qquad \forall(x, Z) \in \Omega \times \mathbb{M}_n^+, \tag{1.5}$$

$$a(x, I) = 0, \qquad \forall x \in \Omega, \tag{1.6}$$

$$Z \cdot \left(\sum_{h,k=1}^n D_{Z_{hk}} a(x, I) Z_{hk} \right) > 0, \qquad \forall(x, Z) \in \bar{\Omega} \times (\text{Sym}_n \backslash \{0\}), \tag{1.7}$$

where $Z_{hk} = Z \cdot (e_h \otimes e_k)$, with $(e_1, \ldots, e_n)$ the canonical base of $\mathbb{R}^n$ and $D_{Z_{hk}}$, denotes the partial derivative operator with respect to Z_{hk}.

If $b(x, y + c) = b(x, y)$ and $\pi(y + c) = \pi(y)$, $\forall(x, y, c) \in \Omega \times \mathbb{R}^n \times \mathbb{R}^n$, namely, if b does not depend on y and π is constant, we will say that the load is *invariant under translations*, while if

$$b(x, Ry) = Rb(x, y), \qquad \forall(x, y, R) \in \Omega \times \mathbb{R}^n \times \mathbb{O}_n^+, \tag{1.8}$$

and

$$\pi(Ry) = \pi(y), \qquad \forall(y, R) \in \Omega \times \mathbb{O}_n^+, \tag{1.9}$$

we will say that the load is *invariant under rotations*. More generally, having given a subgroup $\mathscr{G}$ of the group of all isometries of $\mathbb{R}^n$ (i.e., of all functions from $\mathbb{R}^n$ into itself of the type $y \mapsto c + Ry$ with $c \in \mathbb{R}^n$ and $R \in \mathbb{O}_n^+$), we will say that the load is $\mathscr{G}$-*invariant* when

$$\begin{cases} b(x, \phi(y)) = D\phi(y)b(x, y), & \forall(x, y, \phi) \in \Omega \times \mathbb{R}^n \times \mathscr{G}, \\ \pi(\phi(y)) = \pi(y), & \forall(y, \phi) \in \mathbb{R}^n \times \mathscr{G}. \end{cases} \tag{1.10}$$

* * * * *

Let us introduce a real parameter ε and write (1.2) in the form

$$\begin{cases} \text{div } A(D\phi) + \varepsilon\mu B(\phi) = 0 & \text{in } \Omega, \\ (-A(D\phi) + \varepsilon(\pi \circ \phi) \text{ cof } D\phi)v = 0 & \text{on } \partial\Omega. \end{cases} \tag{1.2}'$$

We recall that (see Chapter I, §1 and §2)

$$(\text{cof } D\phi)(D\phi)^{\mathsf{T}} = (\det D\phi)I, \qquad \text{div(cof } D\phi) = 0, \tag{1.11}$$

provided ϕ is smooth enough. Hence, using the divergence theorem, we easily see that

$$\int_{\partial\Omega} (\pi \circ \phi)(\text{cof } D\phi)v \, d\sigma = \int_{\Omega} (D\pi \circ \phi) \det D\phi \, dx, \tag{1.12}$$

and

$$\int_{\partial\Omega} \chi \otimes ((\pi \circ \phi)(\text{cof } D\phi)v) \, d\sigma = \int_{\Omega} \chi \otimes (D\pi \circ \phi) \det D\phi \, dx$$

$$- \int_{\Omega} (\pi \circ \phi)(\text{cof } D\phi)(D\chi)^{\mathsf{T}} \, dx, \tag{1.13}$$

whenever Ω is regular (see Chapter I, §1) and ϕ, χ are sufficiently smooth $\mathbb{R}^n$-valued functions defined on $\bar{\Omega}$. From (1.12), (1.13) and from (2.10), (2.11) of Chapter V it follows that

$$\int_{\Omega} (\text{div } A(D\phi) + \varepsilon\mu B(\phi)) \, dx + \int_{\partial\Omega} (-A(D\phi) + \varepsilon(\pi \circ \phi) \text{ cof } D\phi)v \, d\sigma$$

$$= \varepsilon \int_{\Omega} (\mu B(\phi) + (D\pi \circ \phi) \det D\phi) \, dx \tag{1.14}$$

and

$$\int_{\Omega} \chi \wedge (\text{div } A(D\phi) + \varepsilon\mu B(\phi)) \, dx + \int_{\partial\Omega} \chi \wedge ((-A(D\phi) + \varepsilon(\pi \circ \phi) \text{ cof } D\phi)v) \, d\sigma$$

$$= \varepsilon \int_{\Omega} \chi \wedge (\mu B(\phi) + D\pi \circ \phi \det D\phi) \, dx$$

$$+ \varepsilon \int_{\Omega} (\pi \circ \phi)(D\chi \text{ cof}(D\phi)^{\mathsf{T}} - \text{cof } D\phi(D\chi)^{\mathsf{T}}) \, dx$$

$$+ \int_{\Omega} (A(D\phi)(D\chi)^{\mathsf{T}} - (D\chi)(A(D\phi))^{\mathsf{T}}) \, dx. \tag{1.15}$$

Note that, in view of $(1.11)_1$, (1.13) gives

$$\int_{\partial\Omega} \phi \wedge ((\pi \circ \phi)(\text{cof } D\phi)v) \, d\sigma = \int_{\Omega} \phi \wedge (D\pi \circ \phi) \det D\phi \, dx. \tag{1.16}$$

We note also that if (1.5) holds, then

$$\int_{\Omega} A(D\phi)(D\phi)^{\mathsf{T}} \, dx \in \text{Sym}_n,$$

and thus (1.15) yields

$$\int_{\Omega} \phi \wedge (\operatorname{div} A(D\phi) + \varepsilon\mu B(\phi)) \, dx + \int_{\partial\Omega} \phi \wedge ((-A(D\phi) + \varepsilon(\pi \circ \phi) \operatorname{cof} D\phi)v) \, d\sigma$$

$$= \varepsilon \int_{\Omega} \phi \wedge (\mu B(\phi) + (D\pi \circ \phi) \det D\phi) \, dx. \quad (1.17)$$

As we did in Chapter V, for any integer $k \geq 1$, any real $p > 1$, any $\lambda \in \,]0, 1]$, and any $R \in \mathbb{O}_n^+$, we set

$$\begin{cases} \mathcal{V}_R^{k,p} = \left\{ v \in W^{k,p}(\Omega, \mathbb{R}^n): \int_{\Omega} v \, dx = 0, R^T \int_{\Omega} Dv \, dx \in \operatorname{Sym}_n \right\}, \\ \mathcal{E}_R^{k,p} = \{(f, g) \in W^{k,p}(\Omega, \mathbb{R}^n) \times W^{m+1-1/p,p}(\partial\Omega, \mathbb{R}^n): (R^T f, R^T g) \text{ is} \\ \quad \text{equilibrated}\}, \end{cases}$$

$$\begin{cases} \mathcal{V}_R^{k,\lambda} = \left\{ v \in C^{k,\lambda}(\overline{\Omega}, \mathbb{R}^n): \int_{\Omega} v \, dx = 0, R^T \int_{\Omega} Dv \, dx \in \operatorname{Sym}_n \right\}, \\ \mathcal{E}_R^{k,\lambda} = \{(f, g) \in C^{k,\lambda}(\overline{\Omega}, \mathbb{R}^n) \times C^{m+1,\lambda}(\partial\Omega, \mathbb{R}^n): (R^T f, R^T g) \text{ is equilibrated}\}. \end{cases}$$

§2. The Case When the Load Is Invariant Under Translations

If the load is translation invariant, problem (1.2)' takes the form

$$\begin{cases} \operatorname{div} A(D\phi) + \varepsilon f = 0 & \text{in } \Omega, \\ (-A(D\phi) + \varepsilon\pi \operatorname{cof} D\phi)v = 0 & \text{on } \partial\Omega, \end{cases} \quad (2.1)$$

where f is a given function of Ω into $\mathbb{R}^n$ and π is a given real number $\neq 0$ (whose physical meaning, in the case $n = 3$, is that of a pressure). Suppose $f \in L^1(\Omega)$. From (1.14) it follows that (for Ω regular and a, ϕ sufficiently smooth) problem (2.1) has a solution for $\varepsilon \neq 0$ only if

$$\int_{\Omega} f \, dx = 0. \quad (2.2)$$

Let ϕ_0 be a rigid deformation of $\overline{\Omega}$ such that

$$\int_{\Omega} \phi_0 \wedge f \, dx = 0. \quad (2.3)$$

Although, here, the load depends on the deformation (in a nontrivial manner), actually the local treatment of problem (2.1) does not involve any substantial difficulty besides those met in studying the traction problem with a dead load (in Chapter V, §6). Note that, in view of (1.12) and (1.16), we have (for Ω regular and ϕ smooth enough)

$$\int_{\partial\Omega} (\operatorname{cof} D\phi)v \, d\sigma = 0, \quad \int_{\partial\Omega} \phi \wedge (\operatorname{cof} D\phi)v \, d\sigma = 0. \quad (2.4)$$

Evidently, from (2.2) and (2.4) it follows that *any constant function of* $\mathbb{R}^n$ *into* $\mathbb{R}^n$ *is critical for the load at any rigid deformation of* $\bar{\Omega}$. Moreover, (2.4) implies that, for any pair $(\mathscr{K}_{\phi_0}(F, G), \mathscr{W}_{\phi_0}(F, G))$ defined in Chapter V, §4, $\mathscr{K}_{\phi_0}(F, G)$ and $\mathscr{W}_{\phi_0}(F, G)$ depend on the load (F, G) only through f: therefore, we will write

$$(\mathscr{K}_{\phi_0}(f), \mathscr{W}_{\phi_0}(f))$$

to mean $((\mathscr{K}_{\phi_0}(F, G), \mathscr{W}_{\phi_0}(F, G))$. Finally, $((2.2), (2.4)_1)$ yields

$$\mathscr{K}_{\phi_0}(f) = 0$$

for any pair $(\mathscr{K}_{\phi_0}(f), \mathscr{W}_{\phi_0}(f))$. We assume that (2.2) holds.

Bearing in mind (2.3), and using arguments analogous to those employed in the proof of Remark 6.1 of Chapter 5 we can prove

Remark 2.1. *Let* $W \in \text{Skew}_n \backslash \{0\}$. *The infinitesimal rigid displacement* $y \mapsto Wy$ *of* $\mathbb{R}^n$ *is critical for the load at* ϕ_0 *if and only if*

$$\int_\Omega W\phi_0 \wedge f \, dx = 0.$$

Because of Remark 2.1, whenever $y \mapsto Wy$ is critical for the load at ϕ_0, we will say that $y \mapsto Wy$ is *critical for* f at ϕ_0.

After Remark 2.1, a procedure quite similar to the one used in proving Lemma 6.4 of Chapter V allows us to prove the following result.

Lemma 2.2. *Let* Π *be a linear subspace of* $\mathbb{R}^n$ *such that* $f(x) \in \Pi$ *for (almost) all* $x \in \Omega$, *let* $(a_1, \ldots, a_k)$ *be an orthonormal base of* Π, *let* $(a_1, \ldots, a_k, \ldots, a_n)$ *be an orthonormal base of* $\mathbb{R}^n$, *let* $\mathscr{W}_\Pi$ *be the linear subspace of* Skew_n *spanned by* $\{a_i \wedge a_j : i = 1, \ldots, k; i < j \leq n\}$, *and let* $\omega_1, \ldots, \omega_k$ *be the eigenvalues of the matrix*

$$\left(\int_\Omega (\phi_0(x) \cdot a_i)(f(x) \cdot a_r) \, dx \right)_{i, r = 1, \ldots, k}. \tag{2.5}$$

For *no infinitesimal rigid displacement of* $\mathbb{R}^n$ *of the type* $y \mapsto Wy$, *with* $0 \neq W \in \mathscr{W}_\Pi$, *to be critical for* f *at* ϕ_0 *it is necessary and sufficient that*

$$\prod_{\substack{i, r = 1, \ldots, k \\ i \neq r}} (\omega_i + \omega_r) \neq 0 \tag{2.6}$$

and, when $k \neq n$, *that matrix (2.5) is nonsingular.*

We are now in a position to state some consequences of Theorem 5.1 of Chapter V. The next theorem follows immediately from Theorem 5.1 of Chapter V and the fact, remarked above, that $\mathscr{K}_{\phi_0}(f) = 0$ for any pair $(\mathscr{K}_{\phi_0}(f), \mathscr{W}_{\phi_0}(f))$.

Theorem 2.3. *Let* m *be an integer* ≥ 0. *Assume that* Ω *is of class* C^{m+2}, *that* $f \in W^{m, p}(\Omega, \mathbb{R}^n)$ *with* $p(m + 1) > n$, *and that* $a \in C^{m+2}(\bar{\Omega} \times \mathbb{M}_n^+, \mathbb{M}_n)$

[*respectively that Ω is of class $C^{m+2,\lambda}$ with $0 < \lambda < 1$, that $f \in C^{m,\lambda}(\bar\Omega, \mathbb{R}^n)$, and that $a \in C^{m+3}(\bar\Omega \times \mathbb{M}_n^+, \mathbb{M}_n)$]. Let ϕ_0 be a rigid deformation of $\bar\Omega$ and let R_0 be the value of $D\phi_0$ (at any $x \in \bar\Omega$). Let $\mathscr{W}$ be a linear subspace of Skew_n containing some $\mathscr{W}_{\phi_0}(f)$ with respect to the space $W^{m+2,p}(\Omega, \mathbb{R}^n)$ [respectively $C^{m+2,\lambda}(\bar\Omega, \mathbb{R}^n)$].† Suppose that (1.4), (1.5), (1.6), (1.7), (2.2), (2.3) hold and that no infinitesimal rigid displacement of $\mathbb{R}^n$ of the type $y \mapsto Wy$, with $0 \ne W \in \mathscr{W}$, is critical for f at ϕ_0. Then, two numbers $\bar\varepsilon > 0$ and $\delta > 0$ exist such that for each $\varepsilon \in [-\bar\varepsilon, \bar\varepsilon] \setminus \{0\}$ there is one and only one deformation‡ ϕ_ε of $\bar\Omega$ belonging to $W^{m+2,p}(\Omega, \mathbb{R}^n)$ [respectively $C^{m+2,\lambda}(\bar\Omega, \mathbb{R}^n)$] and satisfying the following conditions:

$$\begin{cases} \operatorname{div} A(D\phi_\varepsilon) + \varepsilon f = 0 & \text{in } \Omega, \\ (-A(D\phi_\varepsilon) + \varepsilon\pi \operatorname{cof} D\phi_\varepsilon)\nu = 0 & \text{on } \partial\Omega, \end{cases} \tag{2.7}$$

$$\|\phi_\varepsilon - \phi_0\|_{m+2,p} \le \delta \quad [\text{respectively } \|\phi_\varepsilon - \phi_0\|_{m+2,\lambda} \le \delta], \tag{2.8}$$

$$\int_\Omega (\phi_\varepsilon + R_\varepsilon R_0^T \phi_0)\, dx = 0, \tag{2.9}$$

and

$$R_\varepsilon \in \{(\exp W)R_0 : W \in \mathscr{W}\}, \tag{2.10}$$

where R_ε is the element of $\mathbb{O}_n^+$ such that the matrix

$$R_\varepsilon^T \int_\Omega D\phi_\varepsilon\, dx \tag{2.11}$$

is symmetric and positive definite. The mapping $\varepsilon \mapsto \phi_\varepsilon$ (extended to $[-\bar\varepsilon, \bar\varepsilon]$ by assuming that its value at $\varepsilon = 0$ is the given rigid deformation ϕ_0 of $\bar\Omega$) is of class C^1 in a suitable neighborhood of 0. If, in addition, $a \in C^\infty(\bar\Omega \times \mathbb{M}_n^+, \mathbb{M}_n)$ and the functions $(x, Z) \mapsto D_Z^\alpha a(x, Z)$, $|\alpha| \le m + 1$, are analytic in Z, at R_0, uniformly with respect to x, then the mapping $\varepsilon \mapsto \phi_\varepsilon$ is analytic at 0.

If Lemma 2.2 is taken into account, then from Theorem 2.3 we deduce

Theorem 2.4. *Let m be an integer ≥ 0. Assume that Ω is of class C^{m+2}, that $f \in W^{m,p}(\Omega, \mathbb{R}^n)$ with $p(m + 1) > n$, and that $a \in C^{m+2}(\bar\Omega \times \mathbb{M}_n^+, \mathbb{M}_n)$ [respectively that Ω is of class $C^{m+2,\lambda}$ with $0 < \lambda < 1$, that $f \in C^{m,\lambda}(\bar\Omega, \mathbb{R}^n)$, and that $a \in C^{m+3}(\bar\Omega \times \mathbb{M}_n^+, \mathbb{M}_n)$]. Let ϕ_0 be a rigid deformation of $\bar\Omega$ and*

---

† See Chapter V, §4. As $\mathscr{X}_{\phi_0}(f) = 0$, here $\mathscr{W}_{\phi_0}(f)$ is a maximal element of the set of linear subspaces $\mathscr{W}$ of Skew_n such that if, for $W \in \mathscr{W}$, we have $W \cdot \int_\Omega \phi \wedge f\, dx = 0$ for each ϕ belonging to some neighborhood of ϕ_0 in $W^{m+2,p}(\Omega, \mathbb{R}^n)$ [respectively $C^{m+2,\lambda}(\bar\Omega, \mathbb{R}^n)$], then $W = 0$.

‡ We recall that by a deformation of $\bar\Omega$ we mean an one-to-one C^1 function $\phi: \bar\Omega \to \mathbb{R}^n$ such that $\det D\phi > 0$. Thus a deformation of $\bar\Omega$ is an orientation-preserving C^1 diffeomorphism of $\bar\Omega$ onto a subset of $\mathbb{R}^n$. (See Chapter I, §2.)

let R_0 be the value of $D\phi_0$ (at any $x \in \bar{\Omega}$). Let Π be a linear subspace of $\mathbb{R}^n$ such that $f(x) \in \Pi$ for (almost) all $x \in \Omega$, let $(a_1, \ldots, a_k)$ be an orthonormal base of Π, let $(a_1, \ldots, a_k, \ldots, a_n)$ be an orthonormal base of $\mathbb{R}^n$, let $\mathscr{W}_\Pi$ be the linear subspace of Skew_n spanned by $\{a_i \wedge a_j : i = 1, \ldots, k, i < j \leqq n\}$, and let $\omega_1, \ldots, \omega_k$ be the eigenvalues of matrix (2.5). Suppose that (1.4), (1.5), (1.6), (1.7), (2.2), (2.3), (2.6) hold and, if $k \neq n$, matrix (2.5) is nonsingular. Then, two numbers $\bar{\varepsilon} > 0$ and $\delta > 0$ exist such that for each $\varepsilon \in [-\bar{\varepsilon}, \bar{\varepsilon}] \setminus \{0\}$ there is one and only one deformation ϕ_ε of $\bar{\Omega}$ belonging to $W^{m+2, p}(\Omega, \mathbb{R}^n)$ [respectively $C^{m+2, \lambda}(\bar{\Omega}, \mathbb{R}^n)$] and satisfying the conditions (2.7), (2.8), (2.9) and

$$R_\varepsilon \in \{(\exp W)R_0 : W \in \mathscr{W}_\Pi\},$$

where R_ε is the element of $\mathbb{O}_n^+$ such that matrix (2.11) is symmetric and positive definite. The mapping $\varepsilon \mapsto \phi_\varepsilon$ (extended to $[-\bar{\varepsilon}, \bar{\varepsilon}]$ by assuming that its value at $\varepsilon = 0$ is the given rigid deformation ϕ_0 of $\bar{\Omega}$) is of class C^1 in a suitable neighborhood of 0. If, moreover, $a \in C^\infty(\bar{\Omega} \times \mathsf{M}_n^+, \mathsf{M}_n)$ and the functions $(x, Z) \mapsto D_x^\alpha a(x, Z)$, $|\alpha| \leqq m + 1$, are analytic in Z, at R_0, uniformly with respect to x, then the mapping $\varepsilon \mapsto \phi_\varepsilon$ is analytic at 0.

We point out, in the following two corollaries, what Theorem 2.4 yields in the cases when $\dim \Pi = n$ and when $\dim \Pi = 1$.

Corollary 2.5. *Let* Ω, ϕ_0, R_0, *and* m *be as in Theorem 2.4. Assume that* a *belongs to* $C^{m+2}(\bar{\Omega} \times \mathsf{M}_n^+, \mathsf{M}_n)$ *[respectively* $C^{m+3}(\bar{\Omega} \times \mathsf{M}_n^+, \mathsf{M}_n)$*] and satisfies (1.4), (1.5), (1.6), (1.7); further, assume that* f *belongs to* $W^{m, p}(\Omega, \mathbb{R}^n)$ *with* $p(m + 1) > n$ *[respectively to* $C^{m, \lambda}(\bar{\Omega}, \mathbb{R}^n)$ *with* $0 < \lambda < 1$*] and satisfies (2.2), (2.3) and the condition*

$$\prod_{\substack{i, r = 1, \ldots, n \\ i \neq r}} (\omega_i + \omega_r) \neq 0,$$

where $\omega_1, \ldots, \omega_n$ *are the eigenvalues of the matrix*

$$\int_\Omega \phi_0 \otimes f \, dx.$$

Then, two numbers $\bar{\varepsilon} > 0$ *and* $\delta > 0$ *exist such that for each* $\varepsilon \in [-\bar{\varepsilon}, \bar{\varepsilon}] \setminus \{0\}$ *there is one and only one deformation* ϕ_ε *of* $\bar{\Omega}$ *belonging to* $W^{m+2, p}(\Omega, \mathbb{R}^n)$ *[respectively* $C^{m+2, \lambda}(\bar{\Omega}, \mathbb{R}^n)$*] and satisfying conditions (2.7), (2.8), and (2.9). The mapping* $\varepsilon \mapsto \phi_\varepsilon$ *(extended to* $[-\bar{\varepsilon}, \bar{\varepsilon}]$ *by assuming that its value at* $\varepsilon = 0$ *is the given rigid deformation* ϕ_0 *of* $\bar{\Omega}$*) is of class* C^1 *in a suitable neighborhood of* 0. *If, moreover,* $a \in C^\infty(\bar{\Omega} \times \mathsf{M}_n^+, \mathsf{M}_n)$ *and the functions* $(x, Z) \mapsto D_x^\alpha a(x, Z)$, $|\alpha| \leqq m + 1$, *are analytic in* Z, *at* R_0, *uniformly with respect to* x, *then the mapping* $\varepsilon \mapsto \phi_\varepsilon$ *is analytic at* 0.

Corollary 2.6. *Let* Ω, ϕ_0, R_0, *and* m *be as in Theorem 2.4. Assume that* a *belongs to* $C^{m+2}(\bar{\Omega} \times \mathsf{M}_n^+, \mathsf{M}_n)$ *[respectively* $C^{m+3}(\bar{\Omega} \times \mathsf{M}_n^+, \mathsf{M}_n)$*] and satis-*

fies (1.4), (1.5), (1.6), (1.7). *As to* f, *assume that* $f \in W^{m,p}(\Omega, \mathbb{R}^n)$ *with* $p(m + 1) > n$ [*respectively* $f \in C^{m,\lambda}(\bar{\Omega}, \mathbb{R}^n)$ *with* $0 < \lambda < 1$], *that*

$$f(x) \wedge e = 0 \qquad \text{for (almost) all} \quad x \in \Omega,$$

with $e \in \mathbb{R}^n$, $|e| = 1$, *that* (2.2), (2.3) *hold, and that*

$$\int_\Omega \phi_0 \cdot f \, dx \neq 0.$$

Let W_e *be the linear subspace of* Skew_n *spanned by* $(e \wedge a_j)_{j=1,\dots,n}$, *where* $(a_1, \dots, a_n)$ *is an orthonormal base of* $\mathbb{R}^n$ *with* $a_1 = e$. *Then, two numbers* $\bar{\varepsilon} > 0$ *and* $\delta > 0$ *exist such that for each* $\varepsilon \in [-\bar{\varepsilon}, \bar{\varepsilon}] \setminus \{0\}$ *there is one and only one deformation* ϕ_ε *of* $\bar{\Omega}$ *belonging to* $W^{m+2,p}(\Omega, \mathbb{R}^n)$ [*respectively* $C^{m+2,\lambda}(\bar{\Omega}, \mathbb{R}^n)$] *and satisfying the conditions* (2.7), (2.8), (2.9) *and*

$$R_\varepsilon \in \{(\exp W)R_0 : W \in \mathscr{W}_e\},$$

where R_ε *is the element of* $\mathbb{O}_n^+$ *such that matrix* (2.11) *is symmetric and positive definite. The mapping* $\varepsilon \mapsto \phi_\varepsilon$ (*extended to* $[-\bar{\varepsilon}, \bar{\varepsilon}]$ *by assuming that its value at* $\varepsilon = 0$ *is the given rigid deformation* ϕ_0 *of* $\bar{\Omega}$) *is of class* C^1 *in a suitable neighborhood of* 0. *If, moreover,* $a \in C^\infty(\bar{\Omega} \times \mathbb{M}_n^+, \mathbb{M}_n)$ *and the functions* $(x, Z) \mapsto D_x^\alpha a(x, Z)$, $|\alpha| \leq m + 1$, *are analytic in* Z, *at* R_0, *uniformly with respect to* x, *then the mapping* $\varepsilon \mapsto \phi_\varepsilon$ *is analytic at* 0.

§3. The Case When the Load Is Invariant Under Rotations

Remark 3.1. *Let* $n \geq 3$. *The load is invariant under rotations if and only if*

$$\begin{cases} b(x, y) = \varphi(x, |y|)y, & \forall(x, y) \in \Omega \times \mathbb{R}^n, \\ \pi(y) = \eta(|y|), & \forall y \in \mathbb{R}^n, \end{cases} \tag{3.1}$$

with φ *and* η *real-valued functions defined on* $\Omega \times \mathbb{R}$ *and* $\mathbb{R}$, *respectively.*

Proof. Evidently, (3.1) implies ((1.8), (1.9)). We shall prove that (3.1) is a consequence of ((1.8), (1.9)). We first observe that (1.8) implies $b(x, 0) = 0$, $\forall x \in \Omega$. Since for any pair $(y_1, y_2) \in \mathbb{R}^n \times \mathbb{R}^n$ with $|y_1| = |y_2|$ there exists $R \in \mathbb{O}_n^+$ such that $Ry_1 = y_2$, from (1.9) it follows that there is a function $\eta: \mathbb{R} \to \mathbb{R}$ such that $\pi(y) = \eta(|y|)$. We now show that condition (1.8) implies the existence of a function $\varphi: \Omega \times \mathbb{R} \to \mathbb{R}$ such that $b(x, y) = \varphi(x, |y|)y$, $\forall(x, y) \in \Omega \times \mathbb{R}^n$. To this end it suffices to prove that from (1.8) it follows that $b(x, y)$ is parallel to y, $\forall(x, y) \in \Omega \times \mathbb{R}^n$, namely that $b(x, y) = h(x, y)y$, $\forall(x, y) \in \Omega \times \mathbb{R}^n$, with h a real-valued function defined on $\Omega \times \mathbb{R}^n$. Indeed, if this is true, (1.8) immediately yields $h(x, Ry) = Rh(x, y)$, $\forall(x, y, R) \in \Omega \times \mathbb{R}^n \times \mathbb{O}_n^+$, and this implies $h(x, y) = \varphi(x, |y|)$ with φ a function of $\Omega \times \mathbb{R}$ into $\mathbb{R}$. Then let us prove that (1.8) implies that $b(x, y)$ is parallel to y, $\forall(x, y) \in \Omega \times \mathbb{R}^n$. Accordingly, suppose that, for some $(\bar{x}, \bar{y}) \in \Omega \times (\mathbb{R}^n \backslash \{0\})$, $h(\bar{x}, \bar{y})$ is not parallel to $\bar{y}$. Then there are $\lambda \in \mathbb{R}$ and $z \neq 0$ belonging to the subspace of $\mathbb{R}^n$ orthogonal to $\bar{y}$ such that $h(\bar{x}, \bar{y}) = \lambda \bar{y} + z$. As $n \geq 3$, there is an $\bar{R} \in O_n^+$ such that $\bar{R}\bar{y} = \bar{y}$ and $\bar{R}z \neq z$. Hence, from (1.8) and $b(x, y) = \lambda \bar{y} + z$ it follows that $\lambda \bar{y} + z = \bar{R}(\lambda \bar{y} + z) = \lambda \bar{y} + \bar{R}z$, which conflicts with the fact that $\bar{R}z \neq z$. Thus the proof is complete. $\square$

In this section we will study problem (1.2)′ for the case when the load is invariant under rotations and $n \geq 3$. Therefore, throughout this section, we suppose $n \geq 3$. In view of the previous remark there are two functions $\varphi: \Omega \times \mathbb{R} \to \mathbb{R}$ and $\eta: \mathbb{R} \to \mathbb{R}$ such that

$$B(\phi)(x) = \varphi(x, |\phi(x)|)\phi(x), \qquad \pi(\phi(x)) = \eta(|\phi(x)|), \qquad \forall x \in \Omega. \quad (3.2)$$

Suppose that Ω is regular (see Chapter I, §1) and the functions μ, φ, ϕ, η are smooth enough to make sense of what follows. Since, from (3.1)₂,

$$D\pi(y) = \frac{\eta'(|y|)}{|y|}y, \qquad \forall y \neq 0,$$

and in view of (1.14), (1.17), we have

$$\int_\Omega \mu B(\phi)\, dx + \int_{\partial\Omega} (\pi \circ \phi)(\text{cof } D\phi)v\, d\sigma$$

$$= \int_\Omega \left(\mu(x)\varphi(x, |\phi(x)|) + \frac{\eta'(|\phi(x)|)}{|\phi(x)|} \det D\phi(x) \right) \phi(x)\, dx \quad (3.3)$$

and

$$\int_\Omega \phi \wedge \mu B(\phi)\, dx + \int_{\partial\Omega} \phi \wedge (\pi \circ \phi)(\text{cof } D\phi)v\, d\sigma = 0, \qquad (3.4)$$

respectively, provided $\phi(x) \neq 0$, $\forall x \in \bar{\Omega}$.

Let us suppose that a rigid deformation ϕ_0 of $\bar{\Omega}$ exists such that

$$0 \notin \phi_0(\bar{\Omega}) \qquad (3.5)$$

and

$$\int_\Omega \mu B(\phi_0)\, dx + \int_{\partial\Omega} (\pi \circ \phi_0)R_0 v\, d\sigma = 0, \qquad (3.6)$$

where R_0 denotes the value of $D\phi_0$ (at any $x \in \bar{\Omega}$). Note that (3.3) gives, for every $R \in O_n^+$,

$$\int_\Omega \mu B(R\phi)\, dx + \int_{\partial\Omega} (\pi \circ (R\phi))(\text{cof } D(R\phi))v\, d\sigma$$

$$= R\left(\int_\Omega \mu B(\phi)\, dx + \int_{\partial\Omega} (\pi \circ \phi)(\text{cof } D\phi)v\, d\sigma \right);$$

hence, taking into account (3.6), we have, for every $W \in \text{Skew}_n$,

$$\int_\Omega \mu B(\exp(tW))\phi_0) \, dx + \int_{\partial\Omega} (\pi \circ (\exp(tW)\phi_0)(\text{cof } D(\exp(tW)\phi_0))v \, d\sigma = 0,$$

$$\forall t \in \mathbb{R}.$$

Combining this equality with (3.4) and recalling Remark 4.1 of Chapter V we easily realize that *any infinitesimal rigid displacement of $\mathbb{R}^n$ of the type $y \mapsto Wy$ with $W \in \text{Skew}_n$ is critical for the load at ϕ_0.* (See Chapter V, §4.)

We will use the notation $(\mathcal{K}_{\phi_0}(\varphi, \eta), \mathcal{W}_{\phi_0}(\varphi, \eta))$ for $(\mathcal{K}_{\phi_0}(F, G), \mathcal{W}_{\phi_0}(F, G))$. Note that from (3.4) it follows that

$$\mathcal{W}_{\phi_0}(\varphi, \eta) = 0$$

for any pair $(\mathcal{K}_{\phi_0}(\varphi, \eta), \mathcal{W}_{\phi_0}(\varphi, \eta))$. Therefore, the only critical infinitesimal rigid displacements of $\mathbb{R}^n$ for the load at ϕ_0 that we must consider are the constant functions of $\mathbb{R}^n$ into itself. We will say that a unit vector $e \in \mathbb{R}^n$ is a *critical axis at ϕ_0 for the pair (φ, η)* if the constant function $y \mapsto e$, of $\mathbb{R}^n$ into itself, is critical for the load at ϕ_0. (See Chapter V, §4.) In view of Remark 4.1 of Chapter V, and from (3.4), (3.5), and (3.6) it follows that *e is a critical axis at ϕ_0 for (φ, η) if and only if*

$$\lim_{\mathbb{R} \ni \lambda \to 0} \frac{1}{\lambda} \int_\Omega \left(\mu(x)\varphi(x, |\phi_0(x) + \lambda e|) + \frac{\eta'(|\phi_0(x) + \lambda e|)}{|\phi_0(x) + \lambda e|} \right)(\phi_0(x) + \lambda e) \, dx = 0.$$

Remark 3.2. *A critical axis at ϕ_0 for the pair (φ, η) exists if and only if the matrix C defined by*

$$C = \int_\Omega |\phi_0(x)|^{-3}[(\mu(x)|\phi_0(x)|^2 D_2\varphi(x, |\phi_0(x)|) + |\phi_0(x)|\eta''(|\phi_0(x)|)$$

$$- \eta'(|\phi_0(x)|))\phi_0(x) \otimes \phi_0(x) + (\mu(x)|\phi_0(x)|^3\varphi(x, |\phi_0(x)|)$$

$$+ |\phi_0(x)|^2\eta'(|\phi_0(x)|))I] \, dx,$$

is singular. Here $D_2\varphi$ denotes the partial derivative of φ with respect to the second variable.

Proof. Let e be a unit vector of $\mathbb{R}^n$ and let $c: \mathbb{R} \to \mathbb{R}^n$ be the function defined by putting

$$c(\lambda) = \int_\Omega \left(\mu(x)\varphi(x, |\phi_0(x) + \lambda e|) + \frac{\eta'(|\phi_0(x) + \lambda e|)}{|\phi_0(x) + \lambda e|} \right)(\phi_0(x) + \lambda e) \, dx.$$

Vector e is critical at ϕ_0 for (φ, η) if and only if $(dc/d\lambda)(0) = 0$. Then, in order to prove our remark it suffices to observe that

$$\frac{dc}{d\lambda}(0) = Ce. \quad \square$$

Remark 3.3. *No critical axis at ϕ_0 for the pair (φ, η) exists when*

$$\mu(x)(|\phi_0(x)|D_2\varphi(x, |\phi_0(x)|) + \varphi(x, |\phi_0(x)|)) = -\eta''(|\phi_0(x)|), \qquad \forall x \in \Omega \quad (3.7)$$

and the function $x \mapsto \mu(x)|\phi_0(x)|^2 D_2\varphi(x, |\phi_0(x)|) + |\phi_0(x)|\eta''(|\phi_0(x)|) - \eta'(|\phi_0(x)|)$ *is, almost everywhere in Ω, either strictly positive or strictly negative. (Here, again, $D_2\varphi$ denotes the partial derivative of φ with respect to the second variable.)*

Proof. Let the hypotheses on μ, φ, η, and ϕ_0 be satisfied. In view of (3.7) we have

$$C = \int_\Omega q(x)(\phi_0(x) \otimes \phi_0(x) - |\phi_0(x)|^2 I)\, dx,$$

where

$$q(x) = |\phi_0(x)|^{-3}(\mu(x)|\phi_0(x)|^2 D_2\varphi(x, |\phi_0(x)|) + |\phi_0(x)|\eta''(|\phi_0(x)|) - \eta'(|\phi_0(x)|).$$

Observe that, if $\xi \in \mathbb{R}^n$ and $|\xi| = 1$, it is easily seen that

$$(C\xi) \cdot \xi = -\int_\Omega q(x)|\phi_0(x) - (\phi_0(x) \cdot \xi)\xi|^2\, dx;$$

hence, as (by hypothesis) q is, everywhere in Ω, either strictly positive or strictly negative, matrix C is either negative or positive definite; then $\det C \neq 0$ and thus, by Remark 3.2, no critical axis at ϕ_0 for (φ, η) exists. $\square$

Example. A simple, but physically interesting, example of a load invariant under rotations corresponds to the following choice of the functions φ and η:

$$\varphi(x, t) = -\frac{1}{t}, \qquad \eta(t) = \rho(t - r), \qquad (3.8)$$

with ρ and r numbers > 0. With this choice of the pair (φ, η), problem ((1.2), (3.2)), in the case $n = 3$, is the equilibrium problem of an *elastic body when the body force is a central force with center at 0 and is proportional to the density of mass, and a surface pressure is prescribed to be proportional to the distance from a fixed sphere centered at 0.*
 Indeed, if φ and η are defined by (3.8), problem ((1.2), (3.2)) becomes

$$\begin{cases} \operatorname{div} A(D\phi) - \mu\dfrac{\phi}{|\phi|} = 0 & \text{in } \Omega, \\[2mm] A(D\phi)v + \rho(r - |\phi|)(\operatorname{cof} D\phi)v = 0 & \text{on } \partial\Omega, \end{cases}$$

which, by using the notations of §1, derives from problem

$$\begin{cases} \operatorname{div} S_\phi - \mu_\phi\dfrac{\imath_{\phi(\Omega)}}{|\imath_{\phi(\Omega)}|} = 0 & \text{in } \phi(\Omega), \\[2mm] S_\phi v_\phi + \rho(r - |\imath_{\phi(\Omega)}|)v = 0 & \text{on } \partial\Omega, \end{cases}$$

with the functions S_ϕ and μ_ϕ defined by

$$S_\phi(\phi(x)) = s(x, D\phi(x)), \qquad \mu_\phi(\phi(x)) = \frac{\mu(x)}{\det D\phi(x)},$$

where $s: \Omega \times M_n \to M_n$ is the function related to a by $a(x, Z) = s(x, Z) \operatorname{cof} Z$. From Remark 3.3 it follows that, in the case (3.8), *no critical axis at* $\iota_{\bar{\Omega}}$ *for* (φ, η) *exists provided, almost everywhere in* Ω, $\mu - \rho$ *is either strictly positive or strictly negative.*

$$* \quad * \quad * \quad * \quad *$$

Theorem 3.4. *Let m be an integer ≥ 0. Assume that Ω is of class C^{m+2}, that $a \in C^{m+2}(\bar{\Omega} \times M_n^+, M_n)$, that $\varphi \in C^{m+1}(\bar{\Omega} \times \mathbb{R})$, that $\eta \in C^{m+2}(\mathbb{R})$, and that $\mu \in W^{m,p}(\Omega)$ with $p(m + 1) > n$ [respectively Ω of class $C^{m+2,\lambda}$ with $0 < \lambda < 1$, $a \in C^{m+3}(\bar{\Omega} \times M_n^+, M_n)$, $\varphi \in C^{2+\sup(m,1)}(\bar{\Omega}, \mathbb{R})$, $\eta \in C^{3+\sup(m,1)}(\mathbb{R})$, and $\mu \in C^{m,\lambda}(\bar{\Omega})$]. Suppose that there is a rigid deformation ϕ_0 such that (3.5) and (3.6) are satisfied and let R_0 be the value of $D\phi_0$ (at any $x \in \bar{\Omega}$). Let $\mathscr{K}$ be a linear subspace of $\mathbb{R}^n$ containing some $\mathscr{K}_{\phi_0}(\varphi, \eta)$ relative to the space $W^{m+2,p}(\Omega, \mathbb{R}^n)$ [respectively $C^{m+2,\lambda}(\bar{\Omega}, \mathbb{R}^n)$].† Assume that the function a satisfies the conditions (1.4), (1.5), (1.6), (1.7) and that no critical axis at ϕ_0 for (φ, η) belongs to $\mathscr{K}$. Then, two numbers $\bar{\varepsilon} > 0$ and $\delta > 0$ exist such that for each $\varepsilon \in [-\bar{\varepsilon}, \bar{\varepsilon}] \backslash \{0\}$ there is one and only one deformation‡ ϕ_ε of $\bar{\Omega}$ belonging to $W^{m+2,p}(\Omega, \mathbb{R}^n)$ [respectively $C^{m+2,\lambda}(\bar{\Omega}, \mathbb{R}^n)$] and satisfying the following conditions:*

$$\begin{cases} \operatorname{div} A(D\phi_\varepsilon) + \varepsilon\mu B(\phi_\varepsilon) = 0 & \text{in } \Omega, \\ (-A(D\phi_\varepsilon) + \varepsilon(\eta \circ |\phi_\varepsilon|) \operatorname{cof} D\phi_\varepsilon)\nu = 0 & \text{on } \partial\Omega, \end{cases} \tag{3.9}$$

with $B(\phi_\varepsilon)$ the function defined on Ω by putting $B(\phi_\varepsilon)(x) = \varphi(x, |\phi_\varepsilon(x)|)\phi_\varepsilon(x)$, $\forall x \in \Omega$,

$$R_0^T \int_\Omega D\phi_\varepsilon \, dx \text{ is symmetric and positive definite}, \tag{3.10}$$

$$\int_\Omega (\phi_\varepsilon - \phi_0) \, dx \in \mathscr{K},$$

$$\|\phi_\varepsilon - \phi_0\|_{m+2,p} \leq \delta \qquad [\text{respectively } \|\phi_\varepsilon - \phi_0\|_{m+2,\lambda} \leq \delta]. \tag{3.11}$$

The mapping $\varepsilon \mapsto \phi_\varepsilon$ (extended to $[-\bar{\varepsilon}, \bar{\varepsilon}]$ by assuming that its value at $\varepsilon = 0$ is the given rigid deformation ϕ_0 of $\bar{\Omega}$) is of class C^1 in a suitable

---

† As $\mathscr{W}_{\phi_0}(\varphi, \eta) = 0$, here $\mathscr{K}_{\phi_0}(\varphi, \eta)$ is a maximal element of the set of linear subspaces $\mathscr{K}$ of $\mathbb{R}^n$ having the following property: "if for $c \in \mathscr{K}$ we have

$$c \cdot \left(\int_\Omega \mu(x)\varphi(x, |\phi(x)|) + \frac{\eta'(|\phi(x)|)}{|\phi(x)|} \det D\phi(x))\phi(x) \, dx \right) = 0$$

for each ϕ belonging to some neighborhood of ϕ_0 in $W^{m+2,p}(\Omega, \mathbb{R}^n)$ [respectively $C^{m+2,\lambda}(\bar{\Omega}, \mathbb{R}^n)$], then $c = 0$".

‡ See footnote ‡ on p. 138.

neighborhood of 0. *If, in addition, we suppose that* $a \in C^{\infty}(\bar{\Omega} \times M_n^+, M_n)$, *that* $\varphi \in C^{\infty}(\bar{\Omega}, \mathbb{R})$, *that* η *is analytic, that the functions* $(x, Z) \mapsto D_x^{\alpha} a(x, Z)$, $|\alpha| \leqq m + 1$, *are analytic in* Z, *at* R_0, *uniformly with respect to* x, *and that the functions* $(x, t) \mapsto D_x^{\alpha} \varphi(x, t)$, $|\alpha| \leqq m + 1$, *are analytic in* t *uniformly with respect to* x, *then the mapping* $\varepsilon \mapsto \phi_{\varepsilon}$ *is analytic at* 0.

Proof. Let us set†

$$\begin{cases} \mathscr{A}^{m+2,p} = \{\phi \in (W^{m+2,p}(\Omega, \mathbb{R}^n))^+ : |\phi(x)| \neq 0, \ \forall x \in \bar{\Omega}\}, \\ \mathscr{A}^{m+2,\lambda} = \{\phi \in (C^{m+2,\lambda}(\bar{\Omega}, \mathbb{R}^n))^+ : |\phi(x)| \neq 0, \ \forall x \in \bar{\Omega}\}. \end{cases}$$

As regards the definition of $\mathscr{A}^{m+2,p}$ we note that (by the Sobolev embedding theorem) under our hypotheses on Ω, p, and m, $W^{m+2,p}(\Omega)$ can be embedded in $C^1(\Omega)$; of course, the condition $|\phi(x)| \neq 0$, $\forall x \in \bar{\Omega}$ for $\phi \in W^{m+2,p}(\Omega, \mathbb{R}^n)$ must be understood in the sense that it is satisfied by the element of $C^1(\bar{\Omega}, \mathbb{R}^n)$ lying in (the equivalence class) $\phi(x)$.

Using Theorem 4.3 and Corollary 2.3 [respectively Theorem 4.4 and Remark 1.1] of Chapter II and Remark 6.5 of Chapter III, we easily realize that, under the hypotheses made on Ω, μ, η, φ, p, λ, and m in the first part of the statement, the (load) mapping $\phi \mapsto (\mu B(\phi), G(\phi))$, with $B(\phi)$ defined by (3.2) and $G(\phi) = ((\eta \circ |\phi|) \operatorname{cof} D\phi)|_{\partial \Omega} \nu)$, is a C^1 mapping from $\mathscr{A}^{m+2,p}$ into $W^{m,p}(\Omega, \mathbb{R}^n) \times W^{m+1-1/p,p}(\partial \Omega, \mathbb{R}^n)$ [respectively from $\mathscr{A}^{m+2,\lambda}$ into $C^{m,\lambda}(\bar{\Omega}, \mathbb{R}^n) \times C^{m+1,\lambda}(\partial \Omega, \mathbb{R}^n)$]. Furthermore, if we suppose that $\varphi \in C^{\infty}(\bar{\Omega} \times \mathbb{R})$, that η is analytic, and that the functions $(x, t) \mapsto D_x^{\alpha} \varphi(x, t)$, $|\alpha| \leqq m + 1$, are analytic in t uniformly with respect to x, then from Theorem 5.1 [respectively 5.2] of Chapter II and Remark 6.5 of Chapter III if follows that the mapping $\phi \mapsto (\mu B(\phi), G(\phi))$ from $\mathscr{A}^{m+2,p}$ into $W^{m,p}(\Omega, \mathbb{R}^n) \times W^{m+1-1/p,p}(\partial \Omega, \mathbb{R}^n)$ [respectively from $\mathscr{A}^{m+2,\lambda}$ into $C^{m,\lambda}(\bar{\Omega}, \mathbb{R}^n) \times C^{m+1,\lambda}(\partial \Omega, \mathbb{R}^n)$] is analytic at ϕ_0.

After this remark, our theorem can be deduced without difficulty from Theorem 5.1 of Chapter V, bearing in mind that $\mathscr{W}_{\phi_0}(\varphi, \eta) = 0$ for any pair $(\mathscr{K}_{\phi_0}(\varphi, \eta), \mathscr{W}_{\phi_0}(\varphi, \eta))$. $\square$

The following corollary is an obvious consequence of Theorem 3.4.

Corollary 3.5. *Let* Ω, ϕ_0, R_0, μ, m, p, *and* λ *be as in Theorem 3.4. Assume that the function* a *belongs to* $C^{m+2}(\bar{\Omega} \times M_n^+, M_n)$ [*respectively* $C^{m+3}(\bar{\Omega} \times M_n^+, M_n)$] *and satisfies* (1.4), (1.5), (1.6), (1.7); *further, assume that* (φ, η) *belongs to* $C^{m+1}(\bar{\Omega}, \mathbb{R}) \times C^{m+2}(\mathbb{R})$ [*respectively* $C^{2+\sup(m,1)}(\bar{\Omega}, \mathbb{R}) \times C^{3+\sup(m,1)}(\mathbb{R})$] *and that no critical axis at* ϕ_0 *exists for* (φ, η). *Then, numbers* $\bar{\varepsilon} > 0$ *and* $\delta > 0$ *exist such that for each* $\varepsilon \in [-\bar{\varepsilon}, \bar{\varepsilon}] \setminus \{0\}$ *there is one and only one deformation* ϕ_{ε} *of* $\bar{\Omega}$ *belonging to* $W^{m+2,p}(\Omega, \mathbb{R}^n)$ [*respec-*

---

† In accordance with the convention made in Chapter I, §1 $(W^{m+2,p}(\Omega, \mathbb{R}^n))^+ = \{\phi \in W^{m+2,p}(\Omega, \mathbb{R}^n) : \det D\phi > 0\}$ and $(C^{m+2,\lambda}(\bar{\Omega}, \mathbb{R}^n))^+ = \{\phi \in C^{m+2,\lambda}(\bar{\Omega}, \mathbb{R}^n) : \det D\phi > 0\}$.

tively $C^{m+2, \lambda}(\bar{\Omega}, \mathbb{R}^n)$] *and satisfying the conditions* (3.9), (3.10), *and* (3.11). *The mapping* $\varepsilon \mapsto \phi_\varepsilon$ (*extended to* $[-\bar{\varepsilon}, \bar{\varepsilon}]$ *assuming that its value at* $\varepsilon = 0$ *is the given rigid deformation* ϕ_0 *of* $\bar{\Omega}$) *is of class* C^1 *in a suitable neighborhood of* 0. *If, in addition, we suppose that* $a \in C^\infty(\bar{\Omega} \times \mathbb{M}_n^+, \mathbb{M}_n)$, *that* $\varphi \in C^\infty(\bar{\Omega}, \mathbb{R})$, *that* η *is analytic, that the functions* $(x, Z) \mapsto D_x^\alpha a(x, Z)$, $|\alpha| \leq m + 1$, *are analytic in* Z, *at* R_0, *uniformly with respect to* x, *and that the functions* $(x, t) \mapsto D_x^\alpha \varphi(x, t)$, $|\alpha| \leq m + 1$, *are analytic in* t *uniformly with respect to* x, *then the mapping* $\varepsilon \mapsto \phi_\varepsilon$ *is analytic at* 0.

§4. The Case of a Heavy Elastic Body Submerged in a Quiet Heavy Liquid

Let $n \geq 3$ and let e be a fixed unit vector of $\mathbb{R}^n$. Let us denote by $\mathcal{G}_e$ the group of isometries $y \mapsto c + Ry$ of $\mathbb{R}^n$ such that

$$0 \neq c, \qquad c \cdot e = 0, \qquad Re = e. \tag{4.1}$$

Remark 4.1. *The load is* $\mathcal{G}_e$-*invariant* (see §1) *if and only if*

$$b(x, y) = \varphi(x, y \cdot e)e, \qquad \pi(y) = \eta(y \cdot e), \qquad \forall(x, y) \in \Omega \times \mathbb{R}^n, \tag{4.2}$$

with φ *and* η *real-valued functions defined on* $\Omega \times \mathbb{R}$ *and* $\mathbb{R}$, *respectively.*

Proof. It is clear that (4.2) implies

$$b(x, c + Ry) = Rb(x, y), \qquad \pi(c + Ry) = \pi(y) \tag{4.3}$$

for every $(x, y) \in \Omega \times \mathbb{R}^n$ and for every $(c, R) \in \mathbb{R}^n \times \mathbb{O}_n^+$ satisfying (4.1).

Conversely, we assume that (4.3) holds for every $(x, y) \in \Omega \times \mathbb{R}^n$ and for every $(c, R) \in \mathbb{R}^n \times \mathbb{O}_n^+$ satisfying (4.1), and we shall prove that (4.2) holds with suitable functions $\varphi: \Omega \times \mathbb{R} \to \mathbb{R}$ and $\eta: \mathbb{R} \to \mathbb{R}$. If $(4.3)_2$ is true for all $(x, y) \in \Omega \times \mathbb{R}^n$ and for all $(c, R) \in \mathbb{R}^n \times \mathbb{O}_n^+$ satisfying (4.1), then $\pi(y) = \pi((y \cdot e)e)$, $\forall y \in \mathbb{R}^n$ (because we have $y = (y \cdot e)e + c(y)$ with $c(y) \in \mathbb{R}^n$ and $c(y) \cdot e = 0$), which implies $(4.2)_2$, $\forall y \in \mathbb{R}^n$, with $\eta: \mathbb{R} \to \mathbb{R}$ the function defined by $\eta(t) = \pi(te)$, $\forall t \in \mathbb{R}$.

We now prove that $(4.2)_1$ holds $\forall(x, y) \in \Omega \times \mathbb{R}^n$. We begin by showing that (as $(4.3)_1$ has been supposed to be valid for all $(x, y) \in \Omega \times \mathbb{R}^n$ and all $(c, R) \in \mathbb{R}^n \times \mathbb{O}_n^+$ satisfying (4.1)) $b(x, y)$ is parallel to e, $\forall(x, y) \in \Omega \times \mathbb{R}^n$. Accordingly, we suppose that there is an $(\bar{x}, \bar{y}) \in \Omega \times \mathbb{R}^n$ such that $b(\bar{x}, \bar{y}) = \bar{\lambda}e + \bar{c}$ with $\bar{\lambda} \in \mathbb{R}$, $0 \neq \bar{c} \in \mathbb{R}^n$, $c \cdot e = 0$ and let $\bar{R} \in \mathbb{O}_n^+$ be such that $\bar{R}e = e$, $\bar{R}\bar{c} \neq \bar{c}$. Note that $(\bar{y} - \bar{R}\bar{y}) \cdot e = \bar{y} \cdot e - (\bar{R}\bar{y}) \cdot e = \bar{y} \cdot e - \bar{y} \cdot (\bar{R}^T e) = 0$, because the equality $\bar{R}e = e$ implies $\bar{R}^T e = e$; consequently, there is a $\bar{c}_1 \in \mathbb{R}^n$ such that $\bar{c}_1 \cdot e = 0$ and $\bar{y} = \bar{R}\bar{y} + \bar{c}_1$. Then, by $(4.3)_1$, we have $b(x, \bar{y}) = \bar{R}b(x, \bar{y})$, namely $\bar{\lambda}e + \bar{c} = \bar{R}(\bar{\lambda}e + \bar{c})$; this yields $\bar{\lambda}e + \bar{c} = \bar{\lambda}e + \bar{R}\bar{c}$, which conflicts with the assumption $\bar{R}\bar{c} \neq \bar{c}$. Therefore, for each $(x, y) \in \Omega \times \mathbb{R}^n$, $b(x, y)$ is parallel to e, i.e., there is a function $h: \Omega \times \mathbb{R}^n \to \mathbb{R}$ such that $b(x, y) = h(x, y)e$, $\forall(x, y) \in \Omega \times \mathbb{R}^n$. It remains to show that there is a function $\varphi: \Omega \times \mathbb{R} \to \mathbb{R}$ such that $h(x, y) = \varphi(x, y \cdot e)$, $\forall(x, y) \in \Omega \times \mathbb{R}^n$. This is true because the fact that $(4.3)_1$ holds for all $(x, y) \in \Omega \times \mathbb{R}^n$ and for all $(c, R) \in \mathbb{R}^n \times \mathbb{O}_n^+$ satisfying (4.1) implies that $h(x, c + Ry) = h(x, y)$ for all $(x, y) \in \Omega \times \mathbb{R}^n$ and for all $(c, R) \in \mathbb{R}^n \times \mathbb{O}_n^+$ satisfying (4.1), so that $h(x, y) = h(x, (y \cdot e)e)$, $\forall(x, y) \in \Omega \times \mathbb{R}^n$. □

In this section we consider problem $(1.2)'$ with the load $\mathscr{G}_e$-invariant so that (4.2) holds, and we study the particular case in which η *is linear and* φ *is constant*, i.e., we study the problem of finding a deformation $\phi: \bar{\Omega} \to \mathbb{R}^n$ of $\bar{\Omega}$ (see Chapter I, §2) such that

$$\begin{cases} \operatorname{div} A(D\phi) + \varepsilon\mu e = 0 & \text{in } \Omega, \\ (A(D\phi) + \varepsilon\rho(\phi \cdot e) \operatorname{cof} D\phi)\nu = 0 & \text{on } \partial\Omega, \end{cases} \tag{4.4}$$

where e is a fixed element of $\mathbb{R}^n$ with $|e| = 1$ (identified with the constant function $x \mapsto e$ of Ω into $\mathbb{R}^n$), μ is a real-valued function defined on Ω (whose meaning, when $n = 3$, is the density of the elastic body in the reference configuration), ρ is a real number $\neq 0$ (which will be identified with the constant function $x \mapsto \rho$ of Ω into $\mathbb{R}$), $A(D\phi)$ is the $\mathbb{M}_n$-valued function defined on Ω by putting $A(D\phi)(x) = a(x, D\phi(x))$, $\forall x \in \Omega$, with a a given function of $\Omega \times \mathbb{M}_n^+$ into $\mathbb{M}_n$, and ε is real parameter.

Observe that, by using the notations of §1 and taking into account (2.3) and (2.6) of Chapter I, it is easy to recognize that (4.4) derives from

$$\begin{cases} \operatorname{div} S_\phi + \varepsilon\mu_\phi e = 0 & \text{in } \phi(\Omega), \\ S_\phi \nu_\phi + \varepsilon\rho(\iota_{\partial\phi(\Omega)} \cdot e)\nu_\phi = 0 & \text{on } \partial\phi(\Omega) \end{cases}$$

(where $\iota_{\partial\phi(\Omega)}$ denotes the identity of $\partial\phi(\Omega)$ into $\mathbb{R}^n$ and the subscript ϕ is used to indicate a function defined on $\phi(\Omega)$ or $\partial\phi(\Omega)$). Then, on recalling that ν_ϕ denotes the unit outward normal to $\partial\phi(\Omega)$, that μ_ϕ denotes the density of the body in the unknown configuration $\phi(\Omega)$, and that S_ϕ denotes the Cauchy stress corresponding to the unknown deformation ϕ, it is clear that (when $n = 3$, $\mu \geqq 0$, and $\rho > 0$) we are dealing with the equilibrium problem of a heavy elastic body submerged in a quiet liquid of (uniform) density ρ. Indeed, the prescribed traction on the boundary $\partial\phi(\Omega)$ is parallel to ν_ϕ and proportional to the distance from the plane containing the origin and orthogonal to e. Here εe represents the gravitational acceleration.

We shall assume that

$$\int_\Omega (\mu - \rho) \, dx = 0, \qquad \int_\Omega (\mu - \rho)\iota_\Omega \, dx \neq 0 \tag{4.5}$$

(with ι_Ω the identity of Ω into $\mathbb{R}^n$). This implies that μ is not constant. Let $R_0 \in \mathbb{O}_n^+$ be such that

$$e \wedge \int_\Omega (\mu - \rho)R_0\iota_\Omega \, dx = 0. \tag{4.6}$$

In accordance with our convention (see Chapter I, §1) $R_0 \iota_\Omega$ denotes the rigid deformation $x \mapsto R_0 x$ of Ω. Provided Ω is regular (see Chapter I, §1) and ϕ, a, μ are suitably smooth, using (1.14) with $B(\phi) = e$ and $\pi(y) = -\rho y \cdot e$, we obtain

$$\int_\Omega (\operatorname{div} A(D\phi) + \varepsilon\mu e)\, dx - \int_{\partial\Omega} (A(D\phi) + \varepsilon\rho(\phi \cdot e) \operatorname{cof} D\phi)v\, d\sigma$$

$$= \varepsilon \int_\Omega (\mu - \rho \det D\phi)e\, dx, \quad (4.7)$$

while, if (1.5) holds, (1.17) (written for $B(\phi) = e$ and $\pi(y) = -\rho y \cdot e$) gives

$$\int_\Omega \phi \wedge (\operatorname{div} A(D\phi) + \varepsilon\mu e)\, dx - \int_{\partial\Omega} \phi \wedge ((A(D\phi) + \varepsilon\rho(\phi \cdot e) \operatorname{cof} D\phi)v)\, d\sigma$$

$$= \varepsilon \int_\Omega (\mu - \rho \det D\phi)\phi \wedge e\, d\sigma. \quad (4.8)$$

Note that, if $\varepsilon \neq 0$ and (ϕ, ε) satisfies (4.4), in view of (4.7) we have $\operatorname{vol}(\phi(\Omega)) = (1/\rho)\int_\Omega \mu\, dx$, which, when $(4.5)_1$ holds, implies

$$\operatorname{vol}(\phi(\Omega)) = \operatorname{vol}(\Omega).$$

Bearing in mind Remark 1.1, Theorems 4.1, 4.2, 5.1, 5.2 and Corollary 2.3 of Chapter II, and Remark 6.5 of Chapter III we can state

Lemma 4.2. *Let m be an integer ≥ 0. Assume that Ω is of class C^{m+2}, that $a \in C^{m+2}(\bar{\Omega} \times \mathsf{M}_n^+, \mathsf{M}_n)$, and that $\mu \in W^{m,p}(\Omega)$ with $p(m + 1) > n$ [respectively that Ω is of class $C^{m+2,\lambda}$ with $0 < \lambda \leq 1$, that $a \in C^{m+3}(\bar{\Omega} \times \mathsf{M}_n^+, \mathsf{M}_n)$, and that $\mu \in C^{m,\lambda}(\bar{\Omega})$]. Then, putting*

$$\Lambda(\phi, \varepsilon) = (-(\operatorname{div} A(D\phi) + \varepsilon\mu e), (A(D\phi) + \varepsilon\rho(\phi \cdot e) \operatorname{cof} D\phi)|_{\partial\Omega}v), \quad (4.9)$$

$(\phi, \varepsilon) \mapsto \Lambda(\phi, \varepsilon)$ is a C^1 mapping from $(W^{m+2,p}(\Omega, \mathbb{R}^n))^+ \times \mathbb{R}$ into $W^{m,p}(\Omega, \mathbb{R}^n) \times W^{m+1-1/p,p}(\partial\Omega, \mathbb{R}^n)$ [respectively from $(C^{m+2,\lambda}(\bar{\Omega}, \mathbb{R}^n))^+ \times \mathbb{R}$ into $C^{m,\lambda}(\bar{\Omega}, \mathbb{R}^n) \times C^{m+1,\lambda}(\partial\Omega, \mathbb{R}^n)$]† and, for any rigid deformation ϕ_0 of $\bar{\Omega}$, having R_0 as the gradient at any $x \in \bar{\Omega}$, the partial differential $\Lambda'_\phi(\phi_0, 0)$ is the mapping E_{R_0} defined by (2.7) of Chapter V. If, in addition, we suppose that $a \in C^\infty(\bar{\Omega} \times \mathsf{M}_n^+, \mathsf{M}_n)$ and that the functions $(x, Z) \mapsto D_x^\alpha a(x, Z)$, $|\alpha| \leq m + 1$, are analytic in Z, at R_0, uniformly with respect to x, then the mapping $(\phi, \varepsilon) \mapsto \Lambda(\phi, \varepsilon)$ is analytic at $(\phi_0, 0)$.

We note that, if ϕ_0 is a rigid deformation of $\bar{\Omega}$ such that R_0 is the value of $D\phi_0$ (at any $x \in \bar{\Omega}$), then under the hypotheses (1.4), (1.5), (1.6), (1.7) on the function a, we have $\Lambda(\phi_0, 0) = (0, 0)$ and (by Theorems 7.6 and 7.7 of Chapter III) the partial differential $\Lambda'_\phi(\phi_0, 0)$ is a homeomor-

---

† See the footnote on p. 145.

phism of $\mathscr{V}_{R_0}^{m+2,p}$ onto $\mathscr{E}_{R_0}^{m,p}$ with $p(m+1) > n$ [respectively of $\mathscr{V}_{R_0}^{m+2,\lambda}$ onto $\mathscr{E}_{R_0}^{m,\lambda}$ with $0 < \lambda < 1$].

As we remarked in Chapter V, §2, in dealing with the general traction problem, we cannot directly apply the implicit function theorem to the equation $\Lambda(\phi, \varepsilon) = (0, 0)$ in order to express ϕ as a function of ε near $(\phi_0, 0)$, because the elements of $\mathscr{E}_{R_0}^{m,p}$ and $\mathscr{E}_{R_0}^{m,\lambda}$ are equilibrated with respect to ϕ_0 while the values of Λ do not have this property. Furthermore, *not even the abstract method developed in Chapter V for the general traction problem applies to the (load) mapping* $\phi \mapsto (F(\phi), G(\phi))$, when $F(\phi) = \mu e$ and $G(\phi) = -(\rho(\phi \cdot e)\,\text{cof}\,D\phi)|_{\partial\Omega}\nu$. Indeed, by using (4.7) and (4.8), it is not difficult to see that, with such a choice of (F, G),

$$\mathscr{K}_{\phi_0}(F, G) = \mathbb{R}e, \qquad \mathscr{W}_{\phi_0}(F, G) = \{q \wedge e : q \in \mathbb{R}^n\}$$

for any pair $(\mathscr{K}_{\phi_0}(F, G), \mathscr{W}_{\phi_0}(F, G))$ defined in Chapter V, §4, and, *for each* $\lambda \in \mathbb{R}\backslash\{0\}$, *the constant function* $x \mapsto \lambda e$ *of* $\mathbb{R}^n$ *into* $\mathbb{R}^n$ *is critical for the load at* ϕ_0; *therefore, the hypotheses of Lemma 4.2 (and of Theorem 5.1) of Chapter V are not satisfied when* $F(\phi) = \mu e$ *and* $G(\phi) = -(\rho(\phi \cdot e)\,\text{cof}\,D\phi)|_{\partial\Omega}\nu$.

But we will prove that if the hypotheses of the first part of Lemma 4.2 are satisfied and (1.4), (1.5), (1.6), (1.7), (4.5) hold, then a (modified) operator $(\chi, \varepsilon) \mapsto \overline{\Lambda}(\chi, \varepsilon)$ exists with the following three properties:

(i) $\overline{\Lambda}$ is defined on $V_0 \times \mathbb{R}$ with V_0 a suitable neighborhood of $R_0 \iota_{\overline{\Omega}}$ in a certain affine subspace of $W^{m+2,p}(\Omega, \mathbb{R}^n)$ [respectively $C^{m+2,\lambda}(\overline{\Omega}, \mathbb{R}^n)$];

(ii) the implicit function theorem applies to the equation $\overline{\Lambda}(\chi, \varepsilon) = (0, 0)$ at $(R_0 \iota_{\overline{\Omega}}, 0)$;

(iii) there is a rigid deformation ϕ_0 of $\overline{\Omega}$ and a mapping, say $(\chi, \varepsilon) \mapsto \hat{\phi}(\chi, \varepsilon)$, of $V_0 \times \mathbb{R}$ into $W^{m+2,p}(\Omega, \mathbb{R}^n)$ [respectively $C^{m+2,\lambda}(\overline{\Omega}, \mathbb{R}^n)$] such that $\hat{\phi}(R_0 \iota_{\overline{\Omega}}, \varepsilon) = \phi_0$, $\forall \varepsilon \in \mathbb{R}$ and that $(\chi, \varepsilon) \mapsto (\hat{\phi}(\chi, \varepsilon), \varepsilon)$ is a homeomorphism of the set of solutions (χ, ε) of $\overline{\Lambda}(\chi, \varepsilon) = (0, 0)$ belonging to $V_0 \times \mathbb{R}$ onto the set of solutions (ϕ, ε) of $\Lambda(\phi, \varepsilon) = (0, 0)$ with $\varepsilon \in \mathbb{R}$ and ϕ belonging to a suitable neighborhood of ϕ_0 in $W^{m+2,p}(\Omega, \mathbb{R}^n)$ [respectively $C^{m+2,\lambda}(\overline{\Omega}, \mathbb{R}^n)$] and satisfying the conditions

$$e \wedge \left[\int_\Omega \phi\,dx - \frac{1}{n\,\text{vol}(\Omega)} \left(\int_\Omega \text{div}(R_\phi^T \phi)\,dx \right) R_\phi \int_\Omega \iota_\Omega\,dx \right] = 0,$$

$$R_\phi R_0^T \in \{\exp(q \wedge e) : q \in \mathbb{R}^n\},$$

where R_ϕ is the element of $\mathbb{O}_n^+$ such that the matrix $R_\phi^T \int_\Omega D\phi\,dx$ is symmetric and positive definite.

We will obtain such an operator $\overline{\Lambda}$ in three steps, using Lemmas 4.4, 4.5, 4.6, 4.10, 4.13, and 4.15, besides Lemma 3.2 of Chapter V. The starting point is suggested by (4.7), (4.8), and the following remark.

Remark 4.3. *Assume that* (1.4) *holds. Then, for any* $(R, \tau, \varepsilon) \in \mathbb{O}_n^+ \times \mathbb{R} \times (\mathbb{R} \backslash \{0\})$, *a function* $\psi: \bar{\Omega} \to \mathbb{R}^n$ *is a solution of the problem*

$$\begin{cases} \operatorname{div} A(D\psi) + \varepsilon\mu R^T e = 0 & \text{in } \Omega, \\ (A(D\psi) + \rho(\varepsilon\psi \cdot R^T e + \tau)(\operatorname{cof} D\psi))v = 0 & \text{on } \partial\Omega, \end{cases} \tag{4.10}$$

if and only if, putting

$$\phi = R\psi + \frac{\tau}{\varepsilon}e, \tag{4.11}$$

ϕ *is a solution of problem* (4.4).

Proof. Observing that $\operatorname{cof}(RZ) = R \operatorname{cof} Z$, $\forall(R, Z) \in \mathbb{O}_n^+ \times \mathbb{M}_n$, that $(Ry) \cdot e = y \cdot (R^T e)$, $\forall(R, y) \in \mathbb{O}_n^+ \times \mathbb{R}^n$, and that (by (1.4)), $A(R(D\phi)) = RA(D\phi)$, $\forall R \in \mathbb{O}_n^+$, if ϕ and ψ are related by (4.11) we easily obtain

$$\begin{cases} \operatorname{div} A(D\phi) + \varepsilon\mu e = R[\operatorname{div} A(D\psi) + \varepsilon\mu R^T e], \\ A(D\phi) + \varepsilon\rho\phi \cdot e(\operatorname{cof} D\phi) = R[A(D\psi) + \rho(\varepsilon\psi \cdot R^T e + \tau)(\operatorname{cof} D\psi)]; \end{cases}$$

thus Remark 4.3 is justified. $\square$

Putting, in (1.14), $B(\phi) = R^T e$ and $\pi(y) = -(\rho y \cdot R^T e + \tau/\varepsilon)$, we have

$$\int_\Omega (\operatorname{div} A(D\psi) + \varepsilon\mu R^T e) \, dx - \int_{\partial\Omega} (A(D\psi) + \rho(\varepsilon\psi \cdot R^T e + \tau) \operatorname{cof} D\psi)v \, d\sigma$$

$$= \varepsilon\left(\int_\Omega \mu \, dx - \rho \operatorname{vol}(\psi(\Omega))\right) R^T e,$$

which, if $(4.5)_1$ holds, gives

$$\int_\Omega (\operatorname{div} A(D\psi) + \varepsilon\mu R^T e) \, dx - \int_{\partial\Omega} (A(D\psi) + \rho(\varepsilon\psi \cdot R^T e + \tau) \operatorname{cof} D\psi)v \, d\sigma$$

$$= \varepsilon\rho(\operatorname{vol}(\Omega) - \operatorname{vol}(\psi(\Omega)))R^T e.$$

Hence, if ψ is a solution of problem (4.10) and $(4.5)_1$ holds, then (provided Ω is regular, $\mu \in L^1(\Omega)$, and ψ, a are smooth enough) for any $(R, \tau, \varepsilon) \in \mathbb{O}_n^+ \times \mathbb{R} \times (\mathbb{R} \backslash \{0\})$ we have $\operatorname{vol}(\psi(\Omega)) = \operatorname{vol}(\Omega)$. Evidently, this condition is satisfied if $\psi = \gamma(\chi)$ with $\chi \in (C^1(\bar{\Omega}, \mathbb{R}^n))^+$ $(= \{\phi \in C^1(\bar{\Omega}, \mathbb{R}^n): \det D\phi > 0\})$ and with

$$\gamma(\chi) = \left(\frac{\operatorname{vol}(\Omega)}{\operatorname{vol}(\chi(\Omega))}\right)^{1/n} \chi. \tag{4.12}$$

It is not difficult to see that the real-valued mapping

$$\chi \mapsto \left(\frac{\operatorname{vol}(\Omega)}{\operatorname{vol}(\chi(\Omega))}\right)^{1/n}$$

defined on the subset $(C^1(\bar{\Omega}, \mathbb{R}^n))^+$ of $C^1(\bar{\Omega}, \mathbb{R}^n)$ is analytic and its differential at $R_0 \iota_{\bar{\Omega}}$ is the mapping

$$v \mapsto -\frac{1}{n \operatorname{vol}(\Omega)} \int_{\Omega} \operatorname{div}(R_0^T v) \, dx.$$

Then (with the aid of the Sobolev embedding theorem), we can state

Lemma 4.4. *Let Ω be regular,† let $a \in C^1(\bar{\Omega} \times \mathbb{M}_n^+, \mathbb{M}_n)$, let $\mu \in L^1(\Omega)$, and suppose that (4.5) holds. Then, for every $\chi \in (C^1(\bar{\Omega}, \mathbb{R}^n))^+$ and for every $(R, \tau, \varepsilon) \in \mathbb{O}_n^+ \times \mathbb{R} \times \mathbb{R}$ we have*

$$\int_{\Omega} (\operatorname{div} A(D\gamma(\chi)) + \varepsilon\mu Re) \, dx - \int_{\partial\Omega} (A(D(\gamma(\chi)))$$

$$+ \rho(\varepsilon\gamma(\chi)\cdot\operatorname{Re} + \tau) \operatorname{cof} D\gamma(\chi))\nu \, d\sigma = 0$$

with $\gamma(\chi)$ defined by (4.12). The mapping $\chi \mapsto \gamma(\chi)$ is analytic from the open subset $(W^{m+2,p}(\Omega, \mathbb{R}^n))^+$ of $W^{m+2,p}(\Omega, \mathbb{R}^n)$ into itself provided $p(m + 1) > n$ [respectively from the open subset $(C^{m+2,\lambda}(\bar{\Omega}, \mathbb{R}^n))^+$ of $C^{m+2,\lambda}(\bar{\Omega}, \mathbb{R}^n)$ into itself] and we have

$$\gamma'(R_0\iota_\Omega)(v) = v - \left(\frac{1}{n \operatorname{vol}(\Omega)} \int_{\Omega} \operatorname{div}(R_0^T v) \, dx\right) R_0\iota_\Omega \qquad (4.13)$$

for every $v \in W^{m+2,p}(\Omega, \mathbb{R}^n)$ [respectively $v \in C^{m+2,\lambda}(\bar{\Omega}, \mathbb{R}^n)$].

We recall (see Appendix II) that the function $W \mapsto \exp W$ maps Skew_n into $\mathbb{O}_n^+$: it is a diffeomorphism between two suitable neighborhoods of 0 in Skew_n and of I in $\mathbb{O}_n^+$. Let us consider the subgroup $\mathbb{O}_e$ of $\mathbb{O}_n^+$ defined by

$$\mathbb{O}_e = \{\exp(q \wedge e): q \in \mathbb{R}^n\}. \qquad (4.14)$$

Lemma 4.5. *Assume that Ω is of class C^1, that $a \in C^1(\bar{\Omega} \times \mathbb{M}_n^+, \mathbb{M}_n)$, that $\mu \in L^1(\Omega)$, and that (1.4), (1.5), (4.5), (4.6) hold. Then there is an open neighborhood U_0 of $R_0\iota_{\bar{\Omega}}$ in $(C^1(\bar{\Omega}, \mathbb{R}^n))^+$ and a number $\xi > 0$ such that, for each $\psi \in U_0$ there is one and only one $\hat{R}(\psi) \in \mathbb{O}_e$ with $|\hat{R}(\psi) - I| \leq \xi$ and*

$$\int_{\Omega} \psi \wedge (\operatorname{div} A(D\psi) + \varepsilon\mu\hat{R}(\psi)^T e) \, dx$$

$$- \int_{\partial\Omega} \psi \wedge ((A(D\psi) + \rho(\varepsilon\psi\cdot(\hat{R}(\psi)^T e + \tau) \operatorname{cof} D\psi)\nu) \, d\sigma = 0, \quad (4.15)$$

$\forall(\varepsilon, \tau) \in (\mathbb{R}\backslash\{0\}) \times \mathbb{R}$. *Moreover, the mapping $\psi \mapsto \hat{R}(\psi)$, from U_0 into $\mathbb{M}_n$, is analytic at $R_0\iota_{\bar{\Omega}}$.*

---

† See Chapter I, §1.

Proof. From (1.17), written for $B(\phi) = R^T e$ and $\pi(y) = -(\rho y \cdot R^T e + \tau/\varepsilon)$, it follows that if $(R, \tau, \varepsilon) \in \mathbb{O}_n^+ \times \mathbb{R} \times (\mathbb{R}\backslash\{0\})$, then

$$\int_\Omega \psi \wedge (\text{div } A(D\psi) + \varepsilon\mu R^T e) \, dx - \int_{\partial\Omega} \psi \wedge ((A(D\psi)$$

$$+ \rho(\varepsilon\psi \cdot R^T e + \tau) \text{ cof } D\psi)v) \, d\sigma = 0$$

if and only if

$$(R^T e) \wedge \int_\Omega (\mu - \rho \det D\psi)\psi \, dx = 0.$$

Consider the mapping Γ from $\{q \in \mathbb{R}^n : q \cdot e = 0\} \times (C^1(\bar{\Omega}, \mathbb{R}^n))^+$ into $\{y \in \mathbb{R} : y \cdot e = 0\}$ defined by

$$\Gamma(q, \psi) = [((\exp(q \wedge e))e) \wedge \int_\Omega (\mu - \rho \det D\psi)\psi \, dx]e.$$

Note that the following mappings are analytic: $\psi \mapsto \det D\psi$ from $(C^1(\bar{\Omega}, \mathbb{R}^n))^+$ into $C^0(\bar{\Omega})$, $W \mapsto \exp W$ from Skew_n into $\mathbb{O}_n^+$, the multiplication $(f, g) \mapsto fg$ from $L^1(\Omega) \times L^\infty(\Omega)$ into $L^1(\Omega)$, and $(\phi, \psi) \mapsto \int_\Omega \phi \wedge \psi \, dx$ from $L^1(\Omega, \mathbb{R}^n) \times L^\infty(\Omega, \mathbb{R}^n)$ into $\mathbb{M}_n$. Therefore, Γ is analytic because it is defined by composition of analytic mappings: $(q, \psi) \mapsto (\exp(q \wedge e),$ $\mu - \rho \det D\psi) \mapsto ((\exp(q \wedge e)e, (\mu - \rho \det D\psi)) \to \Gamma(q, \psi)$.

Evidently,

$$\Gamma(q, R_0\iota_{\bar{\Omega}}) = \left[(\exp(q \wedge e)e) \wedge \int_\Omega (\mu - \rho)R_0\iota_\Omega \, dx\right]e;$$

hence, in view of (4.6), we have

$$\Gamma(0, R_0\iota_\Omega) = 0.$$

Recalling that the differential at 0 of the function $W \mapsto \exp W$ from Skew_n into $\mathbb{M}_n$ is the identity function from Skew_n into $\mathbb{M}_n$ and that $(y \wedge e)e = y$, $\forall y \in \mathbb{R}^n$ with $y \cdot e = 0$, we easily obtain

$$\Gamma'_q(0, R_0\iota_{\bar{\Omega}})(y) = \left(e \cdot \int_\Omega (\mu - \rho)R_0\iota_\Omega \, dx\right)y,$$

$\forall y \in \mathbb{R}^n$ with $y \cdot e = 0$, where $\Gamma'_q(0, R_0\iota_{\bar{\Omega}})$ denotes the differential at 0 of the mapping $q \mapsto \Gamma(q, R_0\iota_{\bar{\Omega}})$. Therefore, in view of $(4.5)_2$ and (4.6), $\Gamma'_q(0, R_0\iota_{\bar{\Omega}})$ is a bijection from $\{y \in \mathbb{R}^n : y \cdot e = 0\}$ onto itself. Then the implicit function theorem applied to the equation $\Gamma(q, \psi) = 0$ ensures the existence of an open neighborhood U of $R_0\iota_{\bar{\Omega}}$ in $(C^1(\bar{\Omega}, \mathbb{R}^n))^+$ and of a number $\eta > 0$ such that for each $\psi \in U$ there is one and only one element $\hat{q}(\psi)$ of $\mathbb{R}^n$ with $\hat{q}(\psi) \cdot e = 0$ and $\Gamma(\hat{q}(\psi), \psi) = 0$; furthermore, the mapping $\psi \mapsto \hat{q}(\psi)$ from $(C^1(\bar{\Omega}, \mathbb{R}^n))^+$ into $\mathbb{R}^n$ is analytical at $R_0\iota_{\bar{\Omega}}$. Now remark that

$$\Gamma(q, \psi) = \left(e \cdot \int_{\Omega} (\mu - \rho \det D\psi)\psi \, dx \right)(\exp(q \wedge e))e$$

$$- (e \cdot (\exp(q \wedge e))e) \int_{\Omega} (\mu - \rho \det D\psi)\psi \, dx$$

and that $(\exp(q \wedge e))e \neq 0$ for every q belonging to a suitable neighborhood of 0 in $\{q \in \mathbb{R}^n : q \cdot e = 0\}$, thus for every such q we have $\Gamma(q, \psi) = 0$ if and only if

$$((\exp(q \wedge e))e) \wedge \int_{\Omega} (\mu - \rho \det D\psi)\psi \, dx = 0.$$

Remark also that, obviously, $\mathbb{O}_e = \{\exp(q \wedge e) : q \in \mathbb{R}^n, q \cdot e = 0\}$. Therefore, setting

$$\hat{R}(\psi) = \exp(\hat{q}(\psi) \wedge e),$$

$\hat{R}(\psi)$ has the properties required by the statement of the theorem with U_0 a suitable neighborhood of $R_0 \iota_{\bar{\Omega}}$ in $C^1(\bar{\Omega}, \mathbb{R}^n)$ contained in U and with ξ a suitable number > 0.   □

For any $(f, g) \in L^2(\Omega, \mathbb{R}^n) \times L^2(\partial\Omega, \mathbb{R}^n)$ and $R \in \mathbb{O}_n^+$, we denote by

$$\hat{e}_R(f, g)$$

the equilibrated part of (f, g) relative to the rigid deformations of $\bar{\Omega}$ having R as the value of their gradient (at any $x \in \bar{\Omega}$). (See Chapter V, §3.) We point out the fact that, in view of Lemma 3.2 of Chapter V, *there is a neighborhood U of $R_0 \iota_{\bar{\Omega}}$ in $(C^1(\bar{\Omega}, \mathbb{R}^n))^+$ such that if $\psi \in U$ and an element (f, g) of $L^2(\Omega, \mathbb{R}^n) \times L^2(\partial\Omega, \mathbb{R}^n)$ is equilibrated with respect to ψ, then $\hat{e}_{R_0}(f, g) = (0, 0)$ implies $(f, g) = (0, 0)$.*
With reference to Lemma 4.5, let us set

$$\mathscr{A}_{U_0} = \{\chi \in C^1(\bar{\Omega}, \mathbb{R}^n) : \gamma(\chi) \in U_0 \cap U\}$$

and, for $(\chi, \tau, \varepsilon) \in \mathscr{A}_{U_0} \times \mathbb{R} \times \mathbb{R}$,

$$\begin{cases} P(\chi, \varepsilon) = -(\operatorname{div} A(D\gamma(\chi)) + \varepsilon\mu\hat{R}(\gamma(\chi))^{\mathsf{T}}e), \\ Q(\chi, \tau, \varepsilon) = (A(D\gamma(\chi)) + \rho(\varepsilon\gamma(\chi) \cdot (\hat{R}(\gamma(\chi))^{\mathsf{T}}e) + \tau) \operatorname{cof} D\gamma(\chi))|_{\partial\Omega}\nu. \end{cases}$$

Lemma 4.6. *Let m be an integer ≥ 0. Assume that Ω is of class C^{m+2}, that $a \in C^{m+2}(\bar{\Omega} \times \mathbb{M}_n^+, \mathbb{M}_n)$, and that $\mu \in W^{m,p}(\Omega)$ with $p(m + 1) > n$ [respectively that Ω is of class $C^{m+2,\lambda}$ with $0 < \lambda \leq 1$, that $a \in C^{m+3}(\bar{\Omega} \times \mathbb{M}_n^+, \mathbb{M}_n)$, and that $\mu \in C^{m,\lambda}(\bar{\Omega})$]. Assume also that (1.4), (1.5), (1.6), (4.5), (4.6) hold. Then there is an open neighborhood V_0 of $R_0 \iota_{\bar{\Omega}}$ in $W^{m+2,p}(\Omega, \mathbb{R}^n)$*

[respectively $C^{m+2,\lambda}(\bar{\Omega}, \mathbb{R}^n)$] contained in $\mathscr{A}_{U_0}$ and a unique mapping

$$\hat{t}: V_0 \times \mathbb{R} \to \mathbb{R}$$

such that

$$\int_\Omega \operatorname{div}(R_0^T E_{R_0}^{-1}(\hat{e}_{R_0}(P(\chi, \varepsilon), Q(\chi, \hat{t}(\chi, \varepsilon), \varepsilon)))) \, dx = 0, \tag{4.16}$$

$\forall(\chi, \varepsilon) \in V_0 \times \mathbb{R}$, *where E_{R_0} is the linear homeomorphism of $\mathscr{V}_{R_0}^{m+2,p}$ onto $\mathscr{E}_{R_0}^{m,p}$ [respectively of $\mathscr{V}_{R_0}^{m+2,\lambda}$ onto $\mathscr{E}_{R_0}^{m,\lambda}$] defined by (2.7) of Chapter V; $\hat{t}$ is of class C^1 and $\hat{t}(R_0\iota_{\bar{\Omega}}, 0) = 0$. If, furthermore, we suppose that $a \in C^\infty(\bar{\Omega} \times M_n^+, M_n)$ and that the functions $(x, Z) \mapsto D_x^\alpha a(x, Z)$, $|\alpha| \leqq m+1$, are analytic in Z, at R_0, uniformly with respect to x, then V_0 can be chosen such that $\hat{t}$ is analytic.*

Proof. Recall that

$$E_{R_0}(v) = (\operatorname{div} L_{R_0}(Dv), -L_{R_0}(Dv)|_{\partial\Omega}v),$$

$$L_{R_0}(Dv)(x) = \sum_{h,k=1}^n D_{Z_{hk}} a(x, R) D_k v_h(x), \qquad \forall x \in \bar{\Omega}.$$

Let $\chi \in \mathscr{A}_{U_0}$ and let $\tau \in \mathbb{R}$. As $\hat{e}_{R_0}$ and E_{R_0} are linear, the condition

$$\int_\Omega \operatorname{div}(R_0^T E_{R_0}^{-1}(\hat{e}_{R_0}(P(\chi, \varepsilon), Q(\chi, \tau, \varepsilon)))) \, dx = 0$$

takes the form

$$\tau\alpha(\chi) = \beta(\chi, \varepsilon),$$

with

$$\begin{cases} \alpha(\chi) = \int_\Omega \operatorname{div}(R_0^T E_{R_0}^{-1}(\hat{e}_{R_0}(0, \rho \operatorname{cof} D\gamma(\chi)|_{\partial\Omega}v))) \, dx, \\[2mm] \beta(\chi, \varepsilon) = \int_\Omega \operatorname{div}(R_0^T E_{R_0}^{-1}(\hat{e}_{R_0}(\operatorname{div} A(D\gamma(\chi)) + \varepsilon\mu\hat{R}(\gamma(\chi))^T e, \\[2mm] \qquad -(A(D\gamma(\chi)) + \varepsilon\rho\gamma(\chi) \cdot (\hat{R}(\gamma(\chi))^T e) \operatorname{cof} D\gamma(\chi)|_{\partial\Omega}v))) \, dx. \end{cases} \tag{4.17}$$

Since $\gamma(R_0\iota_{\bar{\Omega}}) = R_0\iota_{\bar{\Omega}}$, we have

$$\alpha(R_0\iota_{\bar{\Omega}}) = \int_\Omega \operatorname{div}(R_0^T E_{R_0}^{-1}(\hat{e}_{R_0}(0, \rho R_0 v))) \, dx$$

and thus, as the pair $(0, \rho R_0 v)$ is equilibrated with respect to $R_0\iota_{\bar{\Omega}}$, we have

$$\alpha(R_0\iota_{\bar{\Omega}}) = \int_\Omega \operatorname{div}(R_0^T E_{R_0}^{-1}(0, \rho R_0 v)) \, dx. \tag{4.18}$$

Then, observing that

$$\text{div } L_{R_0}(DE_{R_0}^{-1}(0, \rho R_0 v)) \, dx = 0, \qquad L_{R_0}(DE_{R_0}^{-1}(0, \rho R_0 v)|_{\partial\Omega})v = \rho R_0 v,$$

we obtain (by use of the divergence theorem)

$$
\begin{aligned}
\rho\alpha(R_0 i_{\bar\Omega}) &= \int_{\partial\Omega} \rho(R_0^\mathsf{T} E_{R_0}^{-1}(0, \rho R_0 v)) \cdot v \, d\sigma \\
&= \int_{\partial\Omega} E_{R_0}^{-1}(0, \rho R_0 v) \cdot \rho R_0 v \, d\sigma \\
&= \int_{\partial\Omega} E_{R_0}^{-1}(0, \rho R_0 v) \cdot ((L_{R_0}(DE_{R_0}^{-1}(0, \rho R_0 v)))v) \, d\sigma \\
&= \int_{\Omega} E_{R_0}^{-1}(0, \rho R_0 v) \cdot \text{div } L_{R_0}(DE_{R_0}^{-1}(0, \rho R_0 v)) \, dx \\
&\quad + \int_{\partial\Omega} E_{R_0}^{-1}(0, \rho R_0 v) \cdot ((L_{R_0}(DE_{R_0}^{-1}(0, \rho R_0 v)))v) \, d\sigma \\
&= \int_{\Omega} DE_{R_0}^{-1}(0, \rho R_0 v) \cdot L_{R_0}(DE_{R_0}^{-1}(0, \rho R_0 v)) \, dx,
\end{aligned}
$$

and thus

$$\alpha(R_0 i_{\bar\Omega}) = \frac{1}{\rho}\left(\int_{\Omega} DE_{R_0}^{-1}(0, \rho R_0 v) \cdot L_{R_0}(DE_{R_0}^{-1}(0, \rho R_0 v)) \, dx \right).$$

Therefore

$$\alpha(R_0 i_{\bar\Omega}) \neq 0$$

because $E_{R_0}^{-1}(0, \rho R_0 v) \neq 0$ and because, under the hypotheses made on the function a, we have, for every $v \in W^{1,2}(\Omega, \mathbb{R}^n)$ such that $\int_\Omega v \, dx = 0$ and $\int_\Omega Dv \, dx \in \text{Sym}_n$,

$$\int_\Omega (Dv) \cdot L_{R_0}(Dv) \, dx \geqq c_{R_0} \|v\|_{1,2}^2,$$

with c_{R_0} a number > 0 independent of v (see Chapter III, Corollary 3.4). On the other hand, using Remark 3.1 of Chapter V we see that the mapping $\chi \mapsto \alpha(\chi)$ is continuous from $\mathscr{A}_{U_0}$ into $\mathbb{R}$. Consequently, there is a neighborhood V_1 of $R_0 i_{\bar\Omega}$ in $C^1(\bar\Omega, \mathbb{R}^n)$ such that $V_1 \subseteq \mathscr{A}_{U_0}$ and $\alpha(\chi) \neq 0$, $\forall \chi \in V_1$. Hence if $V_0 = V_1 \cap W^{m+2,p}(\Omega, \mathbb{R}^n)$ [respectively $V_0 = V_1 \cap C^{m+2,\lambda}(\bar\Omega, \mathbb{R}^n)$] then for each $(\chi, \varepsilon) \in V_0 \times \mathbb{R}$ there is one and only one real number $\hat{t}(\chi, \varepsilon)$ satisfying (4.16): this number is given by

$$\hat{t}(\chi, \varepsilon) = \frac{\beta(\chi, \varepsilon)}{\alpha(\chi)}. \tag{4.19}$$

In view of (1.4) and (1.6), we easily see that $\hat{t}(R_0 i_{\bar\Omega}, 0) = 0$. The remaining part of the statement is a consequence of Theorems 4.1, 4.2, 5.1, and 5.2 of Chapter II, combined with Remark 6.5 of Chapter III, Remark 3.1 of Chapter V, and Lemmas 4.4, 4.5 of this section. □

Note that, when the hypotheses of Lemma 4.6 are satisfied, from Remark 3.2 of Chapter III combined with inequality (3.15) of Chapter III, it follows that, for any $x \in \Omega$, the mapping

$$S \mapsto \left((SR_0) \cdot \sum_{k=1}^{n} (R_0)_{sk} D_{Z_{hk}} a(x, R_0) \right)_{h,s=1,\ldots,n}$$

is a (linear) bijection of Sym_n onto itself. Here $(R_0)_{sk}$ denotes the (s, k)th coordinate of R_0. In this case, when the functions $x \mapsto D_{Z_{hk}} a(x, R_0)$ $(h, k = 1, \ldots, n)$ are constant, we will denote by H_0 the element of Sym_n such that, for every $h, s = 1, \ldots, n$,

$$(H_0 R_0) \cdot \sum_{k=1}^{n} (R_0)_{sk} D_{Z_{hk}} a(x, R_0) = I_{hs}. \qquad (4.20)$$

Remark 4.7. *Let the assumptions of Lemma 4.6 be satisfied. If the functions* $x \mapsto D_{Z_{hk}} a(x, R_0)$ $(h, k = 1, \ldots, n)$ *are constant, then*

$$\int_{\Omega} \operatorname{div}(R_0^T E_{R_0}^{-1}(f, g)) \, dx$$

$$= -H_0 \cdot \left(\int_{\Omega} (R_0 \imath_\Omega) \otimes f \, dx + \int_{\partial \Omega} (R_0 \imath_{\partial \Omega}) \otimes g \, d\sigma \right). \qquad (4.21)$$

Proof. Let the functions $x \mapsto D_{Z_{hk}} a(x, R_0)$ $(h, k = 1, \ldots, n)$ be constant. Observe that, by (4.20),

$$\int_{\Omega} \operatorname{div}(R_0^T E_{R_0}^{-1}(f, g)) \, dx = \int_{\Omega} DE_{R_0}^{-1}(f, g)(x) \cdot R_0 \, dx$$

$$= (H_0 R_0) \cdot \int_{\Omega} L_{R_0}(D(E_{R_0}^{-1}(f, g))) \, dx.$$

Hence, setting, for any $x \in \Omega$,

$$v_0(x) = (H_0 R_0)x,$$

we have

$$\int_{\Omega} \operatorname{div}(R_0^T E_{R_0}^{-1}(f, g)) \, dx = \int_{\Omega} Dv_0 \cdot L_{R_0}(D(E_{R_0}^{-1}(f, g))) \, dx.$$

Since $\operatorname{div} L_{R_0}(D(E_{R_0}^{-1}(f, g))) = f$ and $-L_{R_0}(D(E_{R_0}^{-1}(f, g)))|_{\partial\Omega} v = g$, using the divergence theorem we easily obtain

$$-\int_{\Omega} Dv_0 \cdot L_{R_0}(D(R_0^T E_{R_0}^{-1}(f, g))) \, dx = \int_{\Omega} v_0 \cdot f \, dx + \int_{\partial \Omega} v_0 \cdot g \, d\sigma.$$

Then

$$-\int_\Omega \operatorname{div}(R_0^T E_{R_0}^{-1}(f, g))\, dx = \int_\Omega v_0 \cdot f\, dx + \int_{\partial\Omega} v_0 \cdot g\, d\sigma$$

$$= \int_\Omega ((H_0 R_0)\imath_\Omega) \cdot f\, dx + \int_{\partial\Omega} ((H_0 R_0)\imath_{\partial\Omega}) \cdot g\, d\sigma$$

$$= H_0 \cdot \left(\int_\Omega f \otimes (R_0 \imath_\Omega)\, dx + \int_{\partial\Omega} g \otimes (R_0 \imath_{\partial\Omega})\, d\sigma \right)$$

$$= H_0 \cdot \left(\int_\Omega (R_0 \imath_\Omega) \otimes f\, dx + \int_{\partial\Omega} (R_0 \imath_{\partial\Omega}) \otimes g\, d\sigma \right). \qquad \square$$

Remark 4.8. *Let the assumptions of Lemma 4.6 be satisfied. If the functions $x \mapsto D_{z_{hk}} a(x, R_0)$ $(h, k = 1, \dots, n)$ are constant, then the mapping $\hat{t}: V_0 \to \mathbb{R}$ is defined by $\hat{t}(\chi, \varepsilon) = \beta(\chi, \varepsilon)/\alpha(\chi)$, with*

$$\alpha(\chi) = -(H_0 R_0) \cdot \int_\Omega \rho \operatorname{cof} D\gamma(\chi)\, dx$$

$$+ (H_0 R_0) \cdot \left[\int_\Omega \hat{r}_{R_0}(0, \rho \operatorname{cof} D\gamma(\chi)|_{\partial\Omega} v) \otimes \imath_\Omega\, dx \right.$$

$$\left. + \int_{\partial\Omega} \hat{r}_{R_0}(0, \rho \operatorname{cof} D\gamma(\chi)|_{\partial\Omega} v) \otimes \imath_{\partial\Omega}\, d\sigma \right]$$

and

$$\beta(\chi, \varepsilon) = (H_0 R_0) \cdot \left[\int_\Omega A(D\gamma(\chi))\, dx - \varepsilon(\hat{R}(\gamma(\chi))^T e) \otimes \int_\Omega (\mu - \rho \det D\gamma(\chi))\imath_\Omega\, dx \right.$$

$$\left. - \varepsilon\rho \int_\Omega \gamma(\chi) \cdot (\hat{R}(\gamma(\chi))^T e) \operatorname{cof} D\gamma(\chi)\, dx \right]$$

$$+ (H_0 R_0) \cdot \left[\int_\Omega \hat{r}_{R_0}(P(\chi, \varepsilon), Q(\chi, 0, \varepsilon)) \otimes \imath_\Omega\, dx \right.$$

$$\left. + \int_{\partial\Omega} \hat{r}_{R_0}(P(\chi, \varepsilon), Q(\chi, 0, \varepsilon)) \otimes \imath_{\partial\Omega}\, d\sigma \right],$$

where $\hat{r}_{R_0}$ is the (continuous, linear) mapping of $L^2(\Omega, \mathbb{R}^n) \times L^2(\partial\Omega, \mathbb{R}^n)$ into $T_{\imath_{\bar\Omega}}(\mathcal{R}_{\bar\Omega})$ defined by the condition that $\forall (f, g) \in L^2(\Omega, \mathbb{R}^n) \times L^2(\partial\Omega, \mathbb{R}^n)$ the pair $(f - \hat{r}_{R_0}(f, g), g - \hat{r}_{R_0}(f, g)|_{\partial\Omega})$ is equilibrated with respect to $R_0 \imath_{\bar\Omega}$ (see Chapter V, §3).

Proof. Let the functions $x \mapsto D_{z_{hk}} a(x, R_0)$ $(h, k = 1, \dots, n)$ be constant. Taking into account (4.17) and (4.21) we have

$$\alpha(\chi) = -(H_0 R_0) \cdot \int_{\partial\Omega} \rho(\operatorname{cof} D\gamma(\chi)|_{\partial\Omega} v) \otimes \imath_{\partial\Omega}\, d\sigma$$

$$+ (H_0 R_0) \cdot \left[\int_\Omega \hat{r}_{R_0}(0, \rho(\operatorname{cof} D\gamma(\chi))|_{\partial\Omega} v) \otimes \imath_\Omega\, dx \right.$$

$$\left. + \int_{\partial\Omega} \hat{r}_{R_0}(0, \rho(\operatorname{cof} D\gamma(\chi))|_{\partial\Omega} v) \otimes \imath_{\partial\Omega}\, d\sigma \right],$$

and

$$\beta(\chi, \varepsilon) = -(H_0 R_0) \cdot \left[\int_\Omega (\text{div } A(D\gamma(\chi) + \varepsilon\mu(\hat{R}(\gamma(\chi))^T e)) \otimes \iota_\Omega \, dx - \int_{\partial\Omega} (A(D\gamma(\chi)) \right.$$

$$+ \varepsilon\rho\gamma(\chi) \cdot (\hat{R}(\gamma(\chi))^T e) \text{ cof } D\gamma(\chi))|_{\partial\Omega} v) \otimes \iota_{\partial\Omega} \, d\sigma \Big]$$

$$+ (H_0 R_0) \cdot \left[\int_\Omega \hat{r}_{R_0}(P(\chi, \varepsilon), Q(\chi, 0, \varepsilon)) \otimes \iota_\Omega \, dx \right.$$

$$+ \int_{\partial\Omega} \hat{r}_{R_0}(P(\chi, \varepsilon), Q(\chi, 0, \varepsilon))|_{\partial\Omega} \otimes \iota_{\partial\Omega} \, d\sigma \Big].$$

On the other hand, using (1.11) and the divergence theorem, we easily obtain

$$\int_{\partial\Omega} ((\text{cof } D\gamma(\chi))v) \otimes \iota_{\partial\Omega} \, d\sigma = \int_\Omega \text{cof } D\gamma(\chi) \, dx,$$

$$-\int_\Omega (\text{div } A(D\gamma(\chi)) \otimes \iota_\Omega \, dx + \int_{\partial\Omega} ((A(D\gamma(\chi))v) \otimes \iota_{\partial\Omega} \, d\sigma = \int_\Omega A(D\gamma(\chi)) \, dx,$$

and

$$\int_{\partial\Omega} (\gamma(\chi) \cdot (\hat{R}(\gamma(\chi))^T e)(\text{cof } D\gamma(\chi))v) \otimes \iota_{\partial\Omega} \, d\sigma$$

$$= \int_\Omega (\gamma(\chi) \cdot (\hat{R}(\gamma(\chi))^T e) \text{ cof } D\gamma(\chi) \, dx + \int_\Omega \det D\gamma(\chi)(\hat{R}(\gamma(\chi))^T e) \otimes \iota_\Omega \, dx.$$

Thus, to conclude the proof it suffices to recall that $\tau\alpha(\chi) = \beta(\chi, \varepsilon)$, as has been shown in proving Lemma 4.6.   □

Remark 4.9. *If*

$$\begin{cases} \text{div } A(D\gamma(\chi)) + \varepsilon\mu R^T e = 0 & \text{in } \Omega, \\ (A(D\gamma(\chi)) + \rho(\varepsilon\gamma(\chi) \cdot R^T e + \tau) \text{ cof } D\gamma(\chi))v = 0 & \text{on } \partial\Omega \end{cases} \quad (4.22)$$

for $(\chi, R, \varepsilon, \tau) \in \mathscr{A}_{U_0} \times \mathbb{O}_e \times \mathbb{R} \times \mathbb{R}$ *with* $|R - I| \leq \xi$, *where* ξ *is a number* > 0 *as appears in the statement of Lemma 4.5, then* $R = \hat{R}(\gamma(\chi))$ *and* $\tau = \hat{t}(\chi, \varepsilon)$. *Consequently, if* $(\chi, R, \varepsilon, \tau) \in \mathscr{A}_{U_0} \times \mathbb{O}_e \times \mathbb{R} \times \mathbb{R}$ *and* $|R - I| \leq \xi$, *then* (4.22) *holds if and only if* $R = \hat{R}(\gamma(\chi))$, $\tau = \hat{t}(\chi, \varepsilon)$, *and* $\bar{\Lambda}(\chi, \varepsilon) = (0, 0)$, *where*

$$\bar{\Lambda}(\chi, \varepsilon) = \hat{e}_{R_0}(P(\chi, \varepsilon), Q(\chi, \hat{t}(\chi, \varepsilon), \varepsilon)). \quad (4.23)$$

Proof. Let $(\chi, R, \tau, \varepsilon) \in \mathscr{A}_{U_0} \times \mathbb{O}_e \times \mathbb{R} \times \mathbb{R}$ with $|R - I| \leq \delta$ and let (4.22) be satisfied. Then, as $\gamma(\chi) \in U_0$, from Lemma 4.5 it follows that $R = \hat{R}(\gamma(\chi))$. Moreover, in view of (4.22) we have $E_{R_0}^{-1}(\hat{e}_{R_0}(P(\chi, \varepsilon), Q(\chi, \tau, \varepsilon))) =$

0, whence

$$\int_\Omega \operatorname{div} E_{R_0}^{-1}(\hat{e}_{R_0}(P(\chi, \varepsilon), Q(\chi, \tau, \varepsilon))) \, dx = 0,$$

which implies $\tau\alpha(\chi) = \beta(\chi, \varepsilon)$, namely (see (4.19)), $\tau = \hat{t}(\chi, \varepsilon)$. $\square$

Lemma 4.10. *Let the assumptions of the first part of Lemma 4.6 be satisfied. Then the mapping $(\chi, \varepsilon) \mapsto \bar{\Lambda}(\chi, \varepsilon)$, defined by (4.23) on the subset $V_0 \times \mathbb{R}$ of $W^{m+2,p}(\Omega, \mathbb{R}^n) \times \mathbb{R}$ [respectively $C^{m+2,\lambda}(\bar{\Omega}, \mathbb{R}^n) \times \mathbb{R}$], takes its values in the subspace*

$$\left\{ (f, g) \in \mathscr{E}_{R_0}^{m,p} \colon \int_\Omega \operatorname{div}(R_0^{\mathrm{T}} E_{R_0}^{-1}(f, g)) \, dx = 0 \right\}$$

$$\left[\text{respectively } \left\{ (f, g) \in \mathscr{E}_{R_0}^{m,\lambda} \colon \int_\Omega \operatorname{div}(R_0^{\mathrm{T}} E_{R_0}^{-1}(f, g)) \, dx = 0 \right\} \right] \tag{4.24}$$

of $W^{m,p}(\Omega, \mathbb{R}^n) \times W^{m+1-1/p,p}(\partial\Omega, \mathbb{R}^n)$ [respectively $C^{m,\lambda}(\bar{\Omega}, \mathbb{R}^n) \times C^{m+1,\lambda}(\partial\Omega, \mathbb{R}^n)$], where E_{R_0} is the mapping defined by (2.7) of Chapter V. Moreover V_0 can be chosen such that $\bar{\Lambda}$ is of class C^1; the differential at $R_0 I_{\bar{\Omega}}$ of the mapping $\chi \mapsto \bar{\Lambda}(\chi, 0)$ coincides with E_{R_0} on

$$\left\{ v \in \mathscr{V}_{R_0}^{m,p} \colon \int_\Omega \operatorname{div}(R_0^{\mathrm{T}} v) \, dx = 0 \right\}$$

$$\left[\text{respectively } \left\{ v \in \mathscr{V}_{R_0}^{m,\lambda} \colon \int_\Omega \operatorname{div}(R_0^{\mathrm{T}} v) \, dx = 0 \right\} \right]. \tag{4.25}$$

If, furthermore, $a \in C^\infty(\bar{\Omega} \times \mathbb{M}_n^+, \mathbb{M}_n)$ and the functions $(x, Z) \mapsto D_x^\alpha a(x, Z)$, $|\alpha| \leq m + 1$, are analytic in Z, at R_0, uniformly with respect to x, then V_0 can be chosen such that the mapping $(\chi, \varepsilon) \mapsto \bar{\Lambda}(\chi, \varepsilon)$ is analytic.

Proof. From Remark 3.1 of Chapter V it follows that the linear mapping $\hat{e}_{R_0}$ of $L^2(\Omega, \mathbb{R}^n) \times L^2(\partial\Omega, \mathbb{R}^n)$ into itself induces a continuous mapping from $W^{m,p}(\Omega, \mathbb{R}^n) \times W^{m+1-1/p,p}(\partial\Omega, \mathbb{R}^n)$ into itself [respectively from $C^{m,\lambda}(\bar{\Omega}, \mathbb{R}^n) \times C^{m+1,\lambda}(\partial\Omega, \mathbb{R}^n)$ into itself]. Then, in view of Lemmas 4.2, 4.4, and 4.6, $\bar{\Lambda}$ maps $V_0 \times \mathbb{R}$ into (4.24) and V_0 can be chosen such that $\bar{\Lambda}$ is of class C^1; moreover, V_0 can be chosen such that $\bar{\Lambda}$ is analytic provided the function a belongs to $C^\infty(\bar{\Omega} \times \mathbb{M}_n^+, \mathbb{M}_n)$ and the functions $(x, Z) \mapsto D_x^\alpha a(x, Z)$, $|\alpha| \leq m + 1$, are analytic in Z, at R_0, uniformly with respect to x.

The fact that the differential at $R_0 I_{\bar{\Omega}}$ of the mapping $\chi \mapsto \bar{\Lambda}(\chi, 0)$ coincides with E_{R_0} on the subspace (4.25) of $W^{m+2,p}(\Omega, \mathbb{R}^n)$ [respectively $C^{m+2,\lambda}(\bar{\Omega}, \mathbb{R}^n)$] is a consequence of Lemma 4.2 and of the following four remarks.

(i) $\gamma'(R_0 I_{\bar{\Omega}})$ induces the identity on (4.25). (See Lemma 4.4.)
(ii) $\hat{t}(R_0 I_{\bar{\Omega}}, 0) = 0$. (See Lemma 4.6.)

(iii) $\hat{e}_{R_0}$ is a continuous, linear mapping from

$$W^{m,p}(\Omega, \mathbb{R}^n) \times W^{m+1-1/p,p}(\partial\Omega, \mathbb{R}^n)$$

into itself [respectively from $C^{m,\lambda}(\bar{\Omega}, \mathbb{R}^n) \times C^{m+1,\lambda}(\partial\Omega, \mathbb{R}^n)$ into itself].
(iv) The differential at $R_0 \iota_{\bar{\Omega}}$ of the mapping $\chi \mapsto \hat{t}(\chi, 0)$ vanishes on (4.25).

To be convinced of the last assertion it suffices to bear in mind (4.17), (4.18), (4.19) and observe that $\beta(R_0 \iota_{\bar{\Omega}}, 0) = 0$ in view of Remark 3.1 of Chapter III, and that the differential at $R_0 \iota_{\bar{\Omega}}$ of the mapping $\chi \mapsto \beta(\chi, 0)$ vanishes on (4.25). $\square$

Lemma 4.11. *Let the assumptions of the first part of Lemma 4.6 be satisfied and let γ, $\hat{R}$, and $\hat{t}$ be as in the statements of Lemmas 4.4, 4.5, and 4.6. Then two numbers $\varepsilon_0 > 0$ and $\zeta > 0$ exist such that for each $\varepsilon \in [-\varepsilon_0, \varepsilon_0]$ there is one and only one χ_ε belonging to $R_0 \iota_{\bar{\Omega}} + \mathscr{V}_{R_0}^{m+2,p}$ [respectively $R_0 \iota_{\bar{\Omega}} + \mathscr{V}_{R_0}^{m+2,\lambda}$] and satisfying the conditions*

$$\begin{cases} \operatorname{div} A(D\gamma(\chi_\varepsilon)) + \varepsilon\mu\hat{R}(\gamma(\chi_\varepsilon))^{\mathrm{T}}e = 0 & \text{in } \Omega, \\ (A(D\gamma(\chi_\varepsilon)) + \rho(\varepsilon\gamma(\chi_\varepsilon) \cdot \hat{R}(\gamma(\chi_\varepsilon))^{\mathrm{T}}e + \hat{t}(\chi_\varepsilon, \varepsilon)) \operatorname{cof} D\gamma(\chi_\varepsilon))\nu = 0 & \text{on } \partial\Omega, \end{cases}$$
$$(4.26)$$

$$\int_\Omega \operatorname{div}(R_0^{\mathrm{T}}\chi_\varepsilon - \iota_{\bar{\Omega}})\, dx = 0, \tag{4.27}$$

and

$$\|\chi_\varepsilon - R_0\iota_{\bar{\Omega}}\|_{m+2,p} \leqq \zeta \qquad [\text{respectively } \|\chi_\varepsilon - R_0\iota_{\bar{\Omega}}\| \leqq \zeta]. \tag{4.28}$$

The mapping $\varepsilon \mapsto \chi_\varepsilon$ is of class C^1 in a suitable neighborhood of 0 and we have $\chi_0 = R_0\iota_{\bar{\Omega}}$. If, in addition, $a \in C^\infty(\bar{\Omega} \times \mathbb{M}_n^+, \mathbb{M}_n)$ and the functions $(x, Z) \mapsto D_x^\alpha a(x, Z)$, $|\alpha| \leqq m + 1$, are analytic in Z, at R_0, uniformly with respect to x, then the mapping $\varepsilon \mapsto \chi_\varepsilon$ is analytic at 0.

Proof. We deal with the case of Sobolev spaces; in the case of Schauder spaces we can proceed analogously. Consider the open subset

$$\left\{v \in \mathscr{V}_{R_0}^{m+2,p} \cap (V_0 - R_0\iota_{\bar{\Omega}}): \int_\Omega \operatorname{div}(R_0^{\mathrm{T}}v)\, dx = 0\right\} \times \mathbb{R} \tag{4.29}$$

of the Banach space $\{v \in \mathscr{V}_{R_0}^{m+2,p}: \int_\Omega \operatorname{div}(R_0^{\mathrm{T}}v)\, dx = 0\} \times \mathbb{R}$. Let us set, for any (v, ε) belonging to (4.29),

$$\Phi(v, \varepsilon) = \bar{\Lambda}(R_0\iota_{\bar{\Omega}} + v, \varepsilon).$$

In view of Lemma 4.10, $\Phi(v, \varepsilon)$ belongs to the space (4.24); moreover, if V_0 is suitably chosen, then $(v, \varepsilon) \mapsto \Phi(v, \varepsilon)$, regarded as a mapping from (4.29) into the closed subspace (4.24) of $W^{m,p}(\Omega, \mathbb{R}^n) \times W^{m+1-1/p,p}(\partial\Omega, \mathbb{R}^n)$, is of class C^1 and the differential at 0 of the mapping $v \mapsto \Phi(v, 0)$ is E_{R_0}, which is (by Theorem 7.6 of Chapter III) a homeomorphism of $\mathscr{V}_{R_0}^{m+2,p}$ onto $\mathscr{E}_{R_0}^{m,p}$.

Note that $\Phi(0, 0) = (0, 0)$. Then the implicit function theorem applied to the equation $\Phi(v, \varepsilon) = (0, 0)$ assures us that there are two numbers $\varepsilon_0 > 0$ and $\zeta > 0$ such that for each $\varepsilon \in [-\varepsilon_0, \varepsilon_0]$ there is one and only one $\chi_\varepsilon \in R_0 l_{\bar\Omega} + \mathscr{V}_{R_0}^{m+2,p}$ satisfying the conditions $\bar\Lambda(\chi_\varepsilon, \varepsilon) = (0, 0)$, (4.27), and (4.28). To conclude the first part of the proof it suffices to observe that, in view of Lemma 3.2 of Chapter V, the numbers ε_0 and ζ can be selected such that $P(\chi_\varepsilon, \varepsilon) = 0$ and $Q(\chi_\varepsilon, \hat t(\chi_\varepsilon, \varepsilon), \varepsilon) = 0$, i.e., such that (4.26) holds.

With regard to the second part of the statement we recall that (by Lemma 4.10) if $a \in C^\infty(\bar\Omega \times \mathsf{M}_n^+, \mathsf{M}_n)$ and the functions $(x, Z) \mapsto D_x^\alpha a(x, Z)$, $|\alpha| \leq m + 1$, are analytic in Z, at R_0, uniformly with respect to x, then $(v, \varepsilon) \mapsto \Phi(v, \varepsilon)$ is analytic at $(0, 0)$ and thus the (implicit) mapping $\varepsilon \mapsto \chi_\varepsilon$ is analytic at 0. $\square$

Combining Lemma 4.11 with Remark 4.9 we immediately deduce

Corollary 4.12. *Under the assumptions of Lemma 4.6, numbers* $\varepsilon_0 > 0$, $\xi > 0$, *and* $\zeta > 0$ *exist such that for* $|\varepsilon| \leq \varepsilon_0$ *there is one and only one triplet* (χ, R, τ) *belonging to* $(R_0 l_{\bar\Omega} + \mathscr{V}_{R_0}^{m+2,p}) \times \mathbb{O}_e \times \mathbb{R}$ *[respectively* $(R_0 l_{\bar\Omega} + \mathscr{V}_{R_0}^{m+2,\lambda}) \times \mathbb{O}_e \times \mathbb{R}$] *and satisfying the conditions* (4.22), $\int_\Omega \mathrm{div}(R_0^T\chi - l_{\bar\Omega})\, dx = 0$, $|R - I| \leq \xi$, *and* $\|\chi - R_0 l_{\bar\Omega}\|_{m+2,p} \leq \zeta$ *[respectively* $\|\chi - R_0 l_{\bar\Omega}\|_{m+2,\lambda} \leq \zeta$].

The next lemma is crucial in what follows.

Lemma 4.13. *Let the assumptions of Lemma 4.6 be satisfied. Then*†

$$\lim_{\varepsilon \to 0} \frac{\hat t(\chi_\varepsilon, \varepsilon)}{\varepsilon} = \frac{\int_\Omega \mathrm{div}(R_0^T E_{R_0}^{-1}(\mu e, -\rho(e \cdot R_0 l_{\partial\Omega})R_0 v))\, dx}{\int_\Omega \mathrm{div}(R_0^T E_{R_0}^{-1}(0, \rho R_0 v))\, dx}, \qquad (4.30)$$

and, if the function $x \mapsto D_{Z_{hk}} a(x, R_0)$ $(h, k = 1, \ldots, n)$ *are constant,*‡

$$\lim_{\varepsilon \to 0} \frac{\hat t(\chi_\varepsilon, \varepsilon)}{\varepsilon} = \frac{H_0 \cdot (e \otimes R_0 \int_\Omega (\mu - \rho) l_\Omega\, dx)}{\rho\, \mathrm{vol}(\Omega)\, \mathrm{tr}\, H_0} - \frac{1}{\mathrm{vol}(\Omega)} e \cdot \int_\Omega R_0 l_\Omega\, dx, \qquad (4.31)$$

where H_0 *is the element of* Sym_n *defined by* (4.20).

Proof. We will prove the lemma for Sobolev spaces; in the case of Schauder spaces the procedure is exactly the same. Recall that (by Remark 3.1 of Chapter V) $\hat e_{R_0}$ is a (linear and) continuous mapping from $W^{m,p}(\Omega, \mathbb{R}^n) \times W^{m+1-1/p,p}(\partial\Omega, \mathbb{R}^n)$ into itself, that $E_{R_0}^{-1}$ is a (linear and)

---

† Note that the pair $(0, \rho R_0 v)$ is evidently equilibrated with respect to $R_0 l_{\bar\Omega}$ and that, in view of (4.5), (4.6), also the pair $(\mu e, -\rho(e \cdot R_0 l_{\partial\Omega})R_0 v)$ is equilibrated with respect to $R_0 l_{\bar\Omega}$.

‡ Note that $\mathrm{tr}\, H_0 \neq 0$. Indeed, in view of (3.15) of Chapter III, we have $\sum_{h,k=1}^n (H_0 R_0)_{hk}$ $D_{Z_{hk}} a(x, R_0) \cdot (H_0 R_0) > 0$, and, on the other hand, from (4.20) it follows that $\sum_{h,k=1}^n (H_0 R_0)_{hk} D_{Z_{hk}} a(x, R_0) \cdot (H_0 R_0) = \mathrm{tr}\, H_0$.

continuous mapping from a subspace of $W^{m,p}(\Omega, \mathbb{R}^n) \times W^{m+1-1/p,p}(\partial\Omega, \mathbb{R}^n)$ into $W^{m+2,p}(\Omega, \mathbb{R}^n)$, that A is a C^1 mapping from $(W^{m+1,p}(\Omega, \mathbb{R}^n))^+$ into $W^{m+2,p}(\Omega, \mathbb{R}^n)$, that γ is an analytic mapping from $(W^{m+2,p}(\Omega, \mathbb{R}^n))^+$ into itself, that $\hat{R}$ is an analytic mapping from an neighborhood of $R_0 \iota_{\overline{\Omega}}$ in $(W^{m+2,p}(\Omega, \mathbb{R}^n))^+$ into M_n, that $\varepsilon \mapsto \chi_\varepsilon$ is a C^1 mapping from a neighborhood of 0 in $\mathbb{R}$ into $W^{m+2,p}(\Omega, \mathbb{R}^n)$, and that $\chi_0 = R_0 \iota_{\overline{\Omega}}$, $\hat{R}(R_0 \iota_{\overline{\Omega}}) = I$, and $a(x, R_0) = 0$, $\forall x \in \Omega$ (see Remark 3.1 of Chapter III).

A first consequence of these facts is that the real-valued function $\varepsilon \mapsto \alpha(\chi_\varepsilon)$ is continuous and hence

$$\lim_{\varepsilon \to 0} \alpha(\chi_\varepsilon) = \alpha(\chi_0) = \int_\Omega \operatorname{div}(R_0^\mathsf{T} E_{R_0}^{-1}(0, \rho R_0 v))\, dx.$$

Thus, because of (4.19), (4.30) is true provided

$$\lim_{\varepsilon \to 0} \frac{\beta(\chi_\varepsilon, \varepsilon)}{\varepsilon} = \int_\Omega \operatorname{div}(R_0^\mathsf{T} E_{R_0}^{-1}(\mu e, -\rho(e \cdot \chi_0)|_{\partial\Omega} R_0 v))\, dx. \tag{4.32}$$

Observe that, by (4.17),

$$\frac{\beta(\chi_\varepsilon, \varepsilon)}{\varepsilon} = \int_\Omega \operatorname{div}\left(R_0^\mathsf{T} E_{R_0}^{-1}\left(\hat{e}_{R_0}\left(\frac{\operatorname{div} A(D\gamma(\chi_\varepsilon))}{\varepsilon} + \mu \hat{R}(\gamma(\chi_\varepsilon))^\mathsf{T} e, -\left(\frac{A(D\gamma(\chi_\varepsilon))}{\varepsilon} \right)\Big|_{\partial\Omega} v \right) \right. \right.$$
$$\left. \left. - \rho(\gamma(\chi_\varepsilon) \cdot \hat{R}(\gamma(\chi_\varepsilon))^\mathsf{T} e(\operatorname{cof} D\gamma(\chi_\varepsilon)))\Big|_{\partial\Omega} v \right) \right)\, dx. \tag{4.33}$$

In order to prove (4.32) we begin by remarking that, since $A(D\gamma(\chi_\varepsilon)) = 0$, we have

$$\lim_{\varepsilon \to 0} \left\| \frac{A(D\gamma(\chi_\varepsilon))}{\varepsilon} - \left(\frac{d}{d\varepsilon}(\varepsilon \mapsto A(D\gamma(\chi_\varepsilon))) \right)\Big|_{\varepsilon=0} \right\|_{m+1,p} = 0. \tag{4.34}$$

Note that, as

$$\left(\frac{d}{d\varepsilon}(\varepsilon \mapsto A(D\gamma(\chi_\varepsilon))) \right)\Big|_{\varepsilon=0} = (L_{R_0} \circ \gamma'(\chi_0))\left(\frac{d\chi_\varepsilon}{d\varepsilon} \right)\Big|_{\varepsilon=0},$$

in view of (4.13) we have

$$\left(\frac{d}{d\varepsilon}(\varepsilon \mapsto A(D\gamma(\chi_\varepsilon))) \right)\Big|_{\varepsilon=0} = L_{R_0}\left(D\left(\left(\frac{d\chi_\varepsilon}{d\varepsilon} \right)\Big|_{\varepsilon=0} \right) \right)$$
$$- \left(\frac{1}{n\,\operatorname{vol}(\Omega)} \int_\Omega \operatorname{div}\left(R_0^\mathsf{T}\left(\frac{d\chi_\varepsilon}{d\varepsilon} \right)\Big|_{\varepsilon=0} \right)\, dx \right) L_{R_0}(D\chi_0).$$

Note also that, since $\int_\Omega \operatorname{div}(R_0^\mathsf{T}(\chi_\varepsilon - \chi_0))\, dx = 0$, we have

$$\int_\Omega \operatorname{div}\left(R_0^\mathsf{T}\left(\frac{d\chi_\varepsilon}{d\varepsilon} \right)\Big|_{\varepsilon=0} \right)\, dx = 0; \tag{4.35}$$

indeed, $(\chi_\varepsilon - \chi_0)/\varepsilon$ converges to $(d\chi_\varepsilon/d\varepsilon)|_{\varepsilon=0}$ in $W^{m+2,p}(\Omega, \mathbb{R}^n)$ as $\varepsilon \to 0$, hence $\operatorname{div}(R_0^\mathsf{T}((\chi_\varepsilon - \chi_0)/\varepsilon))$ converges to $\operatorname{div}(R_0^\mathsf{T}(d\chi_\varepsilon/d\varepsilon)|_{\varepsilon=0})$ in $W^{m+1,p}(\Omega, \mathbb{R}^n)$

as $\varepsilon \to 0$ and thus

$$\int_\Omega \mathrm{div}\left(R_0^\mathsf{T}\left(\frac{d\chi_\varepsilon}{d\varepsilon}\right)\Big|_{\varepsilon=0}\right) = \lim_{\varepsilon \to 0}\int_\Omega \mathrm{div}\left(R_0^\mathsf{T}\left(\frac{\chi_\varepsilon - \chi_0}{\varepsilon}\right)\right)dx.$$

Then

$$\left(\frac{d}{d\varepsilon}(\varepsilon \mapsto A(D\gamma(\chi_\varepsilon)))\right)\Big|_{\varepsilon=0} = L_{R_0}\left(D\left(\left(\frac{d\chi_\varepsilon}{d\varepsilon}\right)\Big|_{\varepsilon=0}\right)\right)$$

and therefore, by (4.34),

$$\lim_{\varepsilon \to 0}\left\| \frac{A(D\gamma(\chi_\varepsilon)}{\varepsilon} - L_{R_0}\left(D\left(\left(\frac{d\chi_\varepsilon}{d\varepsilon}\right)\Big|_{\varepsilon=0}\right)\right)\right\|_{m+1,p} = 0.$$

It follows that

$$\lim_{\varepsilon \to 0}\left\| \frac{\mathrm{div}\, A(D\gamma(\chi_\varepsilon))}{\varepsilon} - \mathrm{div}\, L_{R_0}\left(D\left(\left(\frac{d\chi_\varepsilon}{d\varepsilon}\right)\Big|_{\varepsilon=0}\right)\right)\right\|_{m,p} = 0 \qquad (4.36)$$

and, in view of Remark 6.5 of Chapter III,

$$\lim_{\varepsilon \to 0}\left\| \frac{1}{\varepsilon} A(D\gamma(\chi_\varepsilon))|_{\partial\Omega} - L_{R_0}\left(D\left(\left(\frac{d\chi_\varepsilon}{d\varepsilon}\right)\Big|_{\varepsilon=0}\right)\right)\Big|_{\partial\Omega}\right\|_{m+1-1/p,p,\partial\Omega} = 0. \qquad (4.37)$$

Since, as $\varepsilon \to 0$, $\gamma(\chi_\varepsilon)\cdot \hat{R}(\gamma(\chi_\varepsilon))e$ converges to $e\cdot\chi_0$ in $W^{m+1,p}(\Omega, \mathbb{R}^n)$ and $\mathrm{cof}\, D\gamma(\chi_\varepsilon))$ converges in $W^{m+1,p}(\Omega, \mathbb{R}^n)$ to the constant function $x \mapsto R_0$, we deduce (recalling that, under our hypotheses, $W^{m+1,p}(\Omega)$ is a Banach algebra) that

$$\lim_{\varepsilon \to 0}\|\gamma(\chi_\varepsilon)\cdot (\hat{R}(\gamma(\chi_\varepsilon))^\mathsf{T}e)(\mathrm{cof}\, D\gamma(\chi_\varepsilon)) - (e\cdot\chi_0)R_0\|_{m+1,p} = 0.$$

Consequently, using Remark 6.5 of Chapter III once more, we obtain

$$\lim_{\varepsilon \to 0}\|(\gamma(\chi_\varepsilon)\cdot (\hat{R}(\gamma(\chi_\varepsilon))^\mathsf{T}e)(\mathrm{cof}\, D\gamma(\chi_\varepsilon))|_{\partial\Omega}v - (e\cdot\chi_0)|_{\partial\Omega}R_0v\|_{m+1-1/p,p,\partial\Omega} = 0.$$
$$(4.38)$$

Now, combining (4.33) with (4.36), (4.37), and (4.38) and recalling that, as $\varepsilon \to 0$, $\hat{R}(\gamma(\chi_\varepsilon))$ converges to I in $\mathbb{M}_n$, we have

$$\lim_{\varepsilon \to 0}\frac{\beta(\chi_\varepsilon)}{\varepsilon} = \int_\Omega \mathrm{div}\left(R_0^\mathsf{T}E_{R_0}^{-1}\left(\hat{e}_{R_0}\left(\mathrm{div}\, L_{R_0}\left(D\left(\left(\frac{d\chi_\varepsilon}{d\varepsilon}\right)\Big|_{\varepsilon=0}\right)\right)\right) + \mu e,\right.\right.$$
$$\left.\left. - L_{R_0}\left(D\left(\left(\frac{d\chi_\varepsilon}{d\varepsilon}\right)\Big|_{\varepsilon=0}\right)\right)\Big|_{\partial\Omega}v - \rho(e\cdot\chi_0)|_{\partial\Omega}R_0v\right)\right)\right)dx.$$

The pairs

$$\left(\mathrm{div}\, L_{R_0}\left(D\left(\left(\frac{d\chi_\varepsilon}{d\varepsilon}\right)\Big|_{\varepsilon=0}\right)\right), -L_{R_0}\left(D\left(\left(\frac{d\chi_\varepsilon}{d\varepsilon}\right)\Big|_{\varepsilon=0}\right)\right)\Big|_{\partial\Omega}v\right) \qquad \text{and}$$

$$(\mu e, -\rho(e\cdot\chi_0)|_{\partial\Omega}R_0v)$$

are equilibrated with respect to χ_0: this is easy to verify by using the divergence theorem and taking into account (4.5), (4.6) and the fact that

(see Remark 3.2 of Chapter III)

$$D_{Z_{hk}} a(x, R_0) R_0^T \in \text{Sym}_n.$$

Hence

$$\lim_{\varepsilon \to 0} \frac{\beta(\chi_\varepsilon)}{\varepsilon} = \int_\Omega \text{div} \left(R_0^T E_{R_0}^{-1} \left(\text{div } L_{R_0} \left(D \left(\left(\frac{d\chi_\varepsilon}{d\varepsilon} \right)\Big|_{\varepsilon=0} \right) \right) \right. \right.$$
$$\left. \left. - L_{R_0} \left(D \left(\left(\frac{d\chi_\varepsilon}{d\varepsilon} \right)\Big|_{\varepsilon=0} \right) \right)\Big|_{\partial\Omega} v \right) \right) dx$$
$$+ \int_\Omega \text{div}(R_0^T E_{R_0}^{-1}(\mu e, -\rho(e \cdot \chi_0)|_{\partial\Omega} R_0 v)) \, dx,$$

which yields (4.32) because of (4.35) and the evident equality

$$\int_\Omega \text{div} \left(R_0^T E_{R_0}^{-1} \left(\text{div } L_{R_0} \left(D \left(\left(\frac{d\chi_\varepsilon}{d\varepsilon} \right)\Big|_{\varepsilon=0} \right) \right), -L_{R_0} \left(D \left(\left(\frac{d\chi_\varepsilon}{d\varepsilon} \right)\Big|_{\varepsilon=0} \right) \right)\Big|_{\partial\Omega} v \right) \right) dx$$
$$= \int_\Omega \text{div} \left(R_0^T \left(\frac{d\chi_\varepsilon}{d\varepsilon} \right)\Big|_{\varepsilon=0} \right) dx.$$

To conclude the proof we note that if the functions $x \mapsto D_{Z_{hk}} a(x, R_0)$ $(h, k = 1, \dots, n)$ are constant, then in view of Remark 4.7 we obtain

$$\int_\Omega \text{div}(R_0^T E_{R_0}^{-1}(\mu e, -\rho(e \cdot R_0 \iota_{\partial\Omega}) R_0 v)) \, dx$$

$$= -H_0 \cdot \left(e \otimes \int_\Omega \mu R_0 \iota_\Omega \, dx - \int_{\partial\Omega} \rho(e \cdot R_0 \iota_{\partial\Omega})(R_0 v) \otimes R_0 \iota_{\partial\Omega} \, d\sigma \right)$$

$$= -H_0 \cdot \left(e \otimes \int_\Omega (\mu - \rho) R_0 \iota_\Omega \, dx - \rho \left(e \cdot \int_\Omega R_0 \iota_\Omega \, dx \right) I \right)$$

and

$$\int_\Omega \text{div}(R_0^T E_{R_0}^{-1}(0, \rho R_0 v)) \, dx = -\rho H_0 \cdot \int_{\partial\Omega} R_0 v \otimes R_0 \iota_{\partial\Omega} \, d\sigma = -\rho \, \text{vol}(\Omega) H_0 \cdot I;$$

thus (4.30) becomes (4.31).   □

Let us set, for $0 < |\varepsilon| \leq \varepsilon_0$,

$$\phi_\varepsilon = \hat{R}(\gamma(\chi_\varepsilon)) \gamma(\chi_\varepsilon) + \frac{1}{\varepsilon} \hat{t}(\chi_\varepsilon, \varepsilon) e \tag{4.39}$$

(where, as always, e is identified with the constant function $x \mapsto e$ of $\bar{\Omega}$ into $\mathbb{R}^n$).

The next corollary follows from Lemma 4.4, 4.5, 4.6, 4.11, 4.13, and Remark 4.3.

Corollary 4.14. *Let the assumptions of the first part of Lemma 4.6 be satisfied. Then, for $0 < |\varepsilon| \leq \varepsilon_0$, the function ϕ_ε is a solution of problem*

(4.4) (i.e.,

$$\begin{cases} \operatorname{div} A(D\phi_\varepsilon) + \varepsilon\mu e = 0 & \text{in } \Omega, \\ (A(D\phi_\varepsilon) + \varepsilon\rho(\phi_\varepsilon \cdot e) \operatorname{cof} D\phi_\varepsilon)\nu = 0 & \text{on } \partial\Omega). \end{cases}$$

Moreover,

$$\lim_{\varepsilon \to 0} \|\phi_\varepsilon - \phi_0\|_{m+2,p} = 0 \quad \left[\text{respectively } \lim_{\varepsilon \to 0} \|\phi_\varepsilon - \phi_0\|_{m+2,\lambda} = 0 \right], \quad (4.40)$$

where ϕ_0 is the $\mathbb{R}^n$-valued function defined on $\bar{\Omega}$ by putting†

$$\phi_0(x) = R_0 x + \frac{\int_\Omega \operatorname{div}(R_0^T E_{R_0}^{-1}(\mu e, -\rho(e \cdot R_0\iota_{\partial\Omega})R_0\nu)) \, dx}{\int_\Omega \operatorname{div}(R_0^T E_{R_0}^{-1}(0, \rho R_0\nu)) \, dx} e. \quad (4.41)$$

The mapping $\varepsilon \mapsto \phi_\varepsilon$ (extended to $[-\varepsilon_0, \varepsilon_0]$ by assuming that its value at $\varepsilon = 0$ is the rigid deformation ϕ_0 of $\bar{\Omega}$ defined by (4.41)) is analytic at 0 provided $a \in C^\infty(\bar{\Omega} \times \mathsf{M}_n^+, \mathsf{M}_n)$ and the functions $(x, Z) \mapsto D_x^\alpha a(x, Z)$, $|\alpha| \le m + 1$, are analytical in Z, at R_0, uniformly with respect to x.

The proof of the (local) uniqueness in Theorem 4.17 is based on

Lemma 4.15. *Let ϕ be an element of $W^{m+2,p}(\Omega, \mathbb{R}^n)$ [respectively $C^{m+2,\lambda}(\Omega, \mathbb{R}^n)$] such that*

$$\det \int_\Omega D\phi \, dx > 0, \qquad \int_\Omega \det D\phi \, dx = \operatorname{vol}(\Omega), \quad (4.42)$$

and

$$e \wedge \left(\int_\Omega \phi \, dx - \frac{1}{n \operatorname{vol}(\Omega)} \left(\int_\Omega \operatorname{div}(R_0^T\phi) \, dx \right) R_\phi \int_\Omega \iota_\Omega \, dx \right) = 0, \quad (4.43)$$

where R_ϕ is the element of $\mathbb{O}_n^+$ such that the matrix

$$R_\phi^T \int_\Omega D\phi \, dx \quad (4.44)$$

is symmetric and positive definite. Then for any $R_0 \in \mathbb{O}_n^+$ there is a unique $(\chi, R, \tau) \in (R_0\iota_{\bar\Omega} + \mathscr{V}_{R_0}^{m+2,p}) \times \mathbb{O}_n^+ \times \mathbb{R}$ [respectively $(\chi, R, \tau) \in (R_0\iota_{\bar\Omega} + \mathscr{V}_{R_0}^{m+2,\lambda}) \times \mathbb{O}_n^+ \times \mathbb{R}$] such that

$$\int_\Omega \operatorname{div}(R_0^T\chi - \iota_\Omega) \, dx = 0, \quad (4.45)$$

$$R_0^T \int_\Omega D\chi \, dx \quad \text{is (symmetric and) positive definite,} \quad (4.46)$$

---

† As e is identified with the constant function $x \mapsto e$ of $\bar{\Omega}$ into $\mathbb{R}^n$, in (4.41) the symbol $e \cdot R_0\iota_{\partial\Omega}$ denotes the function $x \mapsto e \cdot R_0 x$ of $\partial\Omega$ into $\mathbb{R}$.

and

$$\phi(x) = R(\gamma(\chi)(x)) + \tau e, \qquad \forall x \in \Omega, \tag{4.47}$$

where $\gamma(\chi) = (\mathrm{vol}(\Omega)/\mathrm{vol}(\chi(\Omega)))^{1/n}\chi$; R, τ, *and* χ *are related to* ϕ *by*

$$R = R_\phi R_0^{\mathrm{T}}, \quad \tau = \frac{1}{\mathrm{vol}(\Omega)} e \cdot \left(\int_\Omega \phi \, dx - \left(\frac{1}{n \, \mathrm{vol}(\Omega)} \int_\Omega \mathrm{div}(R_0^{\mathrm{T}}\phi) \, dx \right) R_\phi \int_\Omega \iota_\Omega \, dx \right), \tag{4.48}$$

and

$$\chi(x) = \frac{n \, \mathrm{vol}(\Omega)}{\int_\Omega \mathrm{div}(R_\phi^{\mathrm{T}}\phi) \, dx} R_0 \left(R_\phi^{\mathrm{T}}\phi(x) - \frac{1}{\mathrm{vol}(\Omega)} \int_\Omega R_\phi^{\mathrm{T}}\phi \, dx + \frac{1}{\mathrm{vol}(\Omega)} \int_\Omega \iota_\Omega \, dx \right). \tag{4.49}$$

If ϕ *is the function* ϕ_0 *defined by* (4.41), *then* χ *is the function* $R_0 \iota_{\overline{\Omega}}$.

Proof. As a first step of the proof we suppose that (χ, R, τ) exists with the properties of the statement and we prove that (4.48) and (4.49) are true. We can write (4.47) in the form

$$\chi(x) = \left(\frac{\mathrm{vol}(\chi(\Omega))}{\mathrm{vol}(\Omega)} \right)^{1/n} (R^{\mathrm{T}}\phi(x) - \tau R^{\mathrm{T}}e). \tag{4.50}$$

This implies

$$\int_\Omega R_0^{\mathrm{T}} D\chi \, dx = \left(\frac{\mathrm{vol}(\chi(\Omega))}{\mathrm{vol}(\Omega)} \right)^{1/n} R_0^{\mathrm{T}} R^{\mathrm{T}} \int_\Omega D\phi \, dx; \tag{4.51}$$

hence, in view of (4.46), the matrix $R_0^{\mathrm{T}} R^{\mathrm{T}} \int_\Omega D\phi \, dx$ is symmetric and positive definite. Then, by the uniqueness of the (polar) factorization $\int_\Omega D\phi \, dx = QP$ with $Q \in \mathbb{O}_n^+$ and P symmetric and positive definite (see Chapter I, §6), we have $R_0^{\mathrm{T}} R^{\mathrm{T}} = R_\phi^{\mathrm{T}}$, namely $(4.48)_1$.

Note that, since the symmetric matrix (4.44) is positive definite, its trace is > 0, i.e.,

$$\int_\Omega \mathrm{div}(R_\phi^{\mathrm{T}}\phi) \, dx > 0.$$

Combining (4.51) with $(4.48)_1$ we have

$$\int_\Omega \mathrm{div}(R_0^{\mathrm{T}}\chi) \, dx = \left(\frac{\mathrm{vol}(\chi(\Omega))}{\mathrm{vol}(\Omega)} \right)^{1/n} \int_\Omega \mathrm{div}(R_\phi^{\mathrm{T}}\phi) \, dx.$$

Thus, as (by (4.45)) $\int_\Omega \mathrm{div}(R_0^{\mathrm{T}}\chi) \, dx = n \, \mathrm{vol}(\Omega)$, we obtain

$$\left(\frac{\mathrm{vol}(\chi(\Omega))}{\mathrm{vol}(\Omega)} \right)^{1/n} = \frac{n \, \mathrm{vol}(\Omega)}{\int_\Omega \mathrm{div}(R_\phi^{\mathrm{T}}\phi) \, dx}. \tag{4.52}$$

and so (4.50) becomes

$$\chi(x) = \frac{n \, \text{vol}(\Omega)}{\int_\Omega \text{div}(R_\phi^\mathsf{T}\phi) \, dx}(R_0 R_\phi^\mathsf{T}\phi(x) - \tau R_0 R_\phi^\mathsf{T}e). \tag{4.53}$$

Since the condition $\chi \in R_0 l_{\overline{\Omega}} + \mathscr{V}_{R_0}^{m+2,p}$ [respectively $\chi \in R_0 l_{\overline{\Omega}} + \mathscr{V}_{R_0}^{m+2,\lambda}$] requires that $\int_\Omega (\chi - R_0 l_\Omega) \, dx = 0$, namely that $R_0^\mathsf{T} \int_\Omega \chi \, dx = \int_\Omega l_\Omega \, dx$, from (4.53) we easily deduce that

$$\int_\Omega l_\Omega \, dx = \frac{n \, \text{vol}(\Omega)}{\int_\Omega \text{div}(R_\phi^\mathsf{T}\phi) \, dx}\left(R_\phi^\mathsf{T}\int_\Omega \phi \, dx - \tau \, \text{vol}(\Omega)R_\phi^\mathsf{T}e\right),$$

which yields

$$\tau R_\phi^\mathsf{T}e = \frac{1}{\text{vol}(\Omega)}R_\phi^\mathsf{T}\left(\int_\Omega \phi \, dx - \frac{1}{n \, \text{vol}(\Omega)}\left(\int_\Omega \text{div}(R_\phi^\mathsf{T}\phi) \, dx\right)R_\phi\int_\Omega l_\Omega \, dx\right). \tag{4.54}$$

This shows that condition (4.43) is necessary for the existence of a triplet (χ, R, τ) having the properties of the statement. Taking into account (4.43) and (4.54) we get $(4.48)_2$, and so (4.49) follows from (4.53). On the other hand, it is easy to verify that if χ is defined by (4.49), then (4.46) and (4.47) hold and $\chi \in R_0 l_{\overline{\Omega}} + \mathscr{V}_{R_0}^{m+2,p}$ [respectively $\chi \in R_0 l_{\overline{\Omega}} + \mathscr{V}_{R_0}^{m+2,\lambda}$]. Finally, after remarking that $R_{\phi_0} = R_0$, a straightforward calculation shows that $\chi = R_0 l_{\overline{\Omega}}$ when $\phi = \phi_0$. Note that if the center of Ω is the origin O of $\mathbb{R}^n$, i.e., if $\int_\Omega l_\Omega \, dx = 0$, then condition (4.43) becomes

$$e \wedge \int_\Omega \phi \, dx = 0. \quad \square$$

Corollary 4.16. *For any $R_0 \in \mathbb{O}_n^+$ the operator $(\chi, R, \tau) \mapsto \phi$, with ϕ defined by (4.47), is a homeomorphism of the subset $\{(\chi, R, \tau) \in (R_0 l_{\overline{\Omega}} + \mathscr{V}_{R_0}^{m+2,p}) \times \mathbb{O}_e \times \mathbb{R}: (4.45)$ and (4.46) hold$\}$ of $W^{m+2,p}(\Omega, \mathbb{R}^n) \times \mathsf{M}_n \times \mathbb{R}$ onto the subset $\{\phi \in W^{m+2,p}(\Omega, \mathbb{R}^n): (4.42), (4.43)$ hold and $R_\phi R_0^\mathsf{T} \in \mathbb{O}_e\}$ of $W^{m+2,p}(\Omega, \mathbb{R}^n)$, where R_ϕ is the element of $\mathbb{O}_n^+$ such that matrix (4.44) is symmetric and positive definite. The same operator is also a homeomorphism of the subset $\{(\chi, R, \tau) \in (R_0 l_{\overline{\Omega}} + \mathscr{V}_{R_0}^{m+2,\lambda}) \times \mathbb{O}_e \times \mathbb{R}: (4.45)$ and (4.46) hold$\}$ of $C^{m+2,\lambda}(\overline{\Omega}, \mathbb{R}^n) \times \mathsf{M}_n \times \mathbb{R}$ onto the subset $\{\phi \in C^{m+2,\lambda}(\overline{\Omega}, \mathbb{R}^n): (4.42), (4.43)$ hold and $R_\phi R_0^\mathsf{T} \in \mathbb{O}_e\}$ of $C^{m+2,\lambda}(\overline{\Omega}, \mathbb{R}^n)$.*

Proof. We treat only the case of Sobolev spaces, because in the case of Schauder spaces we can proceed in the same way. If, for $(\chi, R, \tau) \in (R_0 l_{\overline{\Omega}} + \mathscr{V}_{R_0}^{m+2,p}) \times \mathbb{O}_e \times \mathbb{R}$, conditions (4.45) and (4.46) are satisfied and ϕ is defined by (4.47), then (as has been shown in the proof of the previous lemma) ϕ satisfies the conditions (4.43) and $R_0 R_\phi^\mathsf{T} = R$, and a straightforward calculation shows that ϕ also satisfies condition (4.42). Therefore, by Lemma 4.15, the mapping $(\chi, R, \tau) \mapsto \phi$, with ϕ defined by (4.47), is a bijection of the set $\{(\chi, R, \tau) \in (R_0 l_{\overline{\Omega}} + \mathscr{V}_{R_0}^{m+2,p}) \times \mathbb{O}_e \times \mathbb{R}: (4.5)$ and

(4.6) hold} onto the set $\{\phi \in W^{m+2,p}(\Omega, \mathbb{R}^n): (4.2), (4.3)$ hold and $R_\phi R_0^T \in \mathbb{O}_e\}$; its inverse is the mappng $\phi \mapsto (\chi, R_0 R_\phi^T, \tau)$ with τ defined by $(4.48)_2$ and χ defined by (4.49).

Evidently, the mapping $(\chi, R, \tau) \mapsto \phi$ (with ϕ defined by (4.47)) from the subset $\{(\chi, R, \tau) \in (R_0 \iota_{\overline{\Omega}} + \mathcal{V}_{R_0}^{m+2,p}) \times \mathbb{O}_e \times \mathbb{R}: (4.45)$ and (4.46) hold} of $W^{m+2,p}(\Omega, \mathbb{R}^n) \times \mathbb{M}_n \times \mathbb{R}$ into $W^{m+2,p}(\Omega, \mathbb{R}^n)$ is continuous. Its inverse is also continuous. Indeed, the mapping $\phi \mapsto \tau$, with τ defined by $(4.48)_2$, is clearly continuous; moreover, as $(R_\phi^T \int_\Omega D\phi \, dx)^2 = (\int_\Omega D\phi \, dx)^T(\int_\Omega D\phi \, dx)$, the mapping $\phi \mapsto R_\phi^T \int_\Omega D\phi \, dx$ is continuous from $\{\phi \in W^{m+2,p}(\Omega, \mathbb{R}^n): \det \int_\Omega D\phi \, dx > 0\}$ into $\mathbb{M}_n$ and hence the mappings $\phi \mapsto R_\phi$ and $\phi \mapsto \chi$ with χ defined by (4.49) are continuous from $\{\phi \in W^{m+2,p}(\Omega, \mathbb{R}^n): \det \int_\Omega D\phi \, dx > 0\}$ into $\mathbb{M}_n$ and from $\{\phi \in W^{m+2,p}(\Omega, \mathbb{R}^n): \det \int_\Omega D\phi \, dx > 0\}$ into itself, respectively. $\quad\square$

We are now in a position to prove the main result of this section, namely

Theorem 4.17. *Let m be an integer ≥ 0. Assume that Ω is of class C^{m+2}, that $a \in C^{m+2}(\overline{\Omega} \times \mathbb{M}_n^+, \mathbb{M}_n)$, and that $\mu \in W^{m,p}(\Omega)$ with $p(m + 1) > n$ [respectively that Ω is of class $C^{m+2,\lambda}$ with $0 < \lambda < 1$, that $a \in C^{m+3}(\overline{\Omega} \times \mathbb{M}_n^+, \mathbb{M}_n)$, and that $\mu \in C^{m,\lambda}(\overline{\Omega})$]. Assume also that (1.4), (1.5), (1.6), (1.7), (4.5) hold and let $R_0 \in \mathbb{O}_n^+$ be such that (4.6) holds. Then, numbers $\overline{\varepsilon} > 0$ and $\delta > 0$ exist such that for each $\varepsilon \in [-\overline{\varepsilon}, \overline{\varepsilon}] \setminus \{0\}$ there is one and only one deformation† ϕ_ε of $\overline{\Omega}$ belonging to $W^{m+2,p}(\Omega, \mathbb{R}^n)$ [respectively $C^{m+2,\lambda}(\overline{\Omega}, \mathbb{R}^n)$] and satisfying the following conditions:*

$$\begin{cases} \operatorname{div} A(D\phi_\varepsilon) + \varepsilon\mu e = 0 & \text{in } \Omega, \\ (A(D\phi_\varepsilon) + \varepsilon\rho(\phi_\varepsilon \cdot e) \operatorname{cof} D\phi_\varepsilon)\nu = 0 & \text{on } \partial\Omega, \end{cases} \tag{4.55}$$

$$\det \int_\Omega D\phi_\varepsilon \, dx > 0, \qquad R_\varepsilon R_0^T \in \mathbb{O}_e, \tag{4.56}$$

$$e \wedge \left[\int_\Omega \phi_\varepsilon \, dx - \frac{1}{n \operatorname{vol}(\Omega)} \left(\int_\Omega \operatorname{div}(R_\varepsilon^T \phi_\varepsilon) \, dx \right) R_\varepsilon \int_\Omega \iota_\Omega \, dx \right] = 0, \tag{4.57}$$

and

$$\|\phi_\varepsilon - \phi_0\|_{m+2,p} \leq \delta \qquad [\text{respectively } \|\phi_\varepsilon - \phi_0\|_{m+2,\lambda} \leq \delta], \tag{4.58}$$

where ϕ_0 is the rigid deformation of Ω defined by (4.41), $\mathbb{O}_e$ is the subgroup of $\mathbb{O}_n^+$ defined by (4.14), and R_ε is the element of $\mathbb{O}_n^+$ such that the matrix

$$R_\varepsilon^T \int_\Omega D\phi_\varepsilon \, dx$$

is symmetric and positive definite. The mapping $\varepsilon \mapsto \phi_\varepsilon$ (extended to $[-\overline{\varepsilon}, \overline{\varepsilon}]$ by assuming that its value at $\varepsilon = 0$ is the rigid deformation ϕ_0 of $\overline{\Omega}$ defined

---

† Cf. footnote ‡ on p. 138.

by (4.41)) *is continuous in a suitable neighborhood of* 0. *If, in addition,*
$a \in C^{\infty}(\bar{\Omega} \times \mathbb{M}_n^+, \mathbb{M}_n)$ *and the functions* $(x, Z) \mapsto D_x^{\alpha} a(x, Z)$, $|\alpha| \leqq m + 1$, *are*
analytic in Z, *at* R_0, *uniformly with respect to* x, *then the mapping* $\varepsilon \mapsto \phi_\varepsilon$
is analytic at 0.

Proof. We consider only the case of Sobolev spaces, because the case of
Schauder spaces can be treated analogously. Let $\hat{R}$, $\hat{t}$, χ_ε, ϕ_0, ε_0, δ, ζ be
as in the statement of Lemmas 4.5, 4.6, 4.11 and Corollaries 4.12, 4.14.
Consider the function $\varepsilon \mapsto \phi_\varepsilon$ defined on $[-\varepsilon_0, \varepsilon_0] \backslash \{0\}$ by (4.39) and ex-
tend it to the whole of $[-\varepsilon_0, \varepsilon_0]$ by assuming that its value at $\varepsilon = 0$ is
the rigid deformation of $\bar{\Omega}$ defined by (4.41). By Corollary 4.14, ϕ_ε is an
element of $W^{m+2,p}(\Omega, \mathbb{R}^n)$ that satisfies (4.55) and (4.40). Recall that γ is
analytic from $(W^{m+2,p}(\Omega, \mathbb{R}^n))^+$ into itself, that $\hat{R}$ is analytic at $R_0 \iota_{\bar{\Omega}}$, that
$\hat{t}$ is of class C^1, and that $\varepsilon \mapsto \chi_\varepsilon$ is of class C^1 in a suitable neighborhood
of 0. Thus $\varepsilon \mapsto \phi_\varepsilon$ is a continuous mapping from a suitable neighborhood
of 0 in $\mathbb{R}$ into $W^{m+2,p}(\Omega, \mathbb{R}^n)$. Furthermore, if $a \in C^{\infty}(\bar{\Omega} \times \mathbb{M}_n^+, \mathbb{M}_n)$ and
the functions $(x, Z) \mapsto D_x^{\alpha} a(x, Z)$, $|\alpha| \leqq m + 1$, are analytic in Z, at R_0,
uniformly with respect to x, then $\hat{t}$ is analytic at $(R_0 \iota_{\bar{\Omega}}, 0)$, and $\varepsilon \mapsto$
χ_ε is analytic at 0, so that the mapping $\varepsilon \mapsto \phi_\varepsilon$, from $[-\varepsilon_0, \varepsilon_0]$ into
$W^{m+2,p}(\Omega, \mathbb{R}^n)$, is analytic at 0.

Let ε_1 be a number with $0 < \varepsilon_1 \leqq \varepsilon_0$ such that for every $\varepsilon \in [-\varepsilon_1, \varepsilon_1]$
the symmetric matrix $R_0^T \int_\Omega D\chi_\varepsilon \, dx$ is positive definite and that

$$\det D\phi_\varepsilon > 0, \qquad \sum_{|\alpha|=1} \sup_{x \in \Omega} |D^{\alpha}(\phi_\varepsilon - \phi_0)(x)| < \frac{1}{c_\Omega},$$

where c_Ω is a number > 0 such that inequality (1.1) of Chapter II holds
(see Remark 1.1 of Chapter II). In order to realize that such an ε_1 exists
it suffices to observe that from (4.40) it follows (by the Sobolev imbedding theorem) that

$$\limsup_{\varepsilon \to 0} \sup_{x \in \Omega} |D^{\alpha}(\phi_\varepsilon - \phi_0)(x)| = 0$$

and to recall that $\chi_0 = R_0 \iota_{\bar{\Omega}}$ (see Lemma 4.11) and that $\varepsilon \mapsto \chi_\varepsilon$ and $\varepsilon \mapsto \phi_\varepsilon$
are continuous from a neighborhood of 0 in $\mathbb{R}$ into $W^{m+2,p}(\Omega, \mathbb{R}^n)$. Note
that, in view of Remark 1.2 of Chapter II, if $|\varepsilon| \leqq \varepsilon_1$ then (the C^1 func-
tion from $\bar{\Omega}$ into $\mathbb{R}^n$ belongs to the equivalence class) ϕ_ε is one-to-one
and hence is a deformation of $\bar{\Omega}$. We recall that χ_ε belongs to $R_0 \iota_{\bar{\Omega}} +$
$\mathcal{V}_{R_0}^{m+2,p}$ and satisfies (4.27); moreover, $R_0^T \int_\Omega D\chi_\varepsilon \, dx$ is positive definite pro-
vided $|\varepsilon| \leqq \varepsilon_1$. Then, using Corollary 4.16 we deduce that, for $|\varepsilon| \leqq \varepsilon_1$,
the deformation ϕ_ε of $\bar{\Omega}$ satisfies conditions (4.56) and (4.57).

With regard to uniqueness we introduce two numbers $\delta_0 > 0$ and
$\zeta_0 > 0$ such that (4.42) holds provided $\|\phi - \phi_0\|_{m+2,p} \leqq \delta_0$ and that (4.46)
holds if $\chi \in R_0 \iota_{\bar{\Omega}} + \mathcal{V}_{R_0}^{m+2,p}$ with $\|\chi - R_0 \iota_{\bar{\Omega}}\|_{m+2,p} \leqq \zeta_0$; by using an argu-
ment of continuity and the fact that $D\phi_0(x) = R_0$, $\forall x \in \Omega$, it is easily seen
that numbers δ_0 and ζ_0 exist. Now let ϕ be a solution, for a fixed $\varepsilon \neq 0$,

of problem (4.4) belonging to $W^{m+2,p}(\Omega, \mathbb{R}^n)$ and satisfying the conditions $\|\phi - \phi_0\|_{m+2,p} \leq \delta_0$, $R_\phi R_0^T \in \mathcal{O}_e$, and (4.43). Since ϕ is a solution of problem (4.4), condition $(4.42)_2$ is also satisfied, as we remarked after Lemma 4.2. Then, in view of Lemma 4.15 and Corollary 4.16, there is one and only one triplet (χ, R, τ) belonging to $(R_0 \iota_{\overline{\Omega}} + \mathcal{V}_{R_0}^{m+2,p}) \times \mathcal{O}_e \times \mathbb{R}$ and satisfying (4.45), (4.46), (4.47); moreover, there is a number δ, with $0 < \delta \leq \delta_0$, such that $\|\phi - \phi_0\|_{m+2,p} \leq \delta$ implies $\|\chi - R_0 \iota_{\overline{\Omega}}\|_{m+2,p} \leq \inf(\zeta, \zeta_0)$ and $|R - I| \leq \xi$, where ζ and ξ are the numbers appearing in the statement of Lemmas 4.5 and 4.11. Then, since, by Remark 4.3,

$$\begin{cases} \operatorname{div} A(D\gamma(\chi)) + \varepsilon\mu R^T e = 0 & \text{in } \Omega, \\ (A(D\gamma(\chi)) + \varepsilon\rho(\gamma(\chi) \cdot R^T e + \tau) \operatorname{cof} D\gamma(\chi))\nu = 0 & \text{on } \partial\Omega, \end{cases}$$

from Lemma 4.11 and Corollary 4.12 it follows that $\chi = \chi_\varepsilon$, $R = \hat{R}(\gamma(\chi_\varepsilon))$, and $\tau = (1/\varepsilon)\hat{t}(\chi_\varepsilon, \varepsilon)$; hence, $\phi = \phi_\varepsilon$, with ϕ_ε as defined in (4.39). Thus, if $\overline{\varepsilon}$ is a number such that $0 < \overline{\varepsilon} \leq \varepsilon_1$ and that $\|\phi_\varepsilon - \phi_0\|_{m+2,p} \leq \delta$ for $0 < |\varepsilon| \leq \overline{\varepsilon}$, we can conclude that for each $\varepsilon \in [-\overline{\varepsilon}, \overline{\varepsilon}]\setminus\{0\}$ there is (one and) only one $\phi_\varepsilon \in W^{m+2,p}(\Omega, \mathbb{R}^n)$ satisfying (4.55), (4.56), (4.57), and (4.58). $\square$

APPENDIX I

On Analytic Mappings Between Banach Spaces. Analytic Implicit Function Theorem

For the proofs of the results quoted here we refer the reader to BOURBAKI [1967b], KRASNOSELSKIJ & VAINIKKO et al. [1972], and PRODI & AMBROSETTI [1973].

Let X and Y be normed (linear) spaces (both over $\mathbb{C}$ or both over $\mathbb{R}$) and let $\mathscr{L}(X, Y)$ be the linear space of continuous linear mappings $\varphi: X \to Y$ equipped with the norm $\|\cdot\|$ defined by $\|\varphi\| = \sup\{\|\varphi(x)\|: \|x\| \leq 1\}$. A basic result is the "open mapping theorem": *if X and Y are Banach spaces and $\varphi \in \mathscr{L}(X, Y)$ is surjective, then φ is open, namely φ maps open sets of X onto open sets of Y.* Consequently, *if X and Y are Banach spaces and $\varphi \in \mathscr{L}(X, Y)$ is a bijection, then its inverse φ^{-1} is continuous.* We also emphasize the well-known fact that *the set of elements of $\mathscr{L}(X, Y)$, that are bijections, is open in $\mathscr{L}(X, Y)$.*

Let U be an open subset of X. A mapping $f: U \to Y$ is *differentiable* at $\bar{x} \in U$ if there is a $\varphi \in \mathscr{L}(X, Y)$ such that

$$\lim_{x \to \bar{x}} \|f(x) - f(\bar{x}) - \varphi(x - \bar{x})\| \, \|x - \bar{x}\|^{-1} = 0.$$

If φ exists, it is unique; it is called the *differential* of f at $\bar{x}$ and it will be denoted by $f'(\bar{x})$. If $f: U \to Y$ is differentiable at any point of U and the mapping $x \mapsto f'(x)$, of U into $\mathscr{L}(X, Y)$, is continuous, then f is said to be of *class C^1* or *continuously differentiable*. Of course, if $f \in \mathscr{L}(X, Y)$, then f is of class C^1 and $f'(x) = f$, $\forall x \in X$. If $f: U \to Y$ is differentiable at $\bar{x}$ and $g: V \to Z$, with V an open subset of Y containing $f(U)$ and Z is a normed space, is differentiable at $f(\bar{x})$, then $g \circ f: U \to Z$ is differentiable at $\bar{x}$ and

$$(g \circ f)'(\bar{x}) = g'(f(\bar{x})) \circ f'(\bar{x}).$$

When $Y = Y_1 \times \cdots \times Y_k$ with $Y_1, \ldots, Y_k$ normed spaces, the mapping $f: U \to Y$ is differentiable at $\bar{x}$ if and only if, for every $i = 1, \ldots, k$, $p_i \circ f: U \to Y_i$ is differentiable at $\bar{x}$, where $p_i: Y \to Y_i$ denotes the ith projection mapping; moreover, $f'(\bar{x})(h) = (f'_i(\bar{x})(h))_{i=1,\ldots,k}$, $\forall h \in X$, where $f_i = p_i \circ f$. When $X = X_1 \times \cdots \times X_k$ with $X_1, \ldots, X_k$ normed spaces, the differential at $\bar{x}_i$ (if it exists) of the mapping

$$x_i \mapsto f(\bar{x}_1, \ldots, \bar{x}_{i-1}, x_i, \bar{x}_{i+1}, \ldots, \bar{x}_k)$$

is called the ith *partial differential* of f at $\bar{x}$ and is denoted by $f'_{x_i}(\bar{x})$ or $f'_i(\bar{x})$; if f is differentiable at $\bar{x}$, then there are partial differentials $f'_i(\bar{x})$ $(i = 1, \ldots, k)$, and

$$f'(\bar{x})(h) = \sum_{i=1}^{k} f'_i(\bar{x})(h_i), \qquad \forall h = (h_i)_{i=1,\ldots,k} \in X;$$

moreover, f is of class C^1 if and only if the partial differentials f'_{x_i} $(i = 1, \ldots, k)$ exist and the mappings $x \mapsto f'_{x_i}(x)$, of U into $\mathscr{L}(X_i, Y)$, are continuous. If $f: U \to Y$ is differentiable at $\bar{x}$, then for any $u \in X$ there is a point in Y, which will be denoted by $D_u f(\bar{x})$, such that

$$\lim_{\mathbb{R} \ni t \to 0} \|t^{-1}(f(\bar{x} + tu) - f(\bar{x})) - D_u f(\bar{x})\| = 0;$$

moreover,

$$f'(\bar{x})(u) = D_u f(\bar{x}).$$

$D_u f(\bar{x})$ is the *derivative* of f at $\bar{x}$ *with respect to* u. If we suppose that $D_u f(x)$ exists $\forall u \in X$ and for every x belonging to a neighborhood of $\bar{x}$, that the map $\gamma_x: u \mapsto D_u f(x)$ is linear and continuous, and that $x \mapsto \gamma_x$ is continuous at $\bar{x}$ as a map with values in $\mathscr{L}(X, Y)$, then f is differentiable at $\bar{x}$ and $f'(\bar{x}) = \gamma_{\bar{x}}$. In the case when $X = \mathbb{R}^n$, the derivatives $D_{e_i} f(\bar{x})$ $(i = 1, \ldots, n)$, where $(e_1, \ldots, e_n)$ is the canonical base of $\mathbb{R}^n$, are called the *partial derivatives* of f at $\bar{x}$ and are denoted by $D_i f(\bar{x})$; for any $h = (h_i)_{i=1,\ldots,n} \in \mathbb{R}^n$ we have $f'(\bar{x})(h) = \sum_{i=1}^{n} D_i f(\bar{x}) h_i$.

Let X_i $(i = 1, \ldots, k)$ and Y be normed spaces. A k-linear mapping (i.e., a mapping which is linear separately in each variable) $\varphi: X_1 \times \cdots \times X_k \to Y$ is continuous if and only if there is a positive number c such that $\|\varphi(x_1, \ldots, x_k)\| \leq c \|x_1\| \ldots \|x_k\|$, $\forall x = (x_1, \ldots, x_k) \in X_1 \times \cdots \times X_k$. When $X_1 = \cdots = X_k = X$ the linear space $\mathscr{L}(X_1, \ldots, X_k; Y)$ of continuous k-linear mappings from $X_1 \times \cdots \times X_k$ into Y will be more briefly denoted by $\mathscr{L}_k(X, Y)$ (and, if $k = 1$, by $\mathscr{L}(X, Y)$). $\mathscr{L}(X_1, \ldots, X_k; Y)$ will be regarded as a normed space with the norm $\|\cdot\|$ defined by

$$\|\varphi\| = \sup_{\|x_1\| \leq 1, \ldots, \|x_k\| \leq 1} \|\varphi(x_1, \ldots, x_k)\|.$$

If X_1, X_2, and Y are normed spaces, the operator which maps any $\varphi \in \mathscr{L}(X_1, X_2; Y)$ to the mapping $X_1 \ni x_1 \mapsto [x_2 \mapsto \varphi(x_1, x_2)]$ is *the canonical (linear) isometry of* $\mathscr{L}(X_1, X_2; Y)$ *onto* $\mathscr{L}(X_1, \mathscr{L}(X_2, Y))$; its inverse maps $\psi \in \mathscr{L}(X_1, \mathscr{L}(X_2, Y))$ to the mapping $X_1 \times X_2 \ni (x_1, x_2) \mapsto \psi(x_1)(x_2)$.

Let X, Y be normed spaces and U an open subset of X. A mapping $f: U \to Y$ is said to be twice differentiable at $\bar{x} \in U$ if f is differentiable in a neighborhood V of $\bar{x}$ and $f': V \to \mathcal{L}(X, Y)$ is differentiable at $\bar{x}$. The differential of f' at $\bar{x}$ is denoted by $f''(\bar{x})$. We have $f''(\bar{x}) \in \mathcal{L}(X, \mathcal{L}(X, Y))$. Because of the canonical isometry of $\mathcal{L}(X, \mathcal{L}(X, Y))$ onto $\mathcal{L}_2(X, Y)$, $f''(\bar{x})$ can be regarded as an element of $\mathcal{L}_2(X, Y)$; i.e., $f''(\bar{x})$ can be identified with the mapping

$$X \times X \ni (h_1, h_2) \mapsto (f''(\bar{x})(h_1))(h_2).$$

By induction we define the kth-order differential $f^{(k)}(\bar{x})$ of f at $\bar{x}$: f is k-times differentiable at $\bar{x}$ if f is $(k - 1)$-times differentiable in a neighborhood V of $\bar{x}$ and $f^{(k-1)}: V \to \mathcal{L}_{k-1}(X, Y)$ is differentiable at $\bar{x}$; the differential at $\bar{x}$ of $f^{(k-1)}$, denoted by $f^{(k)}(\bar{x})$, is said to be the k-*order differential* of f at $\bar{x}$ and can be regarded as an element of $\mathcal{L}_k(X, Y)$. The mapping f is said to be *of class* C^k if f is k-times differentiable at any $x \in U$ and the function $x \mapsto f^{(k)}(x)$ is continuous from U into $\mathcal{L}_k(X, Y)$. f is said to be *of class* C^∞ if it is of class C^k for every positive integer k. A k-linear mapping $\varphi \in \mathcal{L}_k(X, Y)$ is called *symmetric* if it is invariant under all permutations of the indices $1, \ldots, k$. To any $\varphi \in \mathcal{L}_k(X, Y)$ we can associate the symmetric k-linear mapping $\mathrm{Sym}\, \varphi$ defined by

$$\mathrm{Sym}\, \varphi(x_1, \ldots, x_k) = \frac{1}{k!} \sum_{\sigma \in \Sigma_k} \varphi(x_{\sigma(1)}, \ldots, x_{\sigma(k)}),$$

where Σ_k is the set of the permutations of the indices $1, \ldots, k$. If f is k-times differentiable at x, then *the k-linear mapping $f^{(k)}(x) \in \mathcal{L}_k(X, Y)$ is symmetric*. If $f: U \to Y$ is of class C^{k+1} and the line segment $[x, x + h]$ is contained in U, then the following *Taylor formula* holds:

$$f(x + h) = f(x) + f'(x)(h) + \tfrac{1}{2} f''(x)(h)^2 + \cdots + \frac{1}{k!} f^{(k)}(x)(h)^k$$

$$+ \int_0^1 \frac{(1 - t)^k}{k!} f^{(k+1)}(x + th)(h)^{k+1}\, dt,$$

where $f^{(i)}(x)(h)^i = f^{(i)}(x)(h, \ldots, h)$ (the h repeated i times). Any $\varphi \in \mathcal{L}_k(X, Y)$ generates the *monomial* $\tilde{\varphi}: X \to Y$ of degree k: it is defined by $\tilde{\varphi}(x) = \varphi(x, \ldots, x)$ (x repeated k times). f and $\mathrm{Sym}\, f$ generate the same monomial. A monomial of degree k is generated by an unique continuous k-linear symmetric mapping, for if $\varphi \in \mathcal{L}_k(X, Y)$ is symmetric, then $\varphi = \tilde{\varphi}^{(k)}$. For every $\varphi \in \mathcal{L}_k(X, Y)$ we set

$$\|\tilde{\varphi}\| = \sup_{\|x\| \leq 1} \|\tilde{\varphi}(x)\| \qquad \left(= \sup_{x \neq 0} \frac{\|\tilde{\varphi}(x)\|}{\|x\|^k} \right).$$

The function $\tilde{\varphi} \mapsto \|\tilde{\varphi}\|$ is a norm such that

$$\|\tilde{\varphi}\| \leq \|\mathrm{Sym}\, \varphi\| \leq \frac{k^k}{k!} \|\tilde{\varphi}\|. \tag{1.1}$$

Henceforth X and Y shall be (complex or real) Banach spaces and U shall be an open subset of X.

A series in the vector space Y^X is called a *power series* if it is of the type

$$\sum_{k=0}^{\infty} \tilde{\varphi}_k, \tag{1.2}$$

where $\varphi_k \in \mathcal{L}_k(X, Y)$. The *radius of convergence* of (2) is the upper bound, say r, of the set of positive numbers ρ such that $\sum_{k=0}^{\infty} \|\tilde{\varphi}_k\| \rho^k$ is convergent. The series (1.2) is uniformly convergent in any ball $\{x \in X : \|x\| \leq \rho\}$ with $0 < \rho < r$. The upper bound, say r^*, of the set of positive numbers ρ such that $\sum_{k=0}^{\infty} \|\text{Sym } \varphi_k\| \rho^k$ is convergent is called the *strict radius of convergence* of the series (1.2). Since $\lim_{k \to \infty} \sqrt[k]{k!/k^k} = e$, and in view of the Cauchy–Hadamard theorem, from (1.1) it follows that $r/e \leq r^* \leq r$.

A mapping $f : U \to Y$ is said to be *analytic at* $x \in U$ if there is a power series $\sum_{k=0}^{\infty} \tilde{\varphi}_k$ with the radius of convergence $\neq 0$ such that

$$f(x + h) = \sum_{k=0}^{\infty} \tilde{\varphi}_k(h)$$

for $\|h\|$ suitably small. f is said to be *analytic* (in U) if it is analytic at any $x \in U$. If f is analytic at x, then f is of class C^∞ in a neighborhood of x and

$$f(x + h) = \sum_{k=0}^{\infty} \frac{1}{k!} f^{(k)}(x)(h)^k$$

for $\|h\|$ suitably small. We emphasize the following results.

The sum of a power series with the radius of convergence $\neq 0$ is an analytic mapping in the ball $\{x \in U : \|x\| < r^*\}$.
A C^∞ mapping $f : U \to Y$ is analytic at $a \in U$ if and only if there is a neighborhood V_a of a and positive numbers c and ρ such that

$$\|f^{(k)}(x)\| \leq c \frac{k!}{\rho^k}$$

for all $k \in \mathbb{N}$ and all $x \in V_a$.
The composition of analytic mappings is an analytic mapping.

A bijection $f : U \to V$, where U is an open subset of X and V is an open subset of Y is said to be a C^k *diffeomorphism* if f and f^{-1} are of class C^k.

A homeomorphism $f : U \to V$ of class C^k is a C^k diffeomorphism, if and only if for any $x \in U$, $f'(x)$ is a bijection of X onto Y.

Inverse Function Theorem. *Let X, Y be Banach spaces, U an open subset of X, and $x_0 \in U$. If $f : U \to Y$ is of class C^k ($k \geq 1$), and its differential at*

x_0 is a bijection of X onto Y, then there is an open neighborhood U_0 of x_0 in X contained in U and an open neighborhood V_0 of $f(x_0)$ in Y such that f induces a C^k diffeomorphism of U_0 onto V_0. Moreover, if f is analytic at x_0, then its inverse is analytic at $f(x_0)$.

Implicit Function Theorem. *Let X, Y, Z be Banach spaces, U an open subset of $X \times Y$, and let $(x, y) \mapsto f(x, y)$ be a continuous function of U into Z such that the partial differential $f_Y'(x, y)$ exists $\forall (x, y) \in U$ and that $(x, y) \mapsto f_Y'(x, y)$ is continuous in U. Let $(x_0, y_0) \in U$ be such that $f(x_0, y_0) = 0$ and such that $f_Y'(x_0, y_0)$ is a bijection of Y onto Z. Then there is an open neighborhood U_0 of (x_0, y_0) in $X \times Y$, an open neighborhood V_0 of x_0 in X and a continuous function $g: V_0 \to Y$ such that $\{(x, y) \in U_0 : f(x, y) = 0\} = \{(x, y): x \in V_0, y = g(x)\}$. If, furthermore, U_0 is chosen such that $f_Y'(x, y)$ is a bijection of Y onto Z, $\forall (x, y) \in U_0$, then the following facts are true.†If f is of class C^1, then g is of class C^1 and, for $x \in V_0$,*

$$g'(x) = -(f_Y'(x, g(x)))^{-1} \circ (f_x'(x, g(x)));$$

if f is of class C^k ($k \geq 1$), then g is of class C^k; if f is analytic at $(x, g(x))$, then g is analytic at x.

---

† We remark that there is a neighborhood of (x_0, y_0) such that, for each of its points (x, y), $f_Y'(x, y)$ is a bijection of Y onto Z; this assertion is a consequence of the fact that the set of the elements of $\mathscr{L}(X, Y)$ that are bijective is an open subset of $\mathscr{L}(X, Y)$ (see the beginning of this Appendix), and that (by hypotheses) the mapping $(x, y) \mapsto f_Y'(x, y)$ is continuous between U and $\mathscr{L}(Y, Z)$ and $f_Y'(x_0, y_0)$ is a bijection of Y onto Z.

APPENDIX II

On the Representation of Orthogonal Matrices

For $Z \in \mathbb{M}_n$, we set $Z^0 = I$. We recall that the Cauchy–Schwartz inequality $|YZ| \leq |Y||Z|$ holds $\forall Y, Z \in \mathbb{M}_n$, and so

$$|Z^k| \leq |Z|^k, \qquad \forall Z \in \mathbb{M}_n \quad \text{and} \quad \forall k \in \mathbb{N}.$$

Then, since for any $Z \in \mathbb{M}_n$, the series

$$\sum_{k=0}^{\infty} \frac{1}{k!} |Z|^k$$

converges (to $\exp|Z|$), the series

$$\sum_{k=0}^{\infty} \frac{1}{k!} |Z^k|$$

also converges for any $Z \in \mathbb{M}_n$. Therefore the series $\sum_{k=0}^{\infty} (1/k!) Z^k$ converges for any $Z \in \mathbb{M}_n$; we denote its sum by $\exp Z$, i.e., we put

$$\exp Z = \sum_{k=0}^{\infty} \frac{1}{k!} Z^k. \tag{2.1}$$

Thus $Z \mapsto \exp Z$ is an analytic function from $\mathbb{M}_n$ into itself.

It is well known that if $\lambda_1, \ldots, \lambda_n$ are the eigenvalues of $Z \in \mathbb{M}_n$ (each occurring a number of times equal to its multiplicity), then the eigenvalues of $\exp Z$ are $\exp \lambda_1, \ldots, \exp \lambda_n$. Consequently,

$$\det(\exp Z) = \exp(\operatorname{tr} Z). \tag{2.2}$$

We can prove, using Cauchy's multiplication of series, that, for $Y, Z \in M_n$,

$$YZ = ZY \Rightarrow \exp(Y + Z) = (\exp Y)(\exp Z). \tag{2.3}$$

Since, if $W \in \text{Skew}_n$, we have $W + W^T = W^T + W = 0$ and $WW^T = W^TW = (-W^2)$, (2.3) implies

$$(\exp W)(\exp (W^T)) = (\exp(W^T))(\exp W) = I, \qquad \forall W \in \text{Skew}_n;$$

hence, observing that, evidently, $(\exp W)^T = \exp(W^T)$, we obtain

$$(\exp W)(\exp W)^T = (\exp W)^T(\exp W) = I, \qquad \forall W \in \text{Skew}_n. \tag{2.4}$$

Combining (2.2) with (2.4) we get

$$W \in \text{Skew}_n \Rightarrow \exp W \in \mathbb{O}_n^+. \tag{2.5}$$

From definition (2.1) it easily follows that $\exp 0 = I$ and that the differential at 0 of the exponential function $Z \mapsto \exp Z$, of M_n into itself, is the identity function of M_n onto itself. Then, in view of the inverse function theorem and of (2.5), *there is in M_n an open neighborhood U of 0 and an open neighborhood V of I such that the exponential function $Z \mapsto \exp Z$ induces a diffeomorphism of $U \cap \text{Skew}_n$ onto $V \cap \mathbb{O}_n^+$.*

Let $W \in \text{Skew}_n$ with $\det(I - W) \neq 0$. From the equalities $(I - W)(I - W)^{-1} = (I - W)^{-1}(I - W)$ and $(I + W)(I + W)^{-1} = (I + W)^{-1}(I + W)$ it follows that $W(I - W)^{-1} = (I - W)^{-1}W$ and $W(I + W)^{-1} = (I + W)^{-1}W$, respectively, and the latter equalities are equivalent to the following ones:

$$\begin{cases} (I + W)(I - W)^{-1} = (I - W)^{-1}(I + W), \\ (I - W)(I + W)^{-1} = (I + W)^{-1}(I - W). \end{cases} \tag{2.6}$$

Setting

$$\hat{R}(W) = (I - W)^{-1}(I + W), \tag{2.7}$$

we have

$$\hat{R}(W) \in \mathbb{O}_n^+.$$

Indeed

$$\det \hat{R}(W) = \frac{\det(I + W)}{\det(I - W)} = \frac{\det(I - W)^T}{\det(I - W)} = \frac{\det(I - W)}{\det(I - W)} = 1,$$

and, in view of (2.6),

$$\begin{cases} \hat{R}(W)(\hat{R}(W))^T = (I - W)^{-1}(I + W)(I + W)^T((I - W)^T)^{-1} \\ \qquad = (I - W)^{-1}(I + W)(I - W)(I + W)^{-1} = I, \\ (\hat{R}(W))^T\hat{R}(W) = (I + W)^T((I - W)^T)^{-1}(I - W)^{-1}(I + W) \\ \qquad = (I - W)(I + W)^{-1}(I - W)^{-1}(I + W) = I. \end{cases}$$

Note that from (2.7) it follows that $I + \hat{R}(W) = 2(I - W)^{-1}$ and hence

$$W = -(I + \hat{R}(W))^{-1}(I - \hat{R}(W)).$$

Therefore

$$\det(I - W) \neq 0 \Leftrightarrow \det(I + \hat{R}(W)) \neq 0.$$

Note also that it is easy to verify that if $R \in \mathbb{O}_n^+$ and $\det(I + R) \neq 0$, then $(I + R)^{-1}(I + R) \in \mathrm{Skew}_n$. Thus we have shown that *the mapping*

$$W \mapsto (I - W)^{-1}(I + W) \tag{2.8}$$

is a bijection of the set $\{W \in \mathrm{Skew}_n : \det(I - W) \neq 0\}$ *onto the set* $\{R \in \mathbb{O}_n^+ : \det(I + R) \neq 0\}.$†

Henceforth we deal with the case $n = 3$ and we investigate the link between the function (2.8) and the exponential function. For each $W \in \mathrm{Skew}_3$ let us denote by w the (uniquely determined) element (w_1, w_2, w_3) of $\mathbb{R}^3$ such that

$$W = \begin{bmatrix} 0 & w_3 & -w_2 \\ -w_3 & 0 & w_1 \\ w_2 & -w_1 & 0 \end{bmatrix}. \tag{2.9}$$

From the Cayley–Hamilton theorem‡ it follows that

$$W^3 = -|w|^2 W, \qquad \forall W \in \mathrm{Skew}_3,$$

because, for $W \in \mathrm{Skew}_3$, we have $\det(\lambda I - W) = \lambda^3 + |w|^2 \lambda$. Then we easily deduce, for every integer $k \geq 1$ and for every $W \in \mathrm{Skew}_3$,

$$W^{2k+1} = (-1)^k |w|^{2k} W, \qquad W^{2k} = (-1)^{k-1} |w|^{2k-2} W^2.$$

Thus, for every $W \in \mathrm{Skew}_3 \setminus \{0\}$ we have

$$\exp W = I + \sum_{k=0}^{\infty} \frac{1}{(2k+1)!} W^{2k+1} + \sum_{k=1}^{\infty} \frac{1}{(2k)!} W^{2k}$$

$$= I + \left(\sum_{k=0}^{\infty} (-1)^k \frac{|w|^{2k}}{(2k+1)!} \right) W + \left(\sum_{k=1}^{\infty} (-1)^{k-1} \frac{|w|^{2k-2}}{(2k)!} \right) W^2$$

$$= I + \left(\sum_{k=0}^{\infty} (-1)^k \frac{|w|^{2k+1}}{(2k+1)!} \right) \frac{W}{|w|} + \left(\sum_{k=1}^{\infty} (-1)^{k-1} \frac{|w|^{2k}}{(2k)!} \right) \left(\frac{W}{|w|} \right)^2.$$

Hence, for every $W \in \mathrm{Skew}_3$ with $W \neq 0$,

$$\exp W = I + \mathrm{sen}\,|w| \frac{W}{|w|} + (1 - \cos|w|) \left(\frac{W}{|w|} \right)^2. \tag{2.10}$$

---

† A similar result is given in GRÖBNER [1975], p. 113.

‡ The Cayley–Hamilton theorem asserts that *if* $Z \in \mathsf{M}_n$ *and* $P(\lambda)$ *is the polynomial defined by* $P(\lambda) = \det(\lambda I - Z)$, *then* $P(Z) = 0$.

It is interesting to remark that *the right-hand side of (2.10) is the matrix of the rotation of opening* $|w|$ *and axis* $w/|w|$. Putting

$$q = \tan \frac{|w|}{2},$$

we have

$$\operatorname{sen}|w| = \frac{2q}{1 + q^2}, \qquad 1 - \cos|w| = \frac{2q^2}{1 + q^2},$$

and thus

$$\exp W = I + \frac{2}{1 + q^2}\left(q\frac{W}{|w|} + \left(q\frac{W}{|w|}\right)^2\right), \qquad \forall W \in \operatorname{Skew}_3\backslash\{0\}. \quad (2.11)$$

On the other hand, it is not difficult to verify that

$$(I - W)^{-1}(I + W) = I + \frac{2}{1 + |w|^2}(W + W^2). \quad (2.12)$$

for every $W \in \operatorname{Skew}_3$ with $\det(I - W) \neq 0$. Combining (2.11) with (2.12) we get

$$\exp W = \left(I - \left(\frac{1}{|w|}\tan\frac{|w|}{2}\right)W\right)^{-1}\left(I + \left(\frac{1}{|w|}\tan\frac{|w|}{2}\right)W\right)$$

for every $W \in \operatorname{Skew}_3$, with $W \neq 0$, belonging to a suitable neighborhood of 0 in $\mathbb{M}_3$.

Bibliography

ADAMS, R. A. [1975]: *Sobolev Spaces*. Academic Press, New York.

AGMON, S., DOUGLIS, A., & NIRENBERG, L. [1964]: Estimates near the boundary for solutions of elliptic partial differential equations satisfying general boundary conditions, II. *Comm. Pure Appl. Math.* **17**, 35–92.

BALL, J. M. [1977]: Convexity conditions and existence theorems in nonlinear elasticity. *Arch. Rational Mech. Anal.* **63**, 337–403.

BALL, J. M. [1981]: Global invertibility of Sobolev functions and the interpenetration of matter. *Proc. Roy. Soc. Edinburgh* **88A**, 315–328.

BALL, J. M. & SCHAEFFER, D. G. [1983]: Bifurcation and stability of homogeneous equilibrium configurations of an elastic body under dead load traction. *Math. Proc. Cambridge Philos. Soc.* **94**, 315–339.

BERGER, M. S. [1977]: *Nonlinearity and Functional Analysis*. Lectures on Nonlinear Problems in Mathematical Analysis, Vol. 74, Academic Press, New York.

BHARATHA, S. & LEVINSON, M. [1978]: Signorini's perturbation scheme for a general reference configuration in finite elastostatics. *Arch. Rational Mech. Anal.* **67**, 365–394.

BOURBAKI, N. [1967a]: *Eléments de Mathématique: Espaces vectoriels topologiques*, Fasc. XVIII, Livre V. Hermann, Paris.

BOURBAKI, N. [1967b]: *Eléments de Mathématique: Varieté différentielles et analytiques*, Fasc. XXXIII, Livre VI. Hermann, Paris.

BROWDER, F. [1959]: Estimates and existence theorems for elliptic boundary value problems. *Proc. Nat. Acad. Sci. U.S.A.* **20**, 365–372.

CAPRIZ, G. & PODIO-GUIDUGLI, P. [1974]: On Signorini's perturbation method in finite elasticity. *Arch. Rational Mech. Anal.* **57**, 1–30.

CAPRIZ, G. & PODIO-GUIDUGLI, P. [1979]: The role of Fredholm conditions in Signorini's perturbation method. *Arch. Rational Mech. Anal.* **70**, 261–288.

CAPRIZ, G. & PODIO-GUIDUGLI, P. [1981]: *Questions of Uniqueness in Finite Elasticity*. Trends in Applications of Pure Mathematics to Mechanics, Vol. III, pp. 76–95. Pitman, Boston.

CAPRIZ, G. & PODIO-GUIDUGLI, P. [1982a]: Duality and stability questions for the linearized traction problem with live loads in elasticity. In: *Stability in the Mechanics of Continua* (F. H. SCHROEDER, ed.). Springer-Verlag, Berlin.

CAPRIZ, G. & PODIO-GUIDUGLI, P. [1982b]: A generalization of Signorini's perturbation method suggested by two problems of Grioli. *Rend. Sem. Mat. Univ. Padova* **68**, 49–162.

CARTAN, H. [1967]: *Calcul Différentielle*. Hermann, Paris.

CHILLINGWORTH, D. R. J., MARSDEN, J. E., & WAN, Y. H. [1982]: Symmetry and bifurcation in three-dimensional elasticity, Part I. *Arch. Rational Mech. Anal.* **80**, 295–331.

CHILLINGWORTH, D. R. J., MARSDEN, J. E., & WAN, Y. H. [1983]: Symmetry and bifurcation in three-dimensional elasticity, Part II. *Arch. Rational Mech. Anal.* **83**, 362–395.

CIARLET, P. G. [1986]: *Mathematical Elasticity*, Vol. I. North-Holland, Amsterdam.

DIEUDONNE', J. [1970]: *Eléments d'Analyse*, Tome III. Gauthier-Villars, Paris.

FICHERA, G. [1972a]: Existence theorems in elasticity. In: *Handuch der Physik*, Vol. IVa/2. Springer-Verlag, Berlin.

FICHERA, G. [1972b]: *Sulla Propagazione delle onde in un Mezzo Elastico*. Continuum Mechanics and Related Problems of Analysis. Nauka, Moscow, pp. 567–574.

GOBERT, J. [1962]: Une inégalité fondamentale de la théorie de l'élasticité. *Bull. Soc. Roy. Sci. Liège* **3–4**, 182–191.

GRIOLI, G. [1962]: *Mathematical Theory of Elastic Equilibrium (Recent Results)*. Ergebnisse der Angewandte Mathematische, Vol. 7. Springer-Verlag, Berlin.

GRIOLI, G. [1982]: *Statica della Deformazioni Finite. Problemi di Linearizzazione*. Atti del Convegno in memoria di R. Calapso, Ed. V. Veschi, Roma.

GRIOLI, G. [1983a]: On the stress in rigid bodies. *Meccanica* **18**, 3–7.

GRIOLI, G. [1983b]: Mathematical problems in elastic equilibrium with finite deformations. *Applicable Anal.* **15**, 171–186.

GRÖBNER, W. [1975]: Gruppi, Anelli e Algebre di Lie, Poliedro, **20** Ed. Cremonese, Roma.

GURTIN, M. E. [1981]: *An Introduction to Continuum Mechanics*. Academic Press, New York.

GURTIN, M. E. & MARTINS, L. C. [1976]: Cauchy's theorem in classical physics. *Arch. Rational Mech. Anal.* **60**, 305–328.

GURTIN, M. E. & SPECTOR, S. J. [1979]: On stability and uniqueness in finite elasticity. *Arch. Rational Mech. Anal.* **70**, 153–166.

KRASNOSELSKIJ, M. A. *et al.* [1976]: *Integral Operators in Spaces of Summable Functions*. Noordhoff, Leyden.

KRASNOSELSKIJ, M. A., VAINIKKO, G. M. *et al.* [1972]: *Approximate Solutions of Operator Equations*. Wolters–Noordhoff, Groningen.

LANZA DE CRISTOFORIS, M. & VALENT, T. [1982]: On Neumann's problem for a quasi-linear differential system of the finite elastostatics type. Local theorems of existence and uniqueness. *Rend. Sem. Mat. Univ. Padova* **68**, 183–206.

LE DRET, H. [1986]: Structure of the set of equilibrated loads in nonlinear elasticity and applications to existence and nonexistence. *J. Elasticity* (to appear).

LIONS, J. L. & MAGENES, E. [1972]: *Nonhomogeneous Boundary Value Problems and Applications.* Springer-Verlag, New York.

MARSDEN, J. E. & HUGHES, T. J. R. [1978]: *Topics in the Mathematical Foundations of Elasticity.* Heriott–Watt Symposium, Vol. II. Pitman, London, pp. 30–285.

MARSDEN, J. E. & HUGHES, T. J. R. [1983]: *The Mathematical Foundation of Elasticity.* Prentice-Hall, Englewood Cliffs, NJ.

MARSDEN, J. E. & WAN, Y. H. [1983]: Linearization stability and Signorini series for the traction problem in elastostatics. *Proc. Roy. Soc. Edinburgh,* **A95,** 171–180.

MEISTERS, H. & OLECH, C. [1963]: Locally one-to-one mappings and a classical theorem on schlicht functions. *Duke Math. J.* **30,** 63–80.

NEČAS, J. [1967]: *Les Méthodes Directes en Théorie des Equations Elliptiques.* Masson, Paris.

NOLL, W. [1964]: Euclidean geometry and Minkowskian chronometry. *Amer. Math. Monthly* **71,** 129–144.

PODIO-GUIDUGLI, P. & VERGARA CAFFARELLI, G. [1984]: On a class of live traction problems in elasticity. In: *Trends in Applications of Pure Mathematics to Mechanics* (P. G. CIARLET and M. ROSEAU, eds.), Springer-Verlag, Berlin.

PODIO-GUIDUGLI, P., VERGARA CAFFARELLI, G., & VIRGA, E. G. [1984]: Una formula di Green per i problemi di carichi vivi dell'elasticità linearizzata. Atti VII Congresso AIMETA, Trieste.

PODIO-GUIDUGLI, P., VERGARA CAFFARELLI, G., & VIRGA, E. G. [1986]: The role of ellipticity and normality assumptions in formulating live boundary conditions in elasticity. *Quart. Appl. Math.* (to appear).

PRODI, G. & AMBROSETTI, A. [1973]: *Analisi Non Lineare.* Quaderno Scuola Normale Superiore, Pisa.

SEWELL, M. J. [1967]: On configuration-dependent loading. *Arch. Rational Mech. Anal.* **23,** 327–351.

SIGNORINI, A. [1949]: Trasformazioni termoelastiche finite. Memoria II. *Ann. Mat. Pura Appl.,* IV **30,** 1–72.

SIGNORINI, A. [1950]: Un semplice esempio di incompatibilità tra la elastostatica classica e la teoria delle deformazioni elastiche finite. *Rend. Accad. Naz. Lincei* (VIII) **3,** 276–281.

SIMADER, C. G. [1972]: *On Dirichlet's Boundary Value Problems.* Lecture Notes in Mathematics, Vol. 268, Springer-Verlag, Berlin.

SPECTOR, S. J. [1980]: On uniqueness in finite elasticity with general loading. *J. Elasticity* **10,** 145–161.

SPECTOR, S. J. [1982]: On uniqueness for the traction problem in finite elasticity. *J. Elasticity* **12,** 367–383.

STOPPELLI, F. [1954]: Un teorema di esistenza e unicità relativo alle equazioni dell'elastostatica isoterma per deformazioni finite. *Ricerche Mat.* **3,** 247–267.

STOPPELLI, F. [1955]: Sulla sviluppabilità in serie di potenze di un parametro delle soluzioni delle equazioni dell'elastostatica isoterma. *Ricerche Mat.* **4,** 58–73.

STOPPELLI, F. [1957a]: Su un sistema di equazioni integro-differenziali interessante l'elastostatica. *Ricerche Mat.* **6,** 11–26.

STOPPELLI, F. [1957b]: Sull'esistenza di soluzioni delle equazioni dell'elastostatica isoterma nel caso di sollecitazioni dotate di assi di equilibrio I. *Ricerche Mat.* **6,** 241–287.

STOPPELLI, F. [1958]: Sull'esistenza di soluzioni delle equazioni dell'elastostatica isoterma nel caso di sollecitazioni dotate di assi di equilibrio II, III. *Ricerche Mat.* **7**, 71–101, 138–152.

THOMPSON, J. L. [1968]: Some existence theorems for the traction boundary value problems of linearized elastostatics. *Arch. Rational Mech. Anal.* **45**, 369–399.

TOLOTTI, C. [1942]: Orientamenti principali di un corpo elastico rispetto alla sua sollecitazione totale. *Mem. Accad. Ital. Serv.* VII, **13**, 1139–1162.

TRUESDELL, C. [1974]: *Introduction à la Mécanique Rationelle des Milieux Continus.* Masson, Paris.

TRUESDELL, C. [1977]: *A First Course in Rational Continuum Mechanics*, Vol. I. Academic Press, New York.

TRUESDELL, C. & NOLL, W. [1965]: The nonlinear field theories of mechanics. In: *Handbuch der Physik*, Vol. III/3, Springer-Verlag, Berlin.

TRUESDELL, C. & TOUPIN, R. A. [1960]: The classical field theories. In: *Handbuch der Physik*, Vol. III/1, Springer-Verlag, Berlin.

VALENT, T. [1978a]: Sulla differenziabilità dell'operatore di Nemytsky. *Rend. Accad. Naz. Lincei* **65**, 15–26.

VALENT, T. [1978b]: Osservazioni sulla linearizzazione di un operatore differenziale. *Rend. Accad. Naz. Lincei* **65**, 27–37.

VALENT, T. [1979]: Teoremi di esistenza e unicità in elastostatica finita. *Rend. Sem. Mat. Univ. Padova* **60**, 165–181.

VALENT, T. [1981]: Local existence and uniqueness theorems in finite elastostatics. In: *Proc. IUTAM Symposium on Finite Elasticity* (D. E. Carlson and R. T. Shield, eds.). Martinus Nijhoff, The Hague, pp. 401–421.

VALENT, T. [1985a]: Sul problema di posto in elastostatica non lineare. Teoremi di esistenza e di unicità. *Boll. Un. Mat. Ital. Suppl.*, Fisica Matematica, **IV-5** (1), 281–295.

VALENT, T. [1985b]: A property of multiplication in Sobolev spaces. Some applications. *Rend. Sem. Mat. Univ. Padova* **74**, 63–73.

VALENT, T. [1986]: Pressure boundary problems in finite elasticity. Results on local existence, uniqueness and analyticity. In: *Proc. Meeting on Finite Thermoelasticity* (G. GRIOLI, ed.). Accad. Naz. Lincei, Roma, pp. 241–264.

VALENT, T. & ZAMPIERI, G. [1977]: Sulla differenziabilità di un operatore legato a una classe di sistemi differenziali lineari. *Rend. Sem. Mat. Univ. Padova* **57**, 311–322.

VAN BUREN, W. [1968]: On the existence and uniqueness of solutions to boundary value problems in finite elasticity. Thesis. Carnegie–Mellon University; Westinghouse Research Laboratories, Report 68-107-MEKMARI.

WAN, Y. H. & MARSDEN, J. E. [1984]: Symmetry and bifurcation in three-dimensional elasticity, Part III: Stressed reference configurations. *Arch. Rational Mech. Anal.* **84**, 203–233.

WANG, C. & TRUESDELL, C. [1973]: *Introduction to Rational Elasticity.* Noordhoff, Groningen.

Index of Notations

Basic Notations

$\mathbb{N} = \{0, 1, 2, \ldots\}$: set of natural numbers

$\mathbb{Z}$: set of integers

$\mathbb{Q}$: set of rational numbers

$\mathbb{R}$: set of real numbers

Ω: nonempty, bounded, open subset of $\mathbb{R}^n$; $\bar{\Omega}$: closure of Ω; $\partial\Omega$: boundary of Ω

ϕ, ψ, χ: deformations of $\bar{\Omega}$

$\iota_{\mathscr{P}}$: identity mapping of the subset $\mathscr{P}$ of $\mathbb{R}^n$ into $\mathbb{R}^n$

$\mathbb{M}_{m \times n}$: set of real $m \times n$ matrices

$\mathbb{M}_n$: set of real $n \times n$ matrices

I: unit element of the ring $\mathbb{M}_n$

$\det Z$: determinant of $Z \in \mathbb{M}_n$

$\mathbb{M}_n^+ = \{Z \in \mathbb{M}_n: \det Z > 0\}$

Z^T: transpose of $Z \in \mathbb{M}_{m \times n}$

Z^{-1}: inverse of $Z \in \mathbb{M}_n$ with $\det Z \neq 0$

$\operatorname{tr} Z$: trace of Z

$\operatorname{cof} Z$: matrix of cofactors of $Z \in \mathbb{M}_n$, so that $\operatorname{cof} Z = \det Z(Z^{-1})^T$ whenever $\det Z \neq 0$

$\exp Z = \sum\limits_{k=0}^{\infty} \dfrac{1}{k!} Z^k$ for $Z \in \mathbb{M}_n$

Sym_n: set of symmetric, real $n \times n$ matrices

Skew_n: set of skew, real $n \times n$ matrices

$\mathbb{O}_n^+ = \{Z \in \mathbb{M}_n^+: Z^T = Z^{-1}\}$

$$ZW = \left(\sum_{h=1}^{p} Z_{ih} W_{hj} \right)_{i=1,\ldots,m;j=1,\ldots,n} \qquad \text{for } Z = (Z_{ih})_{i=1,\ldots,m;h=1,\ldots,p} \in \mathbb{M}_{m \times p}$$

and $W = (W_{hj})_{h=1,\ldots,p,j=1,\ldots,n} \in \mathbb{M}_{p \times n}$; in particular,

$$Zy = \left(\sum_{j=1}^{n} Z_{ij} y_{j} \right)_{i=1,\ldots,m} \qquad \text{for } Z = (Z_{ij})_{i=1,\ldots,m;j=1,\ldots,n} \in \mathbb{M}_{m \times n}$$

and $y = (y)_{j=1,\ldots,n} \in \mathbb{R}^n$

$Z \cdot W = \displaystyle\sum_{\substack{i=1,\ldots,m \\ j=1,\ldots,n}} Z_{ij} W_{ij}$ for $Z = (Z_{ij})_{i=1,\ldots,m;j=1,\ldots,n}$ and

$W = (W_{ij})_{i=1,\ldots,m;j=1,\ldots,n}$ belonging to $\mathbb{M}_{m \times n}$

$|Z| = (Z \cdot Z)^{1/2}$ for $Z \in \mathbb{M}_{m \times n}$

$y \otimes z = (y_i z_j)_{i,j=1,\ldots,n}$ for $y = (y_i)_{i=1,\ldots,n}$ and $z = (z_j)_{j=1,\ldots,n}$ belonging
to $\mathbb{R}^n$

$y \wedge z = y \otimes z - z \otimes y$

D_j: jth partial derivative operator

$D^\alpha = D_1^{\alpha_1} \ldots D_n^{\alpha_n}$ for $\alpha = (\alpha_1, \ldots, \alpha_n) \in \mathbb{N}^n$

$Dv = (D_j v_i)_{i=1,\ldots,m;j=1,\ldots,n}$ for $v: U \to \mathbb{R}^m$ a mapping with U an open
subset of $\mathbb{R}^n$; here v_i is the ith component of v

$\operatorname{div} S = \left(\sum_{j=1}^{n} D_j S_{ij} \right)_{i=1,\ldots,n}$ for $S: U \to \mathbb{M}_{m \times n}$ a mapping with U an open

subset of $\mathbb{R}^n$; here S_{ij} is the (i, j)th component of S

Special Symbols

$\mathcal{R}$: set of rigid deformations of $\mathbb{R}^n$ 54
$\mathcal{R}_{\bar\Omega}$: set of rigid deformations of $\bar\Omega$ 55, 108
$T_{\rho_0}(\mathcal{R})$: tangent space to $\mathcal{R}$ at $\rho_0 \in \mathcal{R}$ 54, 108
$T_{\rho_0}(\mathcal{R}_{\bar\Omega})$: tangent space to $\mathcal{R}_{\bar\Omega}$ at $\rho_0 \in \mathcal{R}_{\bar\Omega}$ 55, 108
$\hat{e}_\rho(f, g)$: equilibrated part of (f, g) relative to ρ 109
$\mathcal{K}_{\phi_0}(F, G)$, $\mathcal{W}_{\phi_0}(F, G)$ 111
$\mathbb{O}_e$ 151
$\hat{e}_R(f, g)$: equilibrated part of (f, g) relative to the rigid deformations of $\bar\Omega$
having R as the value of their gradient (at any point) 153

Main Operators

D, 2; div, 2; F, 27, 104; F_{x_i}, 27; F_{y_j}, 27; A, 60, 88, 104, 134;
$D_{Z_{hk}}$, 60; L, 64; L_R, 67; Δ, 81; B, 88, 103, 104, 134; $A_{Z_{hk}}$, 93;
G, 103, 104; E_{R_0}, 105; Λ, 105, 148; $\bar\Lambda$, 107, 149; M, 110; γ, 150;
P, Q, 153

Main Spaces of Functions and Distributions. Main Norms

$C^m(\Omega)$, $C^\infty(\Omega)$, $C^m(B)$, $C^\infty(B)$, $C^m(\bar\Omega)$, $C_0^m(\bar\Omega)$, $C^{m,\lambda}(\bar\Omega)$, $\mathcal{D}(\Omega)$, $\mathcal{D}'(\Omega)$ 17
$L^p(\Omega)$, $L^\infty(\Omega)$, $W^{m,p}(\Omega)$, $W_0^{m,p}(\Omega)$, $W^{-m,p}(\Omega)$ 19
$C^{m,\lambda}(\partial\Omega)$, $W^{s,p}(\mathbb{R}^n)$, $W^{s,p}(\partial\Omega)$ 21

Index